Climate, Science and Society

T0177411

Climate, Science and Society: A Primer makes cutting-edge research on climate change accessible to student readers.

The primer consists of 37 short chapters organized within 11 parts written by Science and Technology Studies (STS) and other social science scholars. It covers a range of key topics including communication, justice and inequality, climate policy, and energy transitions, situating each one within the context of STS studies. Each reading translates a focused area of climate change research into short, accessible, and lively prose. Chapter authors open debates where relevant, consider policy implications, critique existing areas of research, and otherwise situate their reading within a larger body of research relevant to climate change courses.

Designed as a jumping-off point for further exploration, this innovative book will be essential reading for students studying climate change, STS, environmental sociology, and environmental sciences.

Zeke Baker is Assistant Professor of Sociology at Sonoma State University, USA. His research investigates how changes in climate knowledge relate to political dynamics, especially over the historical long term, in the United States and comparative contexts.

Tamar Law is a PhD student at Cornell University in Development Studies and holds an MPhil in Human Environmental Geography from the University of Oxford. Her research in the United States and Southeast Asia examines the knowledge and land politics of climate adaptation and mitigation, centering questions of climate justice.

Mark Vardy is a faculty member of the Criminology Department at Kwantlen Polytechnic University, Canada. He is interested in drawing from STS to contribute to discussions of climate justice in green criminology.

Stephen Zehr is Professor Emeritus of Sociology at the University of Southern Indiana, USA. His past research focused on climate change scientific expertise and its representation in the media. He is currently researching maple syrup producers and their adaptation to technological changes, climate change, and labor supply and allegiances.

Climate, Science and Society

A Primer

Edited by
Zeke Baker, Tamar Law, Mark Vardy
and Stephen Zehr

First published 2024
by Routledge
4 Park Square, Milton Park, Abingdon, Oxon OX14 4RN

and by Routledge
605 Third Avenue, New York, NY 10158

Routledge is an imprint of the Taylor & Francis Group, an informa business

British Library Cataloguing-in-Publication Data
A catalogue record for this book is available from the British Library

ISBN: 978-1-032-53016-1 (hbk)
ISBN: 978-1-032-53017-8 (pbk)
ISBN: 978-1-003-40974-8 (ebk)

DOI: 10.4324/9781003409748

Typeset in Times New Roman
by codeMantra

Contents

Contributors

Zeke Baker is Assistant Professor of Sociology at Sonoma State University, USA. His research investigates how changes in climate knowledge relate to political dynamics, especially over the historical long term, in the United States and comparative contexts.

Ankit Bhardwaj is a PhD candidate in Sociology at New York University. He studies the politics of reducing greenhouse gas emissions, focusing on the relations between the state, expertise, and democracy.

Dominic Boyer is a writer and media-maker who teaches at Rice University, where he has recently helped shape the new Rice Sustainability Institute. His most recent book is *No More Fossils* (University of Minnesota Press, 2023). His 2018 film *Not Ok* tells the story of Iceland's first major glacier to fall victim to anthropogenic climate change.

Daniel Breslau is in the Department of Science, Technology, and Society, at Virginia Tech. His research focuses on science, technology, and politics in the formation of markets, with a particular focus on energy markets and climate.

Candis Callison is the Canada Research Chair in Indigenous journalism, media, and public discourse and Associate Professor at the University of British Columbia in the School of Journalism, Writing, and Media and the Institute for Critical Indigenous Studies. Her research focuses on the role of Indigenous and environment-focused journalism and social movements in public discourse, and understanding how climate change becomes meaningful for diverse publics.

Anabela Carvalho is Associate Professor at the Department of Communication Sciences of the University of Minho, Portugal. She has an extensive experience in research on climate change communication and on the climate discourses of various social actors and the media in various countries. In the last few years, her main research interests have been associated with citizens' engagement with climate change and sustainability, especially conditions for political engagement.

Tiago Ribeiro Duarte is Assistant Professor of Sociology at the University of Brasília. He is interested in the science-policy interface with a current focus on post-truth and science denialism.

Adam Fleischmann is a researcher, writer, and teacher with a PhD in anthropology from McGill University. His research and teaching practices bridge anthropological and related approaches to the study of global climate change, science and technology, social movements, the environment, institutions, expertise, knowledge, and ethics. He is currently Senior Research Associate at the School of Public Policy at the University of Calgary and an Affiliate Member of the Centre for Energy Ethics, University of St Andrews.

Allison Ford is Assistant Professor at Sonoma State University, where she researches and teaches on the environment, culture, gender, race, and emotions. Her work asks how people make sense of environmental risk, including climate change, in relation to their social locations and cultural worldviews. Her book-length treatment on prepping as an environmental practice is forthcoming.

Désirée Förster is Assistant Professor at the Department of Media and Culture Studies at Utrecht University. She uses process philosophy, speculative research, and aesthetic practice to explore new phenomenologies of complex issues enabled by new media arts and design.

Tim Forsyth is Professor at the London School of Economics and Political Science. He has conducted research on environmental policy and science in Southeast Asia and is a Lead Author for the Intergovernmental Science-Policy Platform on Biodiversity and Ecosystem Services (IPBES) assessment on transformative change.

Reiner Grundmann is Professor of Science and Technology Studies at the University of Nottingham (UK). His current focus is the relation between knowledge and decision-making. He has written on the nature of expertise in contemporary societies in previous publications. He is Chief Editor for the specialty section Climate and Decision Making in Frontiers in Climate.

Anders Hansson is Senior Associate Professor at Linköping University – Department of Thematic Studies, Sweden. His PhD thesis, published in 2008, was a critical investigation of the political and scientific contexts of carbon capture and storage (CCS). Today he is co-leading an interdisciplinary group that explores geoengineering and the development of methods for carbon dioxide removal.

David J. Hess is Professor of Sociology, the James Thornton Fant Chair in Sustainability Studies at Vanderbilt University, and the Director of Climate and Sustainability Studies. His research and teaching are on the sociology, anthropology, and policy studies of science, technology, health, energy, and the environment.

Cymene Howe is Professor of Anthropology at Rice University and author of *Intimate Activism* (Duke 2013), *Ecologics: Wind and Power in the Anthropocene* (Duke 2019) and co-editor of *Anthropocene Unseen* (Punctum 2020), *Solarities: Elemental Encounters and Refractions* (Punctum 2023) and *The Johns Hopkins Guide to Critical and Cultural Theory*. Her work focuses on the social impacts of climate change and she has conducted research in Mexico, Nicaragua, Iceland, Greenland, Hawai'i, and South Africa.

Shangrila Joshi is a Member of the Faculty in Climate Justice at The Evergreen State College. She is the author of *Climate Change Justice and Global Resource Commons* (Routledge 2021). Her ongoing research seeks to understand how Indigenous ways of being, knowing, and governing are connected, and why this matters for socio-ecological resilience.

Kendra Kintzi is Postdoctoral Fellow in the Atkinson Center for Sustainability at Cornell University. Her research focuses on renewable energy transition, environmental justice, and the possibilities for inclusive, democratic energy governance in Southwest Asia.

Hannah Knox is Professor of Anthropology at University College London. Her research focuses on the intersection of technology, infrastructure, and environment in the UK, Europe, and Peru, with recent work focusing on the relationship between these themes in the context of anthropogenic climate change and energy transitions.

Myanna Lahsen's interdisciplinary scholarship focuses on the politics of knowledge about the global environment and on how to achieve transformations toward sustainability. She has held research positions and professorships in Europe, USA, and Brazil. She is currently Senior Researcher at the Brazilian Institute for Space Research (INPE).

Tamar Law is a PhD student at Cornell University in Development Studies and holds an MPhil in Human Environmental Geography from the University of Oxford. Her research in the United States and Southeast Asia examines the knowledge and land politics of climate adaptation and mitigation, centering questions of climate justice.

Eloisa Beling Loose is in the Department of Communication at Federal University of Rio Grande do Sul, Brazil. Her research focuses on the intersection of environmental journalism with topics such as risks, disasters, and climate change.

Sean Low is a researcher at Aarhus University, and the Copernicus Institute for Sustainable Development, Utrecht University. He explores the politics of knowledge-making in climate and global governance.

Martin Mahony is a lecturer in human geography at the University of East Anglia, UK. His research focuses on the histories and geographies of the environmental sciences, with a particular focus on climate change, colonialism, and the science-policy interface.

Clark A. Miller is Professor in the School for the Future of Innovation in Society and Director of the Center for Energy & Society at Arizona State University. His research focuses on the human design of energy transitions and energy futures.

Robert Næss is Associate Professor of Science and Technology Studies at the Norwegian University of Science and Technology. His research focuses on transportation, mobility, climate adaptation, just transition, and sustainability.

Jeroen Oomen is Assistant Professor at the Urban Futures Studio, where he focuses on the social, cultural, and scientific practices that create societies' conceptions of the future. Coming from an interdisciplinary background, Jeroen's main interest is in how knowledge is made. His main research interests are climate policy, geoengineering, and social theory, specifically where it concerns questions of sustainability.

Roopali Phadke is Professor of Environmental Studies at Macalester College, USA. Working at the interface of political ecology and STS, she investigates issues including hydropower, wind energy, and mining and examines how participatory planning and design techniques build social acceptance, promote justice, and produce locally relevant meanings that root infrastructural technologies to peoples' sense of place.

Marianne Ryghaug is Professor of Science and Technology Studies at the Norwegian University of Science and Technology. Her research focuses on sociotechnical transitions in the areas of energy and transport. Her research interest is in the interface between public engagement, innovation, and climate policy.

Mike S. Schäfer is Professor of Science Communication at the University of Zurich and director of the university's Center for Higher Education and Science Studies. His research focuses on climate change communication in legacy and social media, as well as on public perception of climate change and science-related topics.

Stefan Schäfer investigates the political, ethical, and epistemological dimensions of planetary sciences and technologies. A research group leader at the Research Institute for Sustainability in Potsdam, Germany, he teaches at Humboldt University Berlin. As a member of LiCo collective he makes films, essays, and installations that examine past and future choreographies of mental and environmental life.

Julia Schubert is a postdoctoral researcher at the University of Speyer. She is a sociologist working in the fields of STS and science-policy studies. Her research explores the interrelation of science and politics, with a special focus on notions of expertise.

Tomas Moe Skjølsvold is Professor of Science and Technology Studies at the Norwegian University of Science and Technology. His current interest combines an interest in accelerated and just transitions, including the tensions and controversies that arise from competing goals.

Karolina Sobecka is an artist, designer, and researcher examining social arrangements that exploit, resist, or accommodate technological change. Sobecka has taught at SAIC, RISD, SVA, and NYU, and has been a visiting scholar at Scripps Institution of Oceanography.

Andy Stirling is Professor of Science and Technology Policy in the Science Policy Research Unit at Sussex University where he co-directed the ESRC STEPS Centre, working on politics of uncertainty and diversity in science and society. A fellow of the UK Academy of Social Sciences and former campaigner and board member for Greenpeace International, he's served in many policy advisory and academic evaluation processes.

Marcus Taylor is Professor and Head of Department in Global Development Studies, Queen's University, Canada. He has researched and published widely in the fields of climate change adaptation and rural development with a focus on livelihoods and labor in southern India.

Mark Vardy is a faculty member of the Criminology Department at Kwantlen Polytechnic University, Canada. He is interested in drawing from STS to contribute to discussions of climate justice in green criminology.

Sarah E. Vaughn is Associate Professor of Anthropology at the University of California, Berkeley. She has conducted research and written on climate adaptation on the intersections of technoscience, climate adaptation, and vulnerability throughout the circum-Caribbean.

Xiaoyue Yan is a PhD candidate at the Institute of Communication and Media Research at the University of Zurich. Her research focuses on visual communication of climate change and computational communication science.

Steven Yearley is Professor of Sociology of Scientific Knowledge at the University of Edinburgh in Scotland. He studies the making of environmental knowledge, by scientists, public officials, campaigners, and citizens.

Stephen Zehr is Professor Emeritus of Sociology at the University of Southern Indiana, USA. His past research focused on climate change scientific expertise and its representation in the media. He is currently researching maple syrup producers and their adaptation to technological changes, climate change, and labor supply and allegiances.

Acknowledgments

The editors acknowledge financial support for Open Access publishing from the U.S. National Science Foundation Programs in Science and Technology Studies and Ethical and Responsible Research (Grant # 2145997), University of Southern Indiana Office of the Provost, and Sonoma State University Library.

Zeke Baker acknowledges the helpful feedback from Siobhan Rodriguez and Matthew Bivens-Sommerville for their work as student research and editorial assistants. Baker also acknowledges support from the Sonoma State University School of Social Sciences and the Office of the Provost for their support in allowing me a course release to see this project to completion.

Mark Vardy acknowledges the support of the 0.6% Professional Development Fund from Kwantlen Polytechnic University, which funded a course release to work on this book.

Stephen Zehr acknowledges the helpful feedback from the following student workers: Abigail Burton, Chase Debruyn, and Cole Kneer.

Finally, the editors together acknowledge the hard work of all contributing authors. We appreciate the spirited effort to collectively bring a diverse and wide range of cutting-edge research to non-specialized student audiences. We trust this effort will serve instructors, students, and next-generation critical thinkers and scholars as they work to understand, grapple with, and address climate change.

Introduction

Zeke Baker, Tamar Law, Mark Vardy, and Stephen Zehr

What We All Know About Climate Change

Readers of this primer are already familiar with climate change. Global warming has been a fixture in national news in many nations for three decades. Most readers will have been taught its fundamentals in primary and secondary school years. Most readers will have learned more advanced knowledge at the university level. Except for diehard skeptics, we all know that it is an environmental problem with potentially disastrous consequences. We understand its history is linked to the industrial revolution's turn toward burning fossil fuels as a flexible, movable, storable, and powerful energy source, and the expansion of their use post-WWII in what is sometimes termed the Great Acceleration.

Most of us also understand that climate change is closely linked to global inequality. Per capita greenhouse gas emissions are disproportionately sourced in wealthy nations. We know that inequality from historically differential emission rates is still impactful since greenhouse gases remain in the atmosphere for a long time. But we also know that many nations in the Global South, especially those with high populations and going through rapid development, are catching up and in a few instances surpassing wealthy nations in total annual emissions. We know that while the effects of climate change are felt worldwide, they are particularly impactful to people in the Global South, and poor people everywhere, who face challenges of drought, floods, severe weather, and other climatic impacts with minimal capacity to effectively respond or adapt.

Readers who come to this primer with a natural science or engineering background will know that climate change has been extensively scientifically researched across disciplines, carefully constructed scientific reports have been written, and for several decades scientists have persistently warned of future environmental consequences. So why haven't we heeded those warnings? Most readers will understand the recalcitrance to act largely due to political inertia within and across nations. Our politicians have failed us. Due to skepticism of scientific expertise, ties to the fossil fuel industry, national economies deeply dependent on burning fossil fuel, cross-national challenges between oil-producing and non-oil-producing and richer nations and poorer nations, and so on, our political leaders have been unable to shift us away, quickly enough at least, from fossil fuels to blunt off at least the worst of climate change effects.

In sum, readers are likely well aware of these basic features of climate change, and most will hold deep knowledge of some of them. We are aware of the problem largely thanks to scientific research. We also know that the engineering and design communities have developed technological solutions that would allow us to move away from fossil fuels to renewable energy sources if only there is sufficient political and societal will to implement them. In sum, most readers will be familiar with the argument that the main sticking points lie with the lack of political will, along with a general public that lacks sufficient knowledge or is resistant to making necessary everyday life adjustments.

How Does STS Disrupt What We All Know: Key Questions for the Primer

So why is there a need for yet another academic book on climate change? Our answer is both simple and complex. The editors and authors in this primer approach climate change from an interdisciplinary perspective called **Science and Technology Studies (STS)**, sometimes also referred to as science, technology & society. STS has been around since the 1970s, involving researchers from the social sciences, history, philosophy, and some natural scientists, engineers, and architects and designers. Without too much detail, STS researches the social, political, cultural, and economic shaping of scientific knowledge and technology and how they are communicated and reshape societal infrastructure, institutions, structures, and everyday life. Climate change is only one of many topics of interest, but it receives much STS attention because of its importance as an environmental and societal issue.

What does STS add to what we already know about climate change? The primary and general point, and key motivation for this primer, is that STS opens up important questions that may challenge what we already know about climate change. It challenges and destabilizes some of the truisms contained in the first four paragraphs above. Now some readers might immediately react to this with the thought – "oh, just another book that attacks science and gives fodder to climate change skeptics and the fossil fuel industry." But please read on. Here are some of the key questions and points raised in this primer.

The Social Construction of Climate Change

STS understands climate change as a socially constructed phenomenon that could have been, and still could be, understood and shaped differently. Climate change is not a set, objective entity situated in nature. It is not just *one thing* though it is often represented as such. Rather, climate change evolved from a history of knowledge-making that centered it as *one* global environmental problem, albeit with many diverse effects in different parts of the world. This history of knowledge-making could have been otherwise. Does this mean that climate change as we know it is not "real" or involves "false truths," as some skeptics might claim? No. It means that climate change is open to being interpreted and understood differently and that these alternative narratives are not necessarily wrong. Climate change can mean many different things to many different people. This diversity of thinking about climate change is explored in Parts I and II, and throughout the primer.

Media Representations of Climate Change

STS understands that the information/knowledge flow from knowledge producers (typically scientists) through the media and to the public is far more complex than a simple linear process of accurate communication and careful listening to the scientists. We have all heard that popular request/admonition – "just listen to the science" – as it pertains to climate change and other issues. It makes a great deal of sense. We may feel morally smug because *we* listen to *the* science.

STS unpacks communication complexities that make just "listening to the science" overly simplistic and less meaningful. Climate change itself is heterogeneous and there are many ways that knowledge about it can be framed for different reasons and for different publics. There are diverse types of spokespeople and organizations that present "the science" and they do so in diverse ways, highlighting, downplaying, or ignoring issues depending on their interests and what they think their audience needs to know or wishes to confirm. The result can be mixed messages. The media are not just a standardized filter through which scientific information flows, but diverse sets of agents and technologies for choosing, adjusting, and adapting it for their audiences. Also, different publics bring their own knowledge, values, and interests to climate change, making them not only selective consumers but also active shapers of what climate change is. This complexity is explored in detail in Parts III and IV in this primer, as well as in other locations.

Social Movements and Climate Change

Media are not the only way in which understandings of climate change are shaped. Social movement organizations and environmental non-governmental organizations (NGOs) are also important – indeed vital – when it comes to shaping the public understanding of climate change. Environmental NGOs enact climate change campaigns in ways that borrow from other social movements such as the Civil Rights and Women's Rights, and Gay Liberation movements that emerged in the 1960s and 1970s to demand justice. That is, we often see environmental NGOs staging protest rallies, marches, strikes, and teach-ins to demand action on climate change. At the same time, conservative and libertarian cultural movements, which argue against governmental regulation of the economy, are contributing to alternative understandings of climate change held by many people. Both the conservative and environmental movements profoundly impact public understanding of climate change, and interestingly, both of them often refer to science in a similar way. That is, both movements tend to treat science as a unified body of knowledge, the credibility of which is dependent on adhesion to a single scientific method. For environmentalists, this model of science lends authority to their claims ("just listen to the science"), while for others any deviation from the supposed purity of this model is given as a reason to discredit all of climate science. STS, in contrast, rejects the idea that science is a unified body of knowledge that can be distinguished from other forms of knowledge due to its adherence to a singular scientific method. The chapters in Part IV take this perspective to explore in more detail the relations between environmental NGOs, the conservative and libertarian movements, and public understanding of climate change.

Climate Change Inequality and Justice

STS expands and poses new questions about climate change inequality and justice. There is more complexity to these issues than the historical inequality across nations in greenhouse gas emissions and the fact that low-income nations and poor people in general face more damaging climate change impacts with less resilience to cope. To these valid and important concerns, STS poses questions about inequality in access to and ability to mobilize scientific and other knowledge about climate change. It also examines the justice implications of how climate change is addressed on the ground, through adaptation and mitigation efforts. STS poses questions about the unequal authority of scientific and Indigenous and other local knowledge about climate, its changes, and impacts. STS also opens questions about how we actually conceptualize inequality and justice in the context of climate change. What are the dimensions of environmental inequality and how do we know them when we see them? What forms might "climate change justice" take and which ones could be practically reached? Importantly, STS also opens questions about the means through which climate change inequalities can be addressed by introducing new methods for public engagement and for integrating inequality and justice concerns into imagining climate change futures. Part V of this primer focuses specifically on matters of climate change inequality and justice, but one also finds these issues infused in chapters throughout the primer.

Climate Change Governance and Expertise

The linear model of science for policy informs many representations of climate change. In the linear model, scientists first determine the truth of a given phenomenon, and then politicians and policymakers act upon the knowledge thus generated. This perspective can be seen in the challenges voiced by Greta Thunberg, who many of us admire, when she admonishes world leaders for failing to take immediate action when scientific warnings are so clear. STS challenges the linear model by opening up questions about the structure and operation of advisory bodies, scientific experts, and expertise. Instead of assuming that science and policy exist, or should exist, in two separate and distinct spheres, STS asks questions about how boundaries between scientific advice and political decision-making are drawn in the first place.

In Part VI, chapters explore the complexities of climate governance from the city scale to the international. Different questions crop up at each of these levels. For example, how can cities transform their governance practices from historical concerns, such as economic growth, to becoming climate neutral, or how does the Intergovernmental Panel on Climate Change (IPCC) differentiate between science and national and international governance bodies? What role does trust play in scientific expertise, and why might countries in the Global South mistrust science that is rooted in the Global North?

Climate Change and Sociotechnical Transitions

How do we actually reduce greenhouse gases and adapt to climate change? In social science, natural science, and engineering disciplines, as well as in public and political circles, there are sometimes polarized views about whether technological changes or social and cultural changes should lead the way. With the former, human ingenuity is called upon to develop new technologies and retrofit old ones to wean societies from burning fossil fuels and to build resilient infrastructure that can withstand climate change. There are many examples of efforts in these directions. With the latter, lifestyle and social structural changes are expected to save the day. People, especially in wealthier countries, must adjust their culture, social structure, and societal institutions to radically reduce dependence on fossil fuels, while also shifting resources to more vulnerable people globally, enabling them to develop their societies in environmentally benign ways and adapt to climate changes. For the latter, dependence upon technological development – "technical fixes" to societal problems – may solve one problem but open others due to unanticipated consequences. Technological optimists, on the other hand, assert that it is naïve to think that entrenched human societies and culture can adjust rapidly enough to prevent the worst climate change problems.

STS disrupts this polarity, not by claiming a middle ground, but very simply by analytically combining the technical and the social. For STS scholars, technical changes *are also* social changes and vice versa. For STS they form hybrid arrangements, so it makes little sense to try to separate them out. STS employs the concept "sociotechnical" or one of its variants (used throughout this primer) to refer to how so-called technologies are also social entities with the capacity to act upon, be impacted by, and resist other social and technical forces. STS questions and demonstrates how movement to renewable energy sources, for example, requires inseparable technical, social, political, cultural, economic, and so on, changes that are often resisted, complex, open to failure, have unpredictable consequences, but are potentially transformative. We refer to this process as sociotechnical transition. Different aspects of sociotechnical change involving energy transition are addressed in Part VII of the primer and to some degree in Parts VIII–XI.

Art, Infrastructure, and Design

An enduring impact of STS on other academic disciplines is its insistence that the social sciences and humanities analyze objects and processes once historically reserved for the physical and life sciences. From feminist philosophies of science, for example, we learn that knowledge is always developed through the specific ways in which humans are embodied in the physical world, not through – as much Enlightenment thought presupposed – the ability of humans to cut their ties to it. STS' impact on other disciplines' approaches to climate change can be found in Part IX. The central presupposition shared by these chapters is that climate change challenges humans to rethink how we live, and that art – and critiques of art – are a vital way for reimagining how this can be individually and collectively accomplished. STS is used to engage with other philosophies such as phenomenology to articulate connections among increasing levels of atmospheric carbon dioxide, extreme weather events, and urban stormwater infrastructure designed for outdated climatic norms. But just as importantly, STS connects these elements with the social imaginaries that guided the growth of cities as centers for the accumulation of capital in the first place. Art is a way of depicting these connections visually, and STS is a way of talking about them.

There is an additional element of speculation or mental experiment that chapters in Part IX ask us to engage in. For example, what happens if we think of clouds and atmospheres as media? This may seem strange or counterintuitive at first – after all, aren't media things like movies, newspaper articles, and Instagram feeds? But STS provides a way to see the atmosphere as a giant canvas on which industrialized humanity has inscribed its mark. Once so inscribed, the atmosphere becomes an active agent, authoring events that impact humans. In other words, whereas we normally think of the atmosphere as "out there" or "in nature" separate from humans, and which we can know objectively through science, Part IX encourages us to think how the atmosphere is already a part of who we are.

Climate Engineering (Geoengineering)

What about some of the worst scenarios where we are unable to gravitate away from fossil fuels quickly enough and face calamitous effects of climate change? Are climate engineering projects (also referred to as geoengineering) potential solutions, or are these the worst form of technical fixes noted above? One also finds polarized positions on this issue from social science and natural science scholars. For some (environmentalists and many scholars), even climate engineering *research* is envisioned as creating a loophole for avoiding the more difficult challenge of weaning ourselves from fossil fuels. Moving down this path allows political decision-makers to hesitate, thinking that perhaps some major climate engineering technology will sufficiently remove greenhouse gases from the atmosphere or reduce solar radiation reaching the earth's surface to avoid (expensive) policies that their constituents will oppose. Opposition to climate engineering research also recognizes the likely outcome of unfavorable unanticipated consequences that large technological projects often bring. Those in favor of pursuing climate engineering research, on the other hand, often see it as a Plan B that can be drawn upon should things get out of hand. It's only research after all. We will still hold the option of implementing the technology or not. Relatedly, some scholars argue that it will be impossible to reach targeted concentrations of carbon dioxide equivalent (CO_{2e}) without implementing some climate engineering technologies. Research now will position us with the capacity to choose the most effective technologies with the fewest likely unanticipated consequences.

STS scholars hold a more nuanced perspective on climate engineering. They understand that there is an important history to technological projects to control climate and the weather, dating at least to post-WWII cloud seeding research, and how this history has shaped perceptions of climate engineering research today. STS analyzes the above polarized opinions in light of this history, emphasizing their limitations as discourses. STS addresses the politics of climate engineering, both as an empirical project and to help formulate ethical guidelines for less risky and more just implementation of the technologies should that decision be made. STS researchers analyze the potential implementation of climate engineering technologies, understanding them as sociotechnologies that combine technical, social, political, and cultural components. Along this path, a goal of STS research is to anticipate the unanticipated consequences such that wiser decisions are possible earlier in the development of the technology. Part X of this primer focuses on this research.

Climate Change Futures

Climate change has a past and present, but it also has a future. While much of the past and present are extensively researched with knowledge readily available, climate change futures are more uncertain and pose significant challenges to (especially social science) researchers. From environmentalists and often the scientific community we receive doomsday scenarios, with the caveat that if we act immediately, we can prevent the worst outcomes. Technically, these scenarios take shape in climate change models that predict future warming under different future greenhouse gas emission scenarios. These scenarios range from "business as usual" yielding very stark global warming outcomes to a scenario where strict emission reductions have been implemented in line with the Paris Agreement or other national commitments.

STS critiques climate change modeling research and scenario development for their inability to integrate social and political constraints and opportunities and their singular, global predictions when climate change futures will be experienced locally and differentially by people worldwide. STS emphasizes the need for multiple pathways for building climate change futures and offers a protocol for facilitating and judging them. These pathways typically expand beyond emissions reductions and basic adaptation to address broader sustainable development goals of building more equitable societies, improving human well-being, while also protecting the environment. STS employs the concept of *sociotechnical imaginaries* to describe the methodological process and outcomes of doing this future building work. Rather than a technical procedure like scenario development in climate change modeling, developing sociotechnical imaginaries is more democratic and localized, recognizing that people at local levels necessarily need to be involved in building community resilience to climate change and adjusting away from fossil fuel-dependent daily lives. The emphasis in STS research is on the plural – pathways, rather than one dictated pathway or limited set of future outcomes that emerge from scenario building based on climate change models. STS scholars also have collaborated with artists, designers, and architects in their thinking about designing infrastructure and using art and other aesthetic ideas for building more sustainable futures. These concerns are addressed in Parts VIII, IX, and XI in the primer and are also addressed in chapters in Part VII.

History and Suggested Uses of the Primer

The idea for this primer emerged from discussions at a workshop in June 2022 funded by the U.S. National Science Foundation (Programs in Science and Technology Studies and Ethical and Responsible Research). The workshop, held at the University of Southern Indiana, centered around STS research on climate change. The four editors of this primer took the lead, soliciting chapters from workshop attendees and other STS scholars researching climate change. Funding from NSF (Award #2145997) and from the University of Southern Indiana and Sonoma State University enabled Open Access publication of the full primer.

The model for this primer is based on the American Sociological Association Journal *Contexts*, which publishes cutting-edge research, policy-oriented pieces, and essays written to be accessible to and integrated within sociology undergraduate courses. The editors have aimed for primer chapters that discuss focused STS research projects or that review an area of STS research on climate change. This book is not a textbook and is not intended to provide a comprehensive overview of STS or social science research on climate change. Rather, the primer is more selective and specialized, highlighting key topical areas of research that can be used as depth pieces or examples of broader topics raised in social science or natural science climate change or environmental studies courses. Chapters might be selected and used individually as they relate to specific course topics, or the primer as a whole can be easily integrated into a course.

Chapter authors were asked to minimize their use of references and academic jargon. Readers will therefore find fewer references in chapters than normally found in peer-reviewed academic articles. We did this to improve readability, but also recognize that each chapter owes credit to a larger body of research. To support readers in understanding and defining theoretical ideas, key concepts and theories have been bolded with their definitions close to their initial introduction. The reader will find some repetition across chapters of bolded concepts in order to reinforce them, while also acknowledging and embracing that individual concepts can be put to a range of uses across topical domains. At its conclusion, most chapters contain a selective set of further readings separated from the reference list. These readings are judged by authors to provide the best introduction to research topics contained in the chapter and are a good starting point should the reader wish to pursue their own research project on that topic.

Part I

Climate Change Science as a Social Issue

Introduction

Zeke Baker

Part I of the book is designed to highlight what it means to treat climate science as a social phenomenon. Of course, we know that scientists, engineers, lab technicians, administrative assistants, and others that perform important roles in the scientific process are *people*. We know that these people are trained, that they inhabit certain fields, and that they are shaped in basic ways by their social context, time, and place. When working together, they need to communicate and coordinate their activities in order to do what scientists do to make knowledge. So, clear enough, scientific processes are simultaneously social processes. Thinking further, we know that somehow big ideas (like "natural selection" in biology, "plate tectonics" in geology, or the "greenhouse effect" in climate science) appear to stand *outside* the people that helped create them.

Despite this basic, common understanding about how science works socially, it is nevertheless a common belief that true knowledge can transcend the people and messy processes that made it up. The driving question for Science and Technology Studies (STS) is how to explain the scientific process in terms of its social dynamics, rather than through an undue appeal to the idea that science works by nature – from species to molecules to the climate – revealing *itself*.

It is important to account for and understand the social dynamics of knowledge production for two basic reasons. First is the fact that the history of science, and all areas of contemporary science, are riddled with controversies, open questions, puzzles, and new directions. How do these processes work? STS methods address this question by retooling approaches in the social sciences to study the people, things, and language involved in all this fervor going on amongst scientists and those connected to them. STS tries to explain outcomes in, generally speaking, social terms. It therefore treats critically the idea that science, when done correctly, allows the people, time period and places to fade away.

A second reason for wanting to understand the social dynamics that characterize knowledge production is that science is clearly related to other areas of society, and that these relations are highly consequential. Few aspects of contemporary life are untouched by science/technology. As natural as the products of science may seem, they are not inevitable and must be analyzed as the result of human decisions. For one example, governments and corporations fund science labs and university research centers, and funding decisions are hardly made in a vacuum. They can facilitate or cast aside investment in developments like cancer treatments, atomic weapons, artificial intelligence, or explanations of extreme weather events. As another example, scientists and experts may engage social issues that connect to their expertise, giving credence to a political claim or social position. That position could be about how to structure a health care system that protects public health, how to improve education by knowing how children's brains develop, or what actions might be the cheapest ways to reduce climate change impacts.

DOI: 10.4324/9781003409748-1

When it comes to climate change, it is especially important to account for how society, science, technology, and climate shape one another. The open questions and new directions within climate science may have significant consequences, say, for setting international policy targets aimed at mitigating global warming, or for proposals to geoengineer the planetary atmosphere (or not). Somewhat more abstractly, STS raises questions about the general orientation, paradigms, values, and purposes that characterize climate science in a given time and place. How are not-exactly-scientific concepts, such as sustainability, justice, equity, democracy, accountability, or control, built into climate research and its social impact? Likewise, and more pertinent to this Part, how does the social structure and historical context of current climate science matter for how global warming is known and addressed today?

The three chapters that comprise this Part of the book are designed to introduce readers to ways of thinking critically about these questions at the nexus of science and society. In Chapter 1, Baker sets up a comparison between basic ways of thinking about climate science: is it a body of facts and claims, or instead a "field of social practices?" By drawing upon his empirical research, Baker highlights how tracing the "social life" of the field helps us (1) assess the objectivity of research, (2) explain changing orientations of climate research programs, and (3) question how climate science relates to governmental priorities, especially in the context of anticipated major climate disruptions.

If Baker's chapter provides a way of thinking about the social life of climate science, Mahony's work (Chapter 2) helps situate climate knowledge historically. "Climate change" as we know it today, carries a historical trajectory that pre-dates the entry of global warming into scientific, political, and public consideration and debate. Rather than centering an account of climate change science on concepts like scientific discovery and consensus formation about global warming in particular, Mahony argues that sciences about the global atmosphere stretch farther back in time and connect to a much wider array of scientific and political issues. Of special focus here is the "intertwined histories" of European colonialism and environmental/climate sciences. Mahony not only argues that colonial and imperial formations helped give rise to global climate science, but also that the legacy of this intertwined history remains with us today.

In Chapter 3, Callison neatly follows up on Mahony's argument that contemporary climate knowledge is a fundamentally *historical* and *imperial* formation that risks being reproduced as such. The history of knowledge about climate and the material realities of climate change, Callison shows, must be considered together and centrally acknowledge the colonial context. Callison especially guides us toward a reconsideration of a basic framing of global warming today, namely as a *climate crisis*. Callison builds an Indigenous-centered critique of the notion that global warming, as a crisis, represents a novel break from the past. By engaging with other Indigenous scholars and perspectives, she argues that "climate crisis" may represent less a novel break with history and more a *continuation* of longstanding formations and struggles. Why center colonialism? Callison connects colonization to climate change by pulling out the links between centuries of colonization and the material-economic transformation of the planet that has included global warming. She also examines the social marginalization of Indigenous knowledges (within mainstream science and in society) to suggest that alternative ways of building climate knowledge and action very much remain with us, and indeed may be critical to addressing a climate-impacted future.

1 Future Times and Spaces

Tracing Objectivity, Scale, and Politics in the Social Life of Climate Science

Zeke Baker

Introduction

According to recent survey data (Marlon et al., 2022), an increasing majority of Americans believe the following statement: "global warming is already harming people in the US now or [will] within the next 10 years." The national average on this figure changed from 51 percent to 64 percent over just three years, from 2018 to 2021. It is not difficult to interpret this finding as indicating growing public acceptance of scientific facts. It seems people believe the findings of climate science to be increasingly credible and salient to their own lives.

Climate science, in this interpretation, is primarily a matter of claims, in this case, the claim that the planet is warming and making visible impacts. Such claims make their way to people primarily through textual and visual representations, for example graphs about trends in global temperature and maps about disaster risks associated with climate change.

Rather than consider climate science as a matter of claims and representations, I would like to argue in this chapter that climate science should be understood as a social field of practices. In this view, what comes into focus is the **social life** of climate science rather than what may appear as stand-alone facts (cf. Nelson, 2016). By social life, I mean to highlight how climate science is occupied by people situated in time and place who creatively engage one another, along with technologies and physical phenomena, to help produce what we come to understand as the climate. What might this shift in focus from claims to social life provide?

This chapter lays out three lines of argument to answer this question. First, it interrogates **objectivity** in climate science. By objectivity, I mean the basis upon which science can be evaluated with the criteria of truth and sound knowledge. If knowledge rests upon social interactions (rather than the inherently correct facts revealing themselves), then it stands to reason that the people that make up such interactions matter for how sound knowledge is achieved and evaluated. Second, it helps to explain some recent changes going on in science. Innovation is a poorly equipped concept to account for such changes, and it may be best to focus on explaining how and why climate scientists orient and reorient their work and attention toward (or away from) particular problems. Third, it brings into view the implications of scientific change for other parts of society. I will emphasize the social institution of *the state*.

Strong Objectivity in Climate Science?

A dominant image of climate change is that it is both a global phenomenon and can be known through a universal, objective science. (Even avowed climate skeptics claim to reject mainstream climate science because they allege it reflects "bad science," that is to say, it is not objective.) This dominant image is not an appropriate representation of how things actually

DOI: 10.4324/9781003409748-2

are. The image in STS is a bit more complex. In this section, I use standpoint epistemology to **provincialize** climate science, or recognize the situated practices that make it up. I argue that doing so provides a more robust and realistic version of objectivity, and better acknowledges diverse ways of knowing something as broad as climate. I show how this plays out in the Arctic region and in the case of Indigenous peoples.

Standpoint epistemology is an approach to science that recognizes the socially situated nature of all subjects of knowledge (that is, *knowers*) and the social construction of all objects of knowledge (that is, the things *known*). It holds that objectivity, or what Sandra Harding (2008) calls **strong objectivity**, is maximized when socially marginalized people are systematically valued, incorporated, or even privileged, in the process of making science.

Researchers have employed this approach to challenge basic prevailing social science concepts. Take for example the concept of "modern," "modernity," or "modernization." These concepts can be provincialized (as opposed to universalized) by recognizing how distinctions between modern/non-modern (or civilized versus non-civilized) are invented European categories that took root amid colonialism's global reach. Taking what sociologist Julian Go (2016) has called the "Southern standpoint," composed of marginalized perspectives rooted in experiences and categories of thought among those positioned in, roughly speaking, the global South, can shed light on just how partial, non-objective, and provincial the category of "modern" has been. Only by systematically incorporating marginalized standpoints (for example, colonized or Indigenous groups), the approach contends, can we comprehend global society with stronger objectivity.

Might it be possible to interrogate and build global climate change science in a parallel manner? Provincializing climate science has dangers and opportunities. The danger is the neglect of the planetary scale, which remains significant for projects to measure, explain, regulate, or otherwise address global warming. Does not global warming call forth, and require, a universal planetary standpoint? Are not the monumental achievements of Earth System Modeling a testament not just to the possibility of, but further, the realization, validity, utility, and promise of objectivity through global physical modeling? In a way, yes. However, even global knowledge is achieved in local contexts. So, the opportunity that comes with provincializing global climate knowledge is the deeper recognition that global climate science is at its best because of its *social* achievements—secured through communication, international collaboration, dialogue, and controversy across a range of inherently place-based practices. So, the "God trick" (Haraway, 1988) that would view climate from a universal perspective (or from nowhere) likely remains an overstatement of the nature of global knowledge.

How might climate science, to be strongly objective, further incorporate marginalized standpoints? In part, such incorporation has been realized through non-Western local knowledges, for example, traditional ecological knowledge (TEK, see Callison, this volume); the development of climate change adaptation planning with public input; work in climate risk communication and outreach; and efforts to diversify climate science and climate policy fields along lines of nation, gender, race/ethnicity, culture, and scientific discipline.

Although climate science has broadened in disciplinary and international terms, one disturbing trend is the instrumental use of marginalized standpoints to advance dominant approaches to climate science. Let us consider this concern using the example of climate science in the Arctic, which I know best and is a useful case because the region is warming roughly twice as quickly as mid-latitude regions, thus making climate science and policy immediately relevant. On the one hand, the Arctic Report Card (Moon, Druckenmiller, and Thoman, 2021) is a brilliant achievement of knowledge production. It presents regional data annually in accessible terms, with an emphasis on synthesizing knowledge so that it is usable by a range of stakeholders. It increasingly acknowledges the role of traditional knowledge holders in making climate science, and in making science meaningful to impacted communities. Although in limited ways, it is a scientific text that recognizes diverse voices and speaks back to diverse audiences. On the other hand, within the field of Arctic knowledge, there is not a general problematization of the dominant standpoints, particularly those that divorce scientific concerns from political concerns regarding just what is to be done about global warming impacts. Take for example recent criticism of the U.S. National Science Foundation's major new Navigating the New Arctic program. Some indigenous scholars have criticized the program's call for "co-production," indicating that it may turn indigenous knowledge holders into vessels of TEK, without sufficient incorporation of these communities into the downstream uses of this knowledge and the purposes that it may ultimately serve (Yua et al., 2022). For example, it is not difficult to imagine the utility of indigenous Inupiaq whalers' observations and understanding of Arctic sea ice dynamics to construct physical science models of sea ice, which are then incorporated into global climate change models. (The Arctic ice cap and seasonal ice dynamics play a significant role in global temperature and a range of hemispheric and global geophysical dynamics.) Such models furthermore have a wide range of uses, including for example those of the Arctic transport and fossil fuel drilling sectors seeking to predict the future of an ice-free Arctic for the purposes of financial investment in further oil/gas exploitation. Indigenous livelihoods may be threatened as a result.

To the extent climate science is a field of practices and actors embedded in institutions, it is clearly configured with these economic and geopolitical aims that may or may not be of interest to those invested in indigenous livelihoods. Therefore, in the case of Arctic science, the increasingly mainstream discourse of "**co-production**" (which builds knowledge by scientists partnering with specific, interested communities) presents a mixed bag. Strong objectivity would recognize and interrogate the ways these larger institutions are shaping how knowledge is produced, represented and used. In doing so, it would resist turning non-dominant knowledges into simple instruments of a foreign project, and instead create dialogue in a way that effectively negotiates the social life of Arctic sea ice science. Undoubtedly this would strengthen climate science in the region.

In the Arctic and elsewhere, there is more to accomplish in reconstructing the place-based achievement of global knowledge and some of the possible tensions or contradictions therein. However, this is not the only important direction for STS research, because global knowledge is only part of the social life of climate science. Of special importance is changing orientations toward sub-planetary scales and refined treatment of climate futures across a range of timeframes.

Scientific Change and Scale

In addition to the gains in objectivity from provincializing climate science, there are other reasons to think about climate science at a smaller, non-planetary scale. Given how climate science has developed in the twenty-first century, it is simply inaccurate to consider climate science as invested primarily in planetary-scale phenomena. Downscaling global models, mapping of regional climate change impacts, assessing climate risk and vulnerability, and conducting **event attribution** analyses, which relate specific weather events to global warming—all these developments mean that climate scientists, and the science they produced, are concerned with regional, rather than global, patterns. Here, social studies of climate science bring their strength by following more closely the dynamics of the climate science field: what explains this "rescaling" of climate, and more importantly, what are its effects? In this section, I will answer these questions, again drawing from my work in the Arctic context.

Let us begin with a positive, practical case of how incorporating STS expertise into meteorology (specifically, public weather forecasting and climate services) at local scales can responsibly address social needs and advance equity in the face of climate change impacts. For context, operational meteorology in the U.S. is presently organized in government through the U.S. National Weather Service (NWS). My focus here is on how NWS forecasting can address socially relevant time frames ranging from days to months to decades. Among weather forecasters, timescales like 24-hour, 5-day, and one-month periods are difficult to link together from a meteorological standpoint. A weather forecaster's predictive skill will stop when inter-model comparison makes weather forecasts more like noise than knowledge. Longer range (e.g., seasonal and beyond) outlooks can represent probabilistic trends, but with a questionable sense of whether or not a highly uncertain outlook will be useful to people. (A field of tailored, decision support services, called DSS, has arisen to confront this problem.) In this context, STS scholars can patch together the perspectives that collectively make "future weather" meaningful, thus enriching what sorts of knowledge are possible and desirable.

In the Alaskan Arctic context, where I conducted fieldwork in 2020, of keen interest to many is the fishing season for various commercial and subsistence fisheries. Such a season is determined not only by a calendar but also by regulatory and economic pressures, availability or abundance of the target species, and weather patterns and conditions that render marine activity safe, risky, or impossible. Based on my research with mariners in the Bering Sea, complex rituals structure how mariners evaluate weather forecasts, consider risk, and make decisions. Furthermore, marine-dependent communities express deep uncertainty regarding their livelihood, which may be tied to specific animal species, the populations of which are undergoing rapid shifts because of the changing Bering Sea ecosystem. The timescales of everyday weather, season (broadly construed), and climate thus interact to shape mariners' courses of action. Yet, meteorologists poorly understand the processes through which weather and climate information come to matter for mariner decision-making. These scientists and information providers, indeed, may never have been to the communities they serve. Analysis of the social life of weather and climate science permits us to see scientific information with reference to the range of other factors influencing how people go about anticipating the future. This process of reconstructing what I have called "anticipatory culture" can improve weather forecasting and the value of weather forecasts to people (Baker, 2021). In the case at hand, it will help them fish more safely and think about long-term economic and livelihood strategies. Such an approach to following the rescaling of weather and climate information provides a way to consider climate change impacts across multiple timescales, especially among communities underserved or poorly understood by climate science.

Despite helping communities anticipate the future, are there perhaps some problematic social outcomes regarding the novel capacities to predict and prepare for disasters at smaller geographic scales? To address this question, we need to ask who gets to use this science and who doesn't, and for what purposes. To assess aspects of these questions, meteorologist Friederike Otto and colleagues (2020) performed a study of event attribution in cases of extreme weather events and found stark inequalities in how climate science is rescaled and refined across countries. They identify "discrepancy between where attribution studies are conducted and where the largest damages associated with extreme weather events are," resulting in "a systematic (selection) bias in attribution studies toward focusing on places with lower vulnerability." So, wealthier countries with greater ties between science and government have greater capacity to mobilize attribution science to support forecasting and adaptation efforts. By contrast, poorer countries face greater vulnerability to extreme events *and* have less scientific capacity to forecast and prepare for such events.

What the cases of event attribution science and operational meteorology on Arctic seasonality show is first that climate science is changing, specifically by scaling to national, regional, and local contexts. Scientists are likewise working across timescales. We might call this innovation. However, the key takeaway is that the social life of climate science is necessarily tied up with other social institutions and interests. It may therefore "rescale" in a way that responsively solves the needs of some groups (in the case of refined forecasting techniques), while also generating new forms of social inequality (in the case of unequal access to event attribution and related novel directions for climate science).

How Climate Science Articulates Power

Weather forecasts and climate services form one way in which climate science relates to, or is provided by, the government. These are largely within what sociologist Pierre Bourdieu has called "the left hand" of the state, which invests in citizen well-being by responding to popular demands for welfare and services. We can also follow the social life of rescaled climate science over to the "right hand" of the state. This climate-impacted hand of the state centrally features issues of territoriality, economic and military/geopolitical hegemony, social control, national security, and the political demarcation of what Buxton and Hayes (2015) discuss in the context of climate crisis as "the secure and the dispossessed." How are recent trends in the social life of climate science implicated in state-making projects that aim to govern climate in these terms?

An initial answer begins by recognizing that a major shift in efforts to govern climate change occurred around the middle of the first decade of the twenty-first century, when social actors converged through efforts to know and govern climate change in its *effects* (e.g. by preparing for disasters) and not only in its causes (e.g. by finding ways to reduce greenhouse gas emissions). A governmental logic had come to center on strategic anticipation of and preparation for impending climate risk, perhaps even catastrophe. Although social movements like Extinction Rebellion have more visibly framed climate change with reference to catastrophe, the U.S. national security state is also active on this front. Yet, how did the logic of *national* climate security form, given that environmentalists and scientists had long emphasized climate change as a quintessentially *global* environmental problem? Following the social life of climate science helps put together how scientific developments at the national and regional levels have come to inform a whole new arena for connecting the national state to climate science.

To understand what climate security looks like, let us turn to the rise of climate security experts and the rise of security technologies. **Climate security expertise** comprises the actors and expert organizations that are oriented toward rendering catastrophic climate futures governable in the present. **Security technologies** can be defined as the array of scientific products, surveillance activities, and modeling techniques that represent future risks in order to facilitate strategic action based on anticipated political and military instability or widespread social dislocation.

Climate security experts, who in the U.S. context have arisen out of think tanks, the defense establishment, and the peripheries of climate science and policy, are emerging as one sort of prophet of a governable order amid impending crises. Despite the claims among climate security experts that their activities transcend the polarized context of partisan politics, the way in which future threats are considered, evaluated, and governed is hardly an apolitical exercise. Changes in climate-impacted patterns, like agricultural productivity, the frequency of high-impact disasters, water availability, coastal zone inundation, human displacement and migration flows, and conflicts featuring natural resources are considered in light of governmental priorities and visions of security.

Of special importance is the categorical distinctions made between, on the one hand, those who deserve protection and are to be politically incorporated into the body politic, and on the other hand, those who are to be excluded. Recent history is important to consider how future climate impacts might be governed in part through climate security expertise. Over the first decade of the twenty-first century, a militarized U.S. migration system was fortified along with an international War on Terror, just as a deeply politicized climate science led to climate policy failure. (In 2001, President George W. Bush renounced any intention to abide by the Kyoto Protocol, a treaty that would have required the United States to regulate and reduce carbon emissions.) A growing cadre of experts thus drew upon the security implications of climate change as a powerful framing to spur climate policy action. Given the politics of migration and terrorism, climate-migration and terrorist expertise have gained special relevance to those invested in governing climate change as a real-world threat. Thus, the 2014 U.S. Department of Defense Quadrennial Defense Review emphasized that climate impacts "are threat multipliers that will aggravate stressors," leading to "conditions that can enable terrorist activity and other forms of violence." Climate patterns themselves hardly determine such framings of the climate problem and their uptake in defense and security policy.

Let us turn to security technologies. Several scientific developments are central to how climate security expertise works. First, as introduced in the previous section, rescaled climate models at finer-scale resolution provide an important way for scientists to project how climate change will impact specific areas, resources, populations, and economic sectors. Second, scientists can then use regional modeling to attribute geographically delimited sociopolitical events to global warming. Regional climate modeling techniques, when utilized by climate security experts, can become a powerful technology through which to govern future security threats. Security experts have long recognized that national security threats entail high-risk, low-probability events, like a resource conflict transforming into an international military crisis. Yet, climate security experts have only recently used event attribution studies, adapted from meteorology, to anticipate such events. One example is Kelley et al.'s (2015) study, published in the *Proceedings of the National Academy of Sciences*, which evaluated the impact of global warming on conflict in Syria and subsequent migration. The topic—serving as a case of a "threat-multiplying" climate event—has received widespread attention among climate security experts in recent years. In the cases of the war in Syria or the Arab Spring, studies link drought, urbanization, and armed conflict as events that will likely become more common in the future because of climate change. Particularly significant in these regional climate studies is the relatively novel scientific capacity to isolate the impact of anthropogenic climate change on regional-scaled patterns and events—in this case, regarding Mediterranean drought. We see this elsewhere, too. In the Arctic, anticipated climate security threats include geopolitical tension around what states get to control or regulate Arctic shipping routes that are projected to open with an ice-free Arctic.

Such is the social life of climate security expertise. These developments in climate expertise can be engaged critically by assessing how they matter to various "hands" of the state. Although many changes in science and government are not so simple, it remains possible to ask: is the social life of climate science (1) facilitating situations of inequality and exclusion, or is it (2) building equity in the face of increasing climate risk?

Questions remain about where the field of climate security expertise is headed, and how it might gain prominence as the climate crisis becomes an operating assumption rather than a political challenge to state power. Will developments in climate science be successfully enrolled in climate security projects, or might they reorient to alternative institutions and values? If right-wing political parties disavow climate change denial and weaken their ties to the fossil fuel industry—a situation that would contrast sharply with recent decades—then would a new consensus consider climate security expertise and security technologies as the most important kinds of climate science amid deepening climate crisis? Answering these questions centrally involves following the social life of a fast-changing climate science, with a particular focus on how climate scientists and their tools are embedded in, and dynamically configured with, various hands of the state.

Conclusion

In this chapter, I have argued that climate science can be treated less as a body of facts and claims and more so a dynamic field of practice configured with a range of social institutions. This basic perspective permitted exploration of the objectivity of global climate science, an attention to the changes happening within climate science, and finally unfolding relations between climate knowledge and power. In this account, climate science claims and representations did not carry their own meaning or their own implied actions, and they did not emerge from an invisible hand of scientific rationality or innovation. The climate does not explain itself. Nor does the character of science. Rather, the dynamics of social practices come to shape what climate science "is," what it "does" or performs (and for whom), and how it becomes consequential for social institutions, including, in this case, the state.

This means we can have our cake and eat it too: First, we can advance objectivity—strong objectivity—regarding climate realities, specifically by diversifying the voices within climate science and reflecting upon the presumed values and goals that shape research and information. Second, we can critically engage or question novel directions within climate science while also seeking to improve existing science and its public uses. Finally, we can retain commitments to climate policy goals and social causes like climate justice while also committing to rigorous empirical analysis of how climate science articulates power relations. Such are some of the wide avenues opened up by STS approaches that follow the social life of climate science.

Further Reading

Dalby, S. (2022). *Rethinking Environmental Security*. Cheltenham: Edward Elgar Publishers.

McDonald, M. (2021). *Ecological Security: Climate Change and the Construction of Security*. Cambridge: Cambridge University Press. doi:10.1017/9781009024495

Yua, E., Raymond-Yakoubian, J., Daniel, R. A., and Behe, C. (2022). "A Framework for Co-Production of Knowledge in the Context of Arctic Research." *Ecology and Society* 27(1). doi: 10.5751/ES-12960–270134

References

Baker, Z. (2021). "Anticipatory Culture in the Bering Sea: Weather, Climate, and Temporal Dissonance," *Weather, Climate, and Society* 13(4), pp. 783–95. doi: 10.1175/WCAS-D-21–0066.1

Buxton, N. and Hayes, B., eds. (2015). *The Secure and the Dispossessed: How the Military and Corporations are Shaping a Climate-changed World*. Amsterdam: Pluto Press/TNI.

Go, J. (2016). *Postcolonialism and Social Theory*. London: Oxford University Press.

Haraway, D. (1988). "Situated Knowledges: The Science Question in Feminism and the Privilege of Partial Perspective," *Feminist Studies*, 14(3), pp. 575–599. https://doi.org/10.2307/3178066

Harding, S. (2008). *Sciences from Below: Feminisms, Postcolonialities, and Modernities*. Durham, NC: Duke University Press.

Kelley, C. P., Mohtadi, S., Cane, M. A., Seager, R., and Kushnir, Y. (2015). "Climate change in the Fertile Crescent and implications of the recent Syrian drought," *Proceedings of the National Academy of Sciences*, 112, pp. 3241–3246. https://doi.org/10.1073/pnas.1421533112

Marlon, J., Neyens, L., Jefferson, M., Rosenthal, S., Howe, P., Mildenberger, M. and Leiserowitz, A. (2022). "Perceived Harm from Global Warming Is Becoming More Widespread," Yale Program on Climate Change Communication. Available at: https://climatecommunication.yale.edu/publications/perceived-harm-ycom-2021/

Moon, T. A., Druckenmiller, M. L., and Thoman, R. L., eds. (2021). *Arctic Report Card 2021*. htpp://doi.org/10.25923/5s0f-5163

Nelson, A. (2016). *The Social Life of DNA: Race Reparations and Reconciliation After the Genome*. Boston: Beacon Press.

Otto, F. E. L., Harrington, L., Schmitt, K., Philip, S., Kew, S., van Oldenborgh, G. J., Singh, R., Kimutai, J., and Wolski, P. (2020). "Challenges to Understanding Extreme Weather Changes in Lower Income Countries," *Bulletin of the American Meteorological Society* 101(10), E1851–60. doi: 10.1175/BAMS-D-19–0317.1

2 Meteorology, Climate Science, and Empire

Histories and Legacies

Martin Mahony

Introduction

In an article written for *Time* magazine in 2022, environmental scholar Gaia Vince explored the pressing question of human migration in response to climate change. Vince suggests that everyone on the planet will either need to move to escape the worst impacts of a changing climate or will be involved in welcoming (hopefully) migrant communities to their new homes, communities, and countries.

Vince reasons that on the whole, humanity will have to move northwards and upwards. Areas like the Rocky Mountains in North America or the Alps in Europe will provide refuge from extreme heat and drought, but the article focuses largely on the Arctic – areas like northern Canada, Alaska, Siberia, and Greenland – suggesting that as these regions thaw, they'll become new hubs of settlement, agriculture, and trade. The question of what would become of these regions' current inhabitants, especially indigenous communities and native wildlife, is left largely unanswered. Instead, Vince focuses on how the global community will need to:

> look at the world afresh and develop new plans based on geology, geography, and ecology. In other words, identify where the freshwater resources are, where the safe temperatures are, where gets the most solar or wind energy, and then plan population, food and energy production around that.
>
> (Vince, 2022)

Vince's call for a new way of looking at the world through the lenses of 'geology, geography, and ecology', in order to support a radically new form of political planning based on the global redistribution of human populations and settlements, is certainly radical, and perhaps necessary. Over the last few hundred years the practice of 'planning' has been largely a state-based affair, and one might argue that humanity's collective failure so far to adequately deal with climate change is a product of our most powerful institutions being wedded to the geographical form of the nation-state and being seemingly unable to effectively cooperate beyond borders. Nonetheless, Vince's 'new way of looking' is not entirely unprecedented. In fact, it has curious and – I'll suggest in this chapter – consequential echoes of how things like ecology and climate were thought about and dealt with in earlier colonial contexts, particularly during the period of European 'high imperialism' from the late 19th to the mid-20th centuries. Indeed, we can locate the origins of many of the modern sciences of the environment in this period.

Understanding these histories is vital for making sure that the connections between climate science and decision-making do not reproduce colonial modes of thought and action. We need to understand where climate data comes from, historically, and how their production and use was

DOI: 10.4324/9781003409748-3

shaped by local contexts of power and exploitation. Doing so can help us guard against unjust exercises of power in the name of things like climate change adaptation, and to help redress uneven patterns of participation in climate science today.

An STS Approach to the Problem

Research in science and technology studies (STS) emphasises the power of science in the making of worlds. What does that mean? Well, one way to think about the role of science in making worlds is in relation to long historical processes, like the emergence of nation-states or the rise and fall of empires. Scholarship by historians and STS researchers has shown how scientific practices like geometry and statistics were central to the emergence of the nation-state as a political form in places like Europe and East Asia. Simply put, the nation-state is the conjunction of political power and administrative bureaucracy. To govern effectively, you therefore needed to be able to count, measure, analyse, and perhaps predict things like population changes, agricultural yields, and territorial extent. The similarity of the terms 'state' and 'statistics' is no coincidence. And when we look at the rise of European empires – Portuguese, Spanish, Dutch, French, and British, for example – from the 16th century onwards, other techniques of making robust knowledge to serve powerful interests come to the fore. Sciences like astronomy, hydrography, and meteorology were crucial for maritime navigation; experts in ecology, cartography, geology, and anthropology were crucial for 'taking stock' of new colonial territories and their peoples, environments and resources; and areas of scholarship like tropical and veterinary medicine and 'acclimatisation' were deemed crucial for maintaining the health of transplanted people, animals, and plants in new places. As such, many of the major modern disciplines of the natural, physical, and social sciences were crucial to the expansion and functioning of empires.

In turn, those sciences did very well out of imperialism. Money flowed into them, and careers could be made and fame achieved through feats of exploration, experimentation, and discovery. If science shaped imperial fortunes, imperialism also shaped the development of sciences. Put another way, science and empire were **co-produced**. Research questions and priorities were informed by imperial interests, and colonial ways of seeing the world (e.g. as ripe for exploitation, defined by racial hierarchies, and full of scarily different climates, environments, peoples, and diseases) had profound impacts on how disciplines developed into the forms we see today (Chakrabarti, 2021). Researchers in STS try to understand this two-way traffic between science and empire, asking how they shaped each other, and how the ongoing legacies of imperialism continue to shape how science is done, by whom, on what or who, and to what end.

With these questions in mind, let's turn to the specific contexts in which meteorology and climate science were co-produced with empire.

Science and Empire: Piecing Together the Global Climate?

Arguments about the science and politics of a changing climate have a longer history than you might think. As European adventurers, settlers, and colonists set out for new (to them) corners of the globe from the 16th century onwards, they encountered new climates which, to their frequent disappointment and puzzlement, differed greatly from those at home. European settlers in North America were particularly vexed by the climates of places at similar latitudes to Europe, which nonetheless seemed much more varying and extreme than those of the 'Old World'. For much of this period climate was understood as varying chiefly by latitude. But these new climates raised an intriguing question. Perhaps humans had overridden the natural, latitudinal determinants of climate? What if, over centuries of settlement, deforestation, and agricultural intensification,

European climates had been somehow moderated, their rough edges – as found in the freezing winters and humid summers of New England – smoothed off by human modifications of the landscape? If that was correct, perhaps American climates could be 'tamed' by embarking on a similar programme of landscape transformation – a convenient climatological justification for the rapid colonisation and settlement of new lands. Elsewhere, debate raged about whether deforestation was having more negative impacts on local climates. On islands like St Helena and Mauritius, scholars and administrators worried about whether deforestation had ruined local climates, decreasing the rainfall upon which things like sugar cultivation depended. Forest reserves were instituted in places like Mauritius with the expressed intention of protecting the climate, although scholars have pointed out that economic interests (such as the expansion of sugar plantations) often trumped these early conservationist and climate-protecting efforts, in an interplay of science, politics, and economics that is reminiscent of present-day climate change debates.

During the period of early European colonial expansion from the 16th to early 19th century, climate was seen as something that could make or break imperial fortunes, and as something that was subject to human influence and perhaps even control. But it was also seen as something that determined the enduring characteristics of different people and races. A racial climatology popular in Europe and North America posited that inhabitants of the tropics were inherently less hard-working and productive, and more 'passionate' and sensual, than their distant relatives in the temperate latitudes. This kind of climatological thought, since dubbed *climatic determinism*, was a self-serving, racist means by which Europeans, and people of European descent, sought to give a veneer of scientific legitimacy to their own notions of white superiority, while trying to justify the 'civilising mission' by which white people would spread a more civilised way of life through invasion, colonisation, and settlement. This train of thought travelled well into the 20th century, maintaining that climate was not so much something vulnerable to human influence but rather something unchanging and all-powerful in shaping human fortunes.

Climatic determinism continues to reverberate within more recent scientific research on questions like which temperatures are most amenable to human 'productivity' and, like in Gaia Vince's analysis, which parts of the world in the future are likely to be most economically productive, based on their new climates. While there are undoubtedly physiological limitations to how human bodies and minds can function in extreme meteorological conditions, STS studies of the history of such lines of thought can help us guard against problematic generalisations, and can help us keep in view the diverse adaptations that different communities have made to extreme climates to allow them to live flourishing lives in different times and places.

Towards the end of the 19th century, 'climate' was increasingly seen as something stable – tied to place, and amenable to statistical analysis. Zeke Baker has argued that this 'stabilisation' of climate occurred because of a confluence of scientific findings and economic interests – the idea that climate was fundamentally stable on human timescales played into a new interest in using climatological expertise within government to plan and predict things like agricultural output, particularly in the United States (Baker, 2020). This was also a period when weather and climate were starting to be understood on a global scale. The rapid expansion of European empires towards the end of the 19th century saw a corresponding expansion of systems of meteorological observation. Roving naval and merchant ships, many of which took careful weather measurements, were joined by an increasing number of land-based weather stations, recording things like temperature, pressure, rainfall, and wind direction on hourly or daily bases (see Figure 2.1).

Note how meteorology, as represented here by its weather stations, expands in this 150-year period from being a science largely confined to Europe and North America, to something much

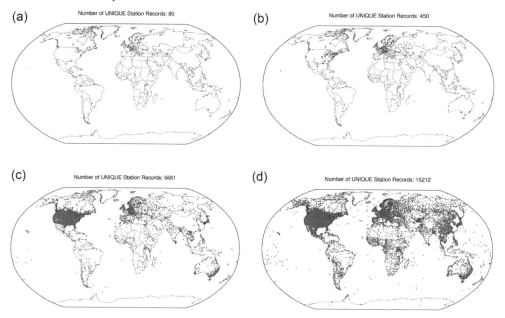

Figure 2.1 The spread of meteorological infrastructure as recorded in counts of station records included in the International Surface Temperature Initiative database: a) 1800–09, b) 1850–59, c) 1900–09, d) 1950–59.

Source: Reproduced with permission from Rennie et al (2014).

more global in scope. The stations mapped in Figure 2.1 featured thermometers for measuring temperature at certain times of day, or for recording daily maximum or minimum temperatures, alongside other instruments for measuring rainfall, atmospheric pressure, and wind. Some would feature instruments that recorded observations onto paper automatically; others, particularly in earlier periods, would be read and maintained by local officials – people like police officers, harbour masters, teachers, priests, and missionaries (Figure 2.2). But a significant number of these weather stations didn't spread in order to further our understanding of the global atmosphere. Rather, they cropped up in service of colonial interests.

Those interests might concern working out which crops could profitably be grown in different environments, delineating areas where European settlement might be encouraged, or even staking a claim to territory and performing cultural superiority. In one disputed area in the northern region of present-day Nigeria, for example, the presence of a British weather station was used in arguments over who could really lay claim to the territory. And the Australian historian Ruth Morgan has shown how efforts to monitor and predict the weather were used politically to showcase the apparent intellectual and cultural superiority of the settler-colonial project (Morgan, 2020). By the 1920s, aviation became a key driver of the expansion of meteorology, with the consequence that the siting of things like barometers (for measuring atmospheric pressure) and anemometers (for measuring wind) was increasingly designed not to serve things like colonial agricultural interests on the ground, but imperial aviation and military interests in the skies. This was much to the annoyance of some colonial scientists who were more interested in doing things like agricultural meteorology than producing on-demand weather forecasts for pilots, the utility of which was largely spent by the time a flight had been completed. Such squabbles aside,

Figure 2.2 Meteorological station in Barombi, German Cameroon, c. 1900.

Source: Bildarchiv der Deutschen Kolonialgesellschaft, Universitätsbibliothek Frankfurt am Main: 043-3026-05. Reproduced with permission. See also Lehmann (2018).

imperial meteorology significantly impacted how climate came to be understood in increasingly global terms.

The expansion of the meteorological infrastructure shown in Figure 2.1 meant that colonial scientists were beginning to piece together pictures of climatic variability across continents. For example, weather records from British India and Australia showed joint fluctuations in barometric pressure, giving colonial scientists an early sighting of the El Niño phenomenon. But we shouldn't just read this science as a precursor to a later, better science of the global atmosphere. Rather, STS approaches urge us to understand how that science was used in practice and must be understood in its local context. Here the work of historian Mike Davis is relevant. In his study of the famines that ravaged India in the late 19th century, he argues that growing scientific understandings of climatic variability in India allowed the British authorities to blame famines on the vagaries of the weather, rather than on things like colonial taxation and traditional agricultural systems being disrupted by the forced introduction of market economies (Davis, 2001). Cutting-edge meteorology helped to 'naturalise' the famines, and helped let the colonial authorities off the hook.

In understanding this spread of meteorological and climatological infrastructure, and the emergence of 'global' understandings of the atmosphere, it's important to look beyond individual imperial systems and beyond state-based scientific institutions. The historian Gregory Cushman has shown how techniques for understanding and forecasting cyclones developed during the middle to late 19th century through a globe-straddling network of observatories and scientists, made up as much by Jesuit missionaries and business interests as by weather-watchers in the pay of colonial governments. Knowledge crossed colonial borders and circulated transnationally, but Cushman has shown that, by the time of the famous Galveston hurricane of 1902, imperial politics were still shaping the making and application of meteorological science. US condescension towards forecasters in Cuba, driven both by ethnic prejudices and an imperial drive to centralise and control, had perverted the development of the science of hurricane prediction with the consequence that opportunities were missed to reduce the deadly impacts of the hurricane on the Gulf coast of Texas (Cushman, 2013).

Imperial expansion enhanced the spatial reach of the sciences of meteorology and climatology. British meteorologists, for example, enthusiastically embraced the capacity for continental-scale analysis that India offered, and hoped that it would allow them to keep up with their North American rivals, equipped as they were with their own continent-sized outdoor laboratory. But what happened when empires began to contract? German meteorologists, on losing their access to colonial field sites after the First World War, became focused on studying regional particularities of climate on smaller scales. Austrian meteorologists, after the collapse of the Habsburg Empire, likewise turned their attention away from continental atmospheric dynamics to things like the microclimates of valleys and mountains (Coen, 2018; Lehmann, 2017). Here we have two clear examples of how changing political structures and institutions, from empires to nation-states, fundamentally altered the capacities, subject matter, and research priorities of the physical sciences.

By this point you might be thinking (if I've done my job!) that 'this is all very interesting as history, but how is it relevant to societies' current struggles to do something about the global climate crisis?' Climatologist Stefan Brönnimann and geographer Jeannine Wintzer argue that users and communicators of climate data need to practice '**climate data empathy**', which they define as a sensitivity to and awareness of the 'political, economic, technological, and cultural' contexts which have shaped the production of climate data over time (Brönnimann and Wintzer, 2019, p. 1). Colonial climate records were often produced with immediate interests in mind, rather than long-term, careful monitoring. As such, many regions of the world still suffer from huge 'data gaps' (look again at Figure 2.1). Data gaps magnify uncertainties about regional climate futures, making it hard for countries to make the case that they are *already* seeing the impacts of climate change, which may be important for accessing things like funding for adaptation projects. Climate data empathy can also emphasise that our vast datasets of meteorological variables are but one way of understanding the atmosphere, rooted in a particular social context. In communicating climate data, scientists can exhibit 'empathy for other observation practices and knowledge traditions', which are increasingly recognised as vital for climate change adaptation but which may themselves have been displaced in the colonial expansion of climate science (Brönnimann and Wintzer, 2019, p. 5).

Conclusion and Next Steps

These arguments for climate data empathy remind us that efforts to 'look at the world afresh and develop new plans based on geology, geography, and ecology … resources … temperature … population, food and energy' (Vince, 2022) will never just be a neutral, academic exercise. As with the colonial past, we must ask questions like: in whose interests are plans being made? Who's sitting at the table, and who isn't? Whose knowledge is being deemed worthy and reliable, and whose is not? These are questions that STS researchers and others are increasingly asking in the context of efforts to mitigate and adapt to climate change – not in order to slow down collective efforts to deal with the problem, but to help make sure that those efforts are just, inclusive, and democratic, and that they don't reproduce colonial ways of seeing and managing the world.

There is more that we still need to know about these colonial histories, as well as their legacies. We still know little, for example, about how sciences like meteorology and climatology interacted with indigenous and vernacular knowledge systems about weather and climate in colonial settings. The historical record at the moment implies that these knowledge systems were largely marginalised and ignored in scientists' efforts to assert their superiority, but much more research needs to be done about, for example, how colonial settlers interacted with or even appropriated the weather knowledges of indigenous peoples, and how the latter served as important informants about local environmental realities. Another research gap concerns the period of European imperial retreat, which coincided with the emergence of a new global environmental consciousness and new worries about climate change. How were these processes linked? How did decolonisation re-shape global science? What role did all those former colonial scientists and experts play in new global institutions like the United Nations?

Alongside these historical questions, pressing questions remain about colonial legacies and enduring inequities of access and participation in climate science; about the links between data coverage, colonial geographies, and climate injustice; and about how diverse knowledge systems can be respectfully brought together to help societies deal with the manifold challenges of a changing climate. These and other questions will keep students and researchers in STS and environmental studies busy for a long time to come, no doubt producing research that can continue to contribute to the conjoined causes of decolonisation and climate justice (Sultana, 2022).

Further Reading

Brönnimann, S. and Wintzer, J. (2019) 'Climate data empathy', *WIREs Climate Change* 10(2): e559. DOI: 10.1002/wcc.559.

Fleetwood, L. (2023) 'Histories of habitability from the oikoumene to the Anthropocene', *Wiley Interdisciplinary Reviews: Climate Change:* e840. DOI: 10.1002/wcc.840.

Mahony, M. and Endfield, G. (2018) 'Climate and colonialism', *Wiley Interdisciplinary Reviews: Climate Change* 9(2): e510. DOI: 10.1002/wcc.510.

Mercer, H. (2022) 'Colonialism: why leading climate scientists have finally acknowledged its link with climate change'. Available at: http://theconversation.com/colonialism-why-leading-climate-scientists-have-finally-acknowledged-its-link-with-climate-change-181642 (accessed 29 March 2023).

Mercer, H. and Simpson, T. (2023) 'Imperialism, colonialism, and climate change science', *WIREs Climate Change*, e851. https://doi.org/10.1002/wcc.851

References

Baker, Z. (2020) 'Agricultural capitalism, climatology and the "stabilization" of climate in the United States, 1850–1920', *British Journal of Sociology* (May), pp. 1–18.

Brönnimann, S. and Wintzer, J. (2019) 'Climate data empathy', *WIREs Climate Change* 10(2): e559. DOI: 10.1002/wcc.559.

Chakrabarti, P. (2021) 'Situating the empire in history of science', in Goss, A. (ed.) *The Routledge Handbook of Science and Empire*. Abingdon: Routledge, pp. 10–20.

Coen, D. R. (2018) *Climate in Motion: Science, Empire, and the Problem of Scale*. Chicago: University of Chicago Press.

Cushman, G. T. (2013) 'The imperial politics of hurricane prediction: from Calcutta and Havana to Manila and Galveston, 1839–1900', in Lawrence, M., Bsumek, E., and Kinkela, D. (eds) *Nation-States and the Global Environment*. Oxford: Oxford University Press, pp. 137–162.

Davis, M. (2001) *Late Victorian Holocausts: El Niño Famines and the Making of the Third World*. London: Verso.

Lehmann, P. (2017) 'Losing the field: Franz Thorbecke and (post-)colonial climatology in Germany', *History of Meteorology* 8, pp. 145–158.

Lehmann, P. (2018) 'Average rainfall and the play of colors: colonial experience and global climate data', *Studies in History and Philosophy of Science Part A* 70, pp. 38–48.

Mahony, M. and Endfield, G. (2018) 'Climate and colonialism', *Wiley Interdisciplinary Reviews: Climate Change* 9(2), p. e510. DOI: 10.1002/wcc.510.

Morgan, R. (2020) 'Prophecy and Prediction: Forecasting Drought and Famine in British India and the Australian Colonies', *Global Environment* 13(1), pp. 95–132.

Sultana, F. (2022) 'The unbearable heaviness of climate coloniality', *Political Geography*, p. 102638. DOI: 10.1016/j.polgeo.2022.102638.

Vince, G. (2022) 'Where We'll End Up Living as the Planet Burns'. Available at: https://time.com/6209432/climate-change-where-we-will-live/ (accessed 29 March 2023).

3 Rethinking Our Histories and Relations with Climate Change

Candis Callison

In the last three decades, the issue of climate change has undergone an enormous transformation. What was once an issue of and for scientists conducting research has more recently become a problem of and for national and transnational policy and public attention. Institutions and systems have emerged to deal with new data, devise new cooperative policy frameworks to decrease global emissions, and address wide-ranging potential impacts. Yet, global temperatures continue to rise such that climate change is now often referred to as a crisis or emergency. And indeed, many of the events we think of as a sign or impact of climate change do constitute a crisis or emergency. Think about massive wildfires in parts of Western North America or Australia. In California, for example, what used to be a seasonal expectation of wildfires now occurs year-round and has led to massive rethinking about fire management and living in fire-prone areas. In Western Canada and Australia in particular, new approaches that involve Indigenous knowledge and burning practices – many of which were outlawed during earlier colonial eras – are finding newly receptive audiences and public attention. Indigenous knowledges and histories of colonialism provide deeper context and resources for living with climate change and its catastrophic impacts, starting with how to think about what this crisis is and how we got here.

"The climate crisis" as a term is the result of the work of many different people and organizations who have spoken out about their climate concerns, including scientists. New reports from the World Meteorological Organization (WMO) in 2023 predict that average global temperatures will exceed the limit of 1.5°C, and already have in polar regions. Further, 1.5°C was the subject of a report by the Intergovernmental Panel on Climate Change (IPCC) in 2018 that warned of catastrophic changes if this global average was exceeded. The IPCC called for massive changes before 2030, essentially providing a 12-year warning in order to avoid a future that took us beyond 1.5°C. Given that, since its inception, climate change has often been described as on the "back burner" with media challenged to report on its slow march toward an uncertain future (Callison, 2014), these reports and the move to calling it a crisis with a timeline for action is, in many ways, an important hard-earned linguistic shift backed up by dire scientific predictions and scenarios. A close cousin of "crisis" is the "climate emergency," but a closer look at this shift and word pairing perhaps raises even more questions. The Oxford English Dictionary defines emergency as "a serious, unexpected, and often dangerous situation requiring immediate action." Certainly, climate change requires immediate action, and is both serious and dangerous, but can we really say it is unexpected?

While calling climate change a crisis does help to both draw public attention to it and communicate the urgent need to do something about it, crisis talk can also contribute to obscuring or ignoring larger questions about history, power, and ecological knowledge (Callison, 2020; Masco, 2017). As scholars have pointed out, when something becomes a crisis, it also often means that questions about "how we got here" get set aside in order to answer "what do we do

DOI: 10.4324/9781003409748-4

about it." Indigenous STS scholars in particular have continually pointed out how deeply connected the past is to both the present and the future, and not too ironically in view of its crisis status, climate change is a standout example of this (Whyte, 2020). What to do about climate change is not only "an environmental problem, but a social one" that involves rethinking human and nonhuman relations as well as built environments and global trade networks that rely on colonialism and industrial capitalism (Callison, 2021a). How we got here to this climate-changed present, with its many possible futures, is absolutely crucial for understanding climate-related impacts *and* devising strategies for what to do about it.

Colonialism and Climate Change

In order to understand how colonialism is a vital part of climate change, the emergence of the term **Anthropocene** is a useful case study. When discussions first emerged about the Anthropocene in the early 2000s, it was because geologists were looking for a way to articulate how profoundly human activity has impacted earth systems (Crutzen and Stoermer, 2000). "Anthropocene" is a combination of the Greek words that mean "human" and "recent" or "new." Geological epochs depend on markers that are reflected in the geological strata, and an epoch lasts for many millennia. The previous epoch, the Holocene, lasted for 10,000 years, and the shift to the Anthropocene in the Geological Time Scale will be visible in rocks and fossils thousands of years from now.

Where to start the clock for the Anthropocene became a major point for discussion among scholars that stretched well over a decade even as use of the term became increasingly commonplace both inside and outside scientific spaces. In their influential 2015 article about this debate, Simon Lewis and Mark Maslin discussed the possibility of starting the newest -cene in either 1964 or 1610. The year 1964 was a top contender because it represented the highest concentration of "the artificial radionuclides spread worldwide by the thermonuclear bomb tests from the early 1950s" (AWG, 2019). Another top contender offered by Paul Crutzen (2002) was the mid-18th century when industrialism became a global force. Lewis and Maslin (2015, p. 177) counter-proposed that the starting point began in 1610 in large part because "colonialism, global trade and coal brought about the Anthropocene." While consideration of geological markers might appear at first glance to be a technical discussion, STS scholarship has made clear in other scientific debates just how much a focus on the technical obscures both the role of culture and contingencies related to history, process, and social structure. The stakes of how to think about the past as a part of the present are evident in Lewis and Maslin's observation that the "formal definition of the Anthropocene makes scientists arbiters, to an extent, of the human–environment relationship, itself an act with consequences beyond geology" p. 171).

A decision was reached in 2019 when 33 geologists who were part of the Anthropocene Working Group (AWG) of the Subcommission on Quaternary Stratigraphy and the International Commission on Stratigraphy chose the mid-20th century (so, 1964-ish) both because it represented "the 'Great Acceleration' of population growth, industrialization and globalization," and the spread of artificial radionuclides. In their statement, the AWG acknowledged that "The 'Anthropocene' has developed a range of meanings among vastly different scholarly communities." Despite the vote and seeming closure, critiques and public discussion have continued, in part because of a study Maslin and Lewis were part of, released the same year as the vote, showing that the deaths of 56 million Indigenous people between 1492 and 1610 resulted in a climatic shift and massive land and vegetation changes (Koch et al., 2019; Maslin and Lewis, 2020). Koch et al. (2019, p. 13) conclude that "the Great Dying of the Indigenous Peoples of the Americas resulted in a human-driven global impact on the Earth System in the two centuries prior to the Industrial Revolution." The findings from "the Great Dying" study add to historical records that postulated even back in the 1600s that Indigenous population declines had a direct impact on the land and vegetation because of their longstanding relations and cultivation practices (Trahant, 2019).

Indigenous STS scholarship has always considered climate change not as a stand-alone crisis or even the first global crisis, but rather as *a continuation* of what began in the late 15th century with colonialism, empire-building, and then industrial capitalism. In their contribution to debates about the Anthropocene, Heather Davis and Zoe Todd (2017) have argued: "Colonialism, especially settler colonialism – which in the Americas simultaneously employed the twinned processes of dispossession and chattel slavery – was always about changing the land, transforming the earth itself, including the creatures, the plants, the soil composition and the atmosphere" (2017, p. 770). And this process wasn't only in the Americas. It was global, as Eyal Weizman (2015) points out:

> Colonial projects from North America through Africa, the Middle East, India and Australia sought to re-engineer the climate. Colonizers did not merely seek to overcome unfamiliar and harsh climatic conditions, but rather to transform them. Native people, who were seen as part of the natural environment, were displaced along with the climate or killed.
>
> (2015, p. 36)

It's for these reasons that Kyle Whyte observes: "Thinking about climate injustice against Indigenous peoples is less about envisioning a new future and more like the experience of *déjà vu*" (2016, p. 12).

This long history of adapting to changing climatic conditions and developing resilience in the face of massive disruptions to their worlds is evident in many Indigenous communities' approaches to addressing and preparing for impacts related to climate change today. Despite "the Great Dying" and the subsequent onslaught of colonial institutions and systems that continue into the present, the rise and fall of colonial empires, and the globalization of industrial capitalism and extractive practices, Indigenous peoples continue to persist, fight for, and work toward their **collective continuance**. Whyte defines collective continuance as "a community's fitness for making adjustments to current or predicted change in ways that contest colonial hardships and embolden comprehensive aims at robust living" (2013, p. 602). New data from the first major assessment of the Intergovernmental Science-Policy Platform on Biodiversity and Ecosystem Services (IPBES), released in 2019, suggests that Indigenous people continue to be highly successful at caring for their lands, waters, and communities. This is significant, as they own, manage, use, or occupy at least a quarter of global land areas, including "approximately 35% of the area that is formally protected, and approximately 35% of all remaining terrestrial areas with very low human intervention." Yet, IPBES also notes that lands under Indigenous control are declining as climate change impacts begin to be felt, but at a much less rapid rate.

Indigenous Knowledge, Relations, and Living Infrastructure

Relations between humans and nonhumans are at the center of discussions about how to understand colonialism and climate change, and they're at the center of what has been termed Indigenous knowledge or Indigenous science. As has been defined by Robin Wall Kimmerer, **Indigenous knowledge** (also called traditional knowledge, TK, or traditional ecological knowledge, TEK) is: "Rational and reliable knowledge that has been developed through generations of intimate contact by native peoples with their lands" (2002, p. 431). Indigenous knowledge can be seen in many forms, both systematic and non-systematic – for example, in the form of highly accurate observations of change made regularly over time (not unlike Western science) or in the form of stories passed down through generations that tell of changes to lands, waters, and nonhumans (all of which are also often seen as having agency). Patricia Cochran (2013) and her co-authors, which include both Indigenous and non-Indigenous scientists, emphasize that "Indigenous understandings of climate change are as diverse as the many environments and cultures in which they are situated," and that there are "common features and differences compared to western science."

Indigenous people are diverse in their social situations, cultures, methods, practices, and relations with nonhumans, lands, and waters. The UN estimates that there are approximately 370 million Indigenous people in 90 countries. The UN Declaration on the Rights of Indigenous People (UNDRIP) was adopted in 2007 after over a decade of discussion involving how to define Indigenous people. The UN considers a range of factors in defining groups as Indigenous: "(1) people whose ancestors were first to occupy their land; (2) self-definition as Indigenous; (3) collective physical and cultural survival based on ancestral claims and distinctive cultural practices related to land; and (4) experiences of subjugation, marginalization, and dispossession" (Callison, 2017; United Nations, 2013). Given this definition, it's not difficult to see how colonialism, land issues, and self-determination are deeply intertwined for Indigenous people, and how climate change forms an additional layer of interlaced new problems, in part because of the vulnerabilities already created by colonialism. Most historical narratives in countries that are built on settler colonialism don't include even a passing mention of either how diverse Indigenous people are or how diverse and deep their relations are with nonhumans (animals, plants), lands, or waters. Many Indigenous people consider nonhumans to be relatives or kin, and Indigenous systems of knowledge are a result of thousands of years of data, experimentation, and adaptation alongside one another.

Indigenous knowledge has in recent years become an area of study and an important contributor to climate discussions, particularly in the Arctic region where observable changes have been happening more rapidly and for a much longer time. The Arctic Climate Impact Assessment released in 2004 was the first formal assessment to integrate Indigenous knowledge with scientific knowledge, involving Indigenous people throughout the process of drafting the report. Inuit knowledge experts worked with over 300 scientists on the assessment, demonstrating both the sensitivity of the Arctic (temperature increases will be much higher compared to mid-latitude regions) and the subsequent impacts for the rest of the world (e.g. glacial melt and sea ice reduction resulting in global sea level rise). In contrast, the first United Nations Framework Convention on Climate Change (UNFCCC) document to mention Indigenous knowledge was the Paris Agreement in 2015, but it still said nothing about *working with* Indigenous people.

Working with Indigenous communities and experts continues to be a challenge for both scientific and political organizations and systems (Ford et al., 2016; Smith and Sharp, 2012), but it is essential if Indigenous knowledge is to become more widely useful in navigating our shared climate futures. Deborah McGregor explains it this way:

> One does TEK [traditional ecological knowledge]; it is not limited to a "body of knowledge." Non-Aboriginal [Non-Indigenous] views of TEK are more concerned with what the knowledge consists of and how it is transmitted. TEK is not just knowledge *about* the relationships with Creation, it *is* the relationship with Creation; it is the way that one relates.
>
> (2004, p. 394, emphasis added).

McGregor's definition pushes against early ideas that Indigenous knowledge could be sup-plemental to Western science in part because "the how you know what you know," the epis-temological basis, is configured differently and stems from a different way of apprehending, being in, and making sense of earth systems and change (Callison, 2014). However, Indigenous knowledge and Western scientific knowledge can be complementary and correlative (e.g. West-ern science has confirmed in many cases what Indigenous knowledge experts have said about both the recent and distant past).

Scholars have continually recommended that instead of seeing differences between Indig-enous and Western scientific knowledge as a barrier, they should be seen as an opportunity for collaboration between scientists and Indigenous communities. One of the earliest and most famous cases of this involves the number of migratory bowhead whales in Alaska. Iñupiat whal-ers, who have long been in close relation with whales (Langlois, 2018; Sakakibara, 2020), suc-cessfully challenged scientific data related to bowhead whales that suggested the population was in steep decline. Iñupiat whalers worked with scientists to help them improve how they collected data and where. This work in turn confirmed Iñupiat knowledge of a much healthier bowhead whale population than previously documented. Subsequently, the moratorium on sub-sistence whaling, put in place in 1977 by the International Whaling Commission (IWC), was lifted (Huntington et al., 2021).

Wildfires and forest management provide a more recent example in which there are hopeful new collaborations. In the U.S., Canada, and Australia, the past decade has seen a rapid increase in the volume, intensity, and reach of wildfires, and there has been a turn to Indigenous knowl-edge in Western Canada and Australia. Kukpi7 Ron Ignace is a leader from the Secwépemc Nation, located in what is now the province of British Columbia in Western Canada where in 2017, a devastating massive wildfire lasting 76 days spread over 192,000 hectares. Ignace di-rectly attributes the devastation to a lack of cultural burning that declined beginning in the 1860s due to the imposition of colonial laws and policies. Ignace's community is part of a three-year recovery program in cooperation with the BC government and eight Secwépemc communities. He described the program as resting on a different foundation: thinking about forests in "a dif-ferent way" as "living infrastructure" (Wood, 2021). The Secwépemc Nation is not alone in their efforts in the region. Just north of their territory, the Tsilhqot'in Nation similarly suffered from massive wildfires in 2017. In turn, they formed a multi-year partnership with Australian Indig-enous fire expert, Victor Steffensen (2020), who has been helping to support and revitalize burn-ing practices that have not been used for many decades due to colonial disruptions. Steffensen described Indigenous burning as not only about preventing wildfires but also about "activating the landscape to look after biodiversity, to improve its health" (Boutsalis, 2020).

Indigenous journalists and scholars have been leading the way in telling these stories that am-plify the efforts of Indigenous experts, who are articulating how these practices are re-emerging in the wake of massive, destructive wildfires, while also acknowledging the long rejection and resistance to Indigenous burning practices (see also Bourke, Atkinson, and Neale, 2020; Krol and Herrera, 2018; Gilio-Whitaker, 2019). Indigenous burning practices prevent large wildfires through culturally and regionally specific use of smaller fires that prevent larger devastating fires. Colonial fire management generally rested on the suppression of all fires, and in some jurisdictions, Indigenous burning practices were criminalized in the 19th and 20th centuries. Clear-cut forest logging and an overall decline in the diversity and age of forests have also created more vulnerabilities for forests. In writing about the boreal forest in North America, Christianson et al. (2022) stress the need for scientists to collaborate with Indigenous communi-ties, most of whom live near or among large forested areas, and to rewrite historical narratives such that they take into account Indigenous knowledge. "Indigenous peoples in the boreal have applied fire on their landscapes for a multitude of reasons. They understand fire as an active, alive agent. As an agent, fire is capable of movement, destruction, and creation, acting on the landscape to create order, within a living, connected environment" (2022, p. 271).

As "the Great Dying" study and many historical records confirm, Indigenous people in close relation with their lands, waters, and nonhuman relatives had a long history of cultivation and management practices that were "unrecognizable to settlers upon their arrival throughout the Americas and the Pacific" (Callison, 2021b). Indeed, the notion of forests as a "living infrastructure" remains a leap for many, even now. Nevertheless, these kinds of interventions, practices, and approaches to relations with forests, lands, and other nonhumans may most meaningfully influence climate adaptation planning (and potentially also climate mitigation). Vulnerability, a much-discussed aspect of assessing risks related to climate change, is not a natural state for Indigenous people. Much of the vulnerability related to climate change facing Indigenous people, whether it be wildfire risks or the location of communities in high-risk areas are due to colonial policies. Whyte (2013, p. 521) suggests that "the ecological challenges of climate change are entangled, or coupled, with political obstructions" and that societal institutions can either create more constraints or opportunities for collective continuance. In order to create more opportunities for navigating climate change, it is essential to see Indigenous communities as navigating often burdensome colonial histories and systems that have emerged from settler-colonial frameworks. Furthermore, it is critical to acknowledge that Indigenous peoples also offer distinctive knowledge and approaches to being in good relations with both human and nonhuman worlds.

Conclusion

Climate change is what anthropologists might call "lively." How we talk about it shifts with cultural changes, and that shifting has stakes and consequences for the politics, systems, and institutions associated with climate change (Fischer, 2009; Rajan, 2012). When I initially began to study climate change in the early 2000s, I found that how climate change got talked about reflected social concerns – that facts had "communal lives" and that climate change was a "form of life" that evolved culturally and socially rather than being a fixed, scientific issue (Callison, 2014). In corporate social responsibility discourse, for example, climate change became "climate risk" in order to tap into concerns about value and investment that are already embedded in financial markets. For American evangelicals active at that time, it was a matter of "caring for creation," and climate facts required a "blessing" from "trusted messengers" that made climate change a real and actionable issue for evangelicals. For Inuit leaders, climate change was already a direct experience. Therefore, advocating for the Arctic and their communities required both "putting a human face" on climate change and navigating colonialism, science, and varying kinds of political institutions and legal systems.

In many ways, the emergence of "crisis" and "emergency" as ways to talk about climate change reflect widening cultural anxieties, as assessments and reports mount with dire predictions and we watch as some of those predictions, like massive and devastating wildfires, become reality. For many who see climate change as the first epic crisis for humanity, the past is seen as a paragon of stability. Yet, our understanding of both stability and change are culturally specific and have much to do with how we see humans in relation to nonhumans and the level of awareness of the destruction and devastation wrought by colonialism. In an Indigenous knowledge framework, stability is something that must be continually assessed and maintained; it is a result of being in good relations with nonhuman relatives and kin that include lands, waters, forests, animals, and others. There are reciprocal obligations and responsibilities that are part of these relationships – many of which have been disrupted by colonialism. As the work related to Indigenous burning practices shows, moving toward reciprocity and relational thinking, based on Indigenous knowledge and expertise, provides a path forward through devastating impacts related to climate change.

By situating a term like the Anthropocene that has been widely used as a means for understanding and articulating the present as deeply affected by humans – and in crisis – within broader discussions about colonialism, the aim of this chapter has been to show how climate change is a crisis, but not necessarily a new one. Indigenous scholars, who have looked closely at Indigenous knowledge and lifeways alongside scientific rationales and methods, recognize the past and ongoing impact of colonialism, and the ways in which climate change amplifies vulnerabilities created by colonial impositions. In this sense, climate change is indeed a continuation of the crises begun with colonialism and empire-building that span the last 500 years. Yet, Indigenous communities, along with their care for lands, waters, and nonhumans, have persisted and offer a wealth of approaches, knowledge, and expertise about how to navigate and adapt to a climate-changed future.

References

Anthropocene Working Group, Subcommission on Quaternary Stratigraphy (2019), 21 May. http://quaternary.stratigraphy.org/working-groups/anthropocene/ (Accessed: August 28, 2023).

Boutsalis, K. (2020) "The art of fire: reviving the Indigenous craft of cultural burning." *The Narwhal.* https://thenarwhal.ca/indigenous-cultural-burning/ (Accessed: August 28, 2020).

Bourke, M., Atkinson A., and Neale, T. (2020) "Putting country back together: A conversation about collaboration and Aboriginal fire management." *Postcolonial Studies* 23(4), pp. 546–551. https://doi.org/10.1080/13688790.2020.1751909

Callison, C. (2021a) "What COVID-19 and climate change teach us about 'syndemics'." *Policy Options* 3. https://policyoptions.irpp.org/magazines/march-2021/what-covid-19-and-climate-change-teach-us-about-syndemics/ (Accessed: June 9, 2023).

Callison, C. (2021b) "Refusing more empire: Utility, colonialism, and Indigenous knowing." *Climatic Change* 167, p. 58. https://doi.org/10.1007/s10584-021-03188-9

Callison, C. (2020) "The twelve-year warning." *Isis* 111(1), pp. 129–137.

Callison, C. (2017) "Climate change communication and Indigenous publics." *Oxford Research Encyclopedia of Climate Science.* DOI: 10.1093/acrefore/9780190228620.013.411

Callison, C. (2014) *How climate change comes to matter: The communal life of facts.* Durham, NC: Duke University Press.

Christianson, A. C., Sutherland, C. R., Moola, F., Gonzalez Bautista, N., Young, D., and MacDonald, H. (2022) "Centering Indigenous voices: The role of fire in the Boreal Forest of North America." *Current Forestry Reports* 8(3), pp. 257–276.

Cochran, P., Huntington, O. H., Pungowiyi, C., Tom, S., Chapin F. S., Huntington, H. P., Maynard N. G., and Trainor, S. F. (2013) "Indigenous Frameworks for Observing and Responding to Climate Change in Alaska." *Climatic Change* 120(3), pp. 557–567.

Crutzen P. J. and Stoermer, E. F. (2000) "The 'Anthropocene'." *Global Change Newsletter*, May, pp. 17–18.

Davis, H. and Todd, Z. (2017) "On the importance of a date: Or, decolonizing the Anthropocene." *ACME: An International E-Journal for Critical Geographies* 16(4), pp. 761–780.

Ford, J. D., Cameron, L., Rubis, J., Maillet, M., Nakashima, D., Willox, A. C., and Pearce, T. (2016) "Including indigenous knowledge and experience in IPCC assessment reports." *Nature Climate Change* 6(4), pp. 349–353.

Gilio-Whitaker, D. (2019) *As long as grass grows: The Indigenous fight for environmental justice, from colonization to standing rock.* Boston: Beacon Press.

Fischer, Michael M. J. (2009) *Anthropological futures.* Durham, NC: Duke University Press.

Huntington, H. P., Ferguson, S. H., George, J. C., Noongwook, G., Quakenbush, L., and Thewissen, J. G. M. (2021) "Indigenous knowledge in research and management." In: *The bowhead whale*, edited by J. C. George and J. G. M. Thewissen. Cambridge, MA: Academic Press, pp. 549–564.

IPBES (2019) "Media release: Nature's dangerous decline 'unprecedented': Species Extinction Rates 'Accelerating'." Intergovernmental Science-Policy Platform on Biodiversity and Ecosystem Services (IPBES). https://ipbes.net/news/Media-Release-Global-Assessment (Accessed: May 8, 2019).

IPCC (2018) "Global Warming of 1.5°C: An IPCC Special Report on the impacts of global warming of 1.5°C above pre-industrial levels and related global greenhouse gas emission pathways, in the context of strengthening the global response to the threat of climate change, sustainable development, and efforts to eradicate poverty." Geneva: World Meteorological Organization.

Koch, A., Brierley, C., Maslin, M. A., and Lewis, S. L. (2019) "Earth System Impacts of the European Arrival and Great Dying in the Americas after 1492." *Quaternary Science Reviews* 207, pp. 13–36.

Krol, D. U. and Herrera, A. (2018) "California wildfires weren't always this destructive." *High Country News*, November 15. www.hcn.org/articles/tribal-affairs-california-wildfires-werent-always-this-destructive/print_view

Langlois, K. (2018) "When whales and humans talk." *Hakai Magazine*, April 3, https://hakaimagazine.com/features/when-whales-and-humans-talk/

Lewis, S. L. and Maslin M. A. (2015) "Defining the Anthropocene." *Nature* 519 (7542), pp. 171–180.

Maslin, M. and Lewis, S. (2020) "Why the Anthropocene began with European colonisation, mass slavery and the 'great dying' of the 16th century." *The Conversation*, June 25. https://theconversation.com/why-the-anthropocene-began-with-european-colonisation-mass-slavery-and-the-great-dying-of-the-16th-century-140661

Masco, J. (2017) "The crisis in crisis." *Current Anthropology* 58(15), pp. 65–76.

McGregor, D. (2004) "Coming full circle: Indigenous knowledge, environment, and our future." *American Indian Quarterly* 28(3), pp. 385–410.

PBES (2019) "Global assessment report on biodiversity and ecosystem services of the Intergovernmental Science-Policy Platform on Biodiversity and Ecosystem Services." IPBES Secretariat, Bonn.

Rajan, K. S. (Ed.). (2012) *Lively capital: Biotechnologies, ethics, and governance in global markets*. Durham, NC: Duke University Press.

Sakakibara, C. (2020) *Whale snow: Iñupiat, climate change, and multispecies resilience in Arctic Alaska*. Tucson: University of Arizona Press.

Smith H. A. and Sharp K. (2012) "Indigenous climate knowledges." *WIREs Climate Change* 3, pp. 467–476.

Steffensen, V. (2020) *Fire country: How Indigenous fire management could help save Australia*. Richmond, Australia: Hardie Grant Publishing.

Trahant, M. (2019) "Documenting colonialism: Millions killed, a changed climate, and Europe gets rich." *Indian Country Today*, February 4, 2019. https://newsmaven.io/indiancountrytoday/news/documenting-colonialism-millions-killed-a-changed-climate-europe-gets-rich-GF4REDdir0-SBj10c_KZTg/

United Nations (2013) "The United Nations Declaration on the Rights of Indigenous Peoples: A manual for national human rights institutions." Geneva: Asia Pacific Forum of National Human Rights Institutions and the Office of the United Nations High Commissioner for Human Rights.

Weizman, E. 2015. *The conflict shoreline*. Göttingen: Steidl.

Whyte, K. P. (2013) "Justice forward: Tribes, climate adaptation and responsibility." *Climatic Change*, 120(3), pp. 517–530.

Whyte. K. P. (2018) "Indigenous science (fiction) for the Anthropocene: Ancestral dystopias and fantasies of climate change crises." *Environment and Planning E: Nature and Space* 1(1–2), pp. 224–242.

Whyte, K. (2020) "Against crisis epistemology." In: Hokowhitu, B., Moreton-Robinson, A., Tuhiwai-Smith, L., Larkin, S., and Andersen, C. (eds) *Handbook of critical indigenous studies*. New York: Routledge.

Wood, S. (2021) "After the fire: The long road to recovery." *The narwhal*. https://thenarwhal.ca/bc-forest-fires-restoration-secwepemc/ (Accessed: January 19, 2021).

World Meteorological Organization (2023) "WMO global annual to decadal climate update (Target years: 2023–2027)." World Meteorological Organization. https://library.wmo.int/index.php?lvl=notice_display&id=22272 (Accessed: June 25, 2023).

Part II

Theorizing Climate, Science, and Society

Introduction

Stephen Zehr

A distinct section on theory might seem expendable in a primer on climate change. Why bother when so many other issues need attention? Why not cut to the chase, avoid academic posturing, and limit theoretical points to essential locations within topical chapters? It is the editors' view, however, that understanding some key Science and Technology Studies (STS) and social science theories is essential for a deeper and more meaningful understanding of climate change.

Why theory? What does it do? Who is it for? In general, STS and social science theories provide guidelines for seeing the world – in this case, how to see climate change and its relationship to human beings and their knowledge systems and technologies. The social and natural worlds are very complex, and even more so when integrated within problems like climate change. *Where* does one look and *how* does one look at climate change? What questions need to be posed? This is where theory provides guideposts. A theory makes certain questions important and indicates where and how one should look for answers. In more technical words, theory provides an ontology and epistemology.

Ontology refers to the nature of what one is studying. On its surface, this might seem obvious in the case of climate change. Are we not just studying the buildup of greenhouse gasses and their impacts on climate, terrestrial and aquatic systems, and human societies? In STS and social science research this is too simplistic. Climate change is also about global power arrangements and inequality, human perceptions and knowledge of the environment, technologies that cause the problem and potentially provide solutions, and so on. A theory guides you into seeing whether *this* or *that* is an essential substance of climate change.

Epistemology refers to how one understands and can develop knowledge about climate change. It includes research methods but much more. In the case of climate change, should one proceed by ignoring what people think about the issue because people are often wrong, and focus instead on more "objective" data gathered by scientists or governmental agencies? Or should we focus on people's perceptions and beliefs about climate change because they significantly impact their reactions to it? Should one research the natural environment and social arrangements, scientific knowledge, and technologies as separate entities or combine them together in some way?

Collectively, the ontology and epistemology of social theory inform us of the substance of issues we are researching and how to obtain knowledge about them. Myanna Lahsen introduces and emphasizes the importance of social constructionist (sometimes referred to as constructivist) theory in climate change research. Social constructionism has a long history in STS research and an even longer history in sociology and other social sciences. The basic idea is simple. We study how humans construct the world around them through their culture-producing activities.

DOI: 10.4324/9781003409748-5

We might study common discourses, framing, ideologies, everyday actions, or other symbolic practices that produce joint understandings and order in the world.

Where social constructionism becomes more provocative and sometimes controversial is when it is directed to scientists' research and truth-claims about the natural world. Historically, it was generally assumed that through application of the scientific method scientists possessed the ability to develop truthful, reliable, and universal knowledge, resulting over time in a one-to-one correspondence between nature and what scientists said about it. While social constructionists acknowledge that scientific knowledge is generally reliable and useful, they emphasize its development as a socially and culturally contingent activity, opening the door that it could be otherwise in different social and cultural circumstances. Thus, social constructionists view the ontology of climate change as a varying but sometimes stable set of truth-claims across scientific communities, political actors, publics, environmentalists, fossil fuel industry spokespeople, media, and so on. Epistemologically, social constructionists research the procedures these social actors have used to build stable claims.

Lahsen explores opposition to social constructionist theory in the study of climate change. Lahsen argues that some influential social scientists are fearful that social constructionist approaches may empower climate change skeptics, open the door to relativism (i.e., everyone's claims about climate change are equally legitimate), and hinder progress towards mitigation and adaptation. Lahsen then discusses the utility of social constructionist theory, noting how it opens new and important questions about climate change as a singular global problem and the relationship between scientific knowledge and political action.

Zeke Baker takes a historical approach to clarify the close linkages between industrial fossil-fuel-dependent technologies and capitalism's drive towards ever-growing profit to locate causes of the climate crisis experienced today. Baker shows how these close linkages form the backdrop to a dominant political-economy theory in the social sciences often called "treadmill of production theory". This theory emphasizes the capitalist economic forces that historically and currently drive a fossil-fuel-dependent world that has as one consequence a changing climate. Baker questions the role and position of both science and technology in this theoretical framework. Are they clear saviors and villains or is their positioning much more complex?

Baker also introduces the social science theory of ecological modernization, which emphasizes that societies are now going through reindustrialization, introducing technologies that are more environmentally benign. Within this theory, capitalism is flexibly open to different types of environmental consequences from damaging to beneficial. Its emphasis is on researching the importance of environment and environmentalism in contemporary societies and how it impacts the scientific knowledge and technologies we produce.

Baker also introduces empirical studies that comment on the capabilities of the treadmill of production versus ecological modernization theories to explain the current relationship among science, technology, capitalism, and climate change. Baker argues that STS research on climate change must always be cognizant of and address the economic interests that underpin climate-impactful behavior. Baker demonstrates how it can be done.

4 We Cannot Afford *Not* to Perform Constructionist Studies of Mainstream Climate Science

Myanna Lahsen

Introduction

Reflecting skepticism about modern cultural conceptions of science, technology, and rationality as engines of emancipation and progress, **social constructionism**, with its method of **deconstruction**, became one of the most influential currents in social science and humanities during the 1970s and 1980s, and a central theoretical orientation in Science and Technology Studies (STS) (Fuchs and Ward, 1994). Social constructionism seeks to understand the role of social contexts, processes, power dynamics, and cultural beliefs and values in shaping the development, interpretation, and impact of scientific and technological knowledge. It recognizes that scientific knowledge and technological artifacts are developed within specific social, political, and historical contexts, and that they reflect the perspectives, biases, priorities, and power structures prevailing among the individuals and communities involved in their creation. The method of deconstruction involves "opening up" (critical examination) of knowledge claims to identify social and cultural meanings and influences that make them convincing, and in that sense stable, in particular contexts.

The notion that human understanding of the world is determined not solely by natural or inherent properties but also by power-inflected human interpretation, language, and social interactions challenges the premise of a singular objective reality. The implication of these social filters is that scientific and technological knowledge is not merely discovered or revealed but also shaped ("constructed") through human perceptions, activities, and social negotiations. STS researchers vary in the relative weight they grant to social and subjective versus more objective factors, but most, if not all, recognize that objective reality significantly restrains what can be plausibly presented as truth. In other words, they do not subscribe to **ontological relativism** (that is, the view that reality *only* exists as constructions, lacking a material objectivity), despite common misunderstandings to that effect (Jasanoff, 1996).

However, STS researchers working within a social constructionist tradition do generally subscribe to **methodological relativism**. This methodological approach requires the researcher to remain agnostic about the truth value of scientists' claims, attending instead to how reality is subject to multiple interests, perspectives, and interpretations, mediated through interactions, language, and cultural contexts. Constructionist analyses "open up" science and technology in the sense of revealing assumptions, interests, and social and cultural forces that underpin the very stability of knowledge claims. They ask critical questions: Who is the knower? On what basis do they claim to have authority and speak the truth? Where do they see from, with what limits to their vision, and with what political consequences?

Adopting social constructionism in the 1980s, the **Strong Programme** in the sociology of science also urged "symmetrical" sociological inquiry and explanation in scientific controversies.

DOI: 10.4324/9781003409748-6

Regardless of scientists' or societies' perceptions of which side was correct, all sides of the controversy would be equally analyzed. This approach reacted against ("weak") sociological and historical studies of science and technology that sought to explain "failed" technologies or "false" theories in terms of extra-scientific factors, such as scientific subcultural particularities or broader socio-political biases and interests, while assuming that "successful" technologies and "correct" scientific theories were consistent with natural forces and captured a singular reality perceivable by all (Barnes and Bloor, 1982).

Thus, for example, social constructionist studies of the acid rain controversy in the 1980s did not start with the assumption that acid rain exists and has damaging environmental consequences because mainstream scientists tell us so. Rather, these studies remained agnostic about the reality of acid rain. Instead, they sought to explain both mainstream and non-mainstream scientific claims in terms of contextual social, cultural, economic, or political forces. In other words, these studies took a symmetrical approach to each side of the controversy, as did social constructionist research on social problems such as drug addiction, homelessness, or violent crime. Importantly, this was a methodological tool to ensure rigorous sociological analysis. It did not imply skepticism or denial that acid rain existed.

Apprehension About Constructionist Studies of Environmental Science and Technology

With the rise of the anti-environmental movement in the 1990s, some scholars increasingly weighed the value of constructionist research against concerns to protect science and support environmental protection. They feared that revelations of science as a human and political enterprise complicated appeals to hard scientific authority to justify desired policies. This response grew in a context of intensifying, politically motivated attacks on environmental science as variously corrupt and insufficiently certain to merit environmental protective policies (Lahsen, 2005b).

The utility of deconstructions of environmental science thus grew less immediately obvious, especially among staunch science defenders. STS scholars continued to stress the importance of constructionist studies of science, but they increasingly adopted the idiom of **co-production**. In constructionist STS, the term co-production emphasizes the mutual influences and restraints that science and political structures impose on each other. Science restrains what can be plausibly and legitimately said and done in politics, but politics also shape and restrain what passes as scientifically true and worthy of study. As such, work in science and engineering not only produces knowledge and technologies; it also shapes social and political arrangements around them. The term "co-production" sharpened conceptualizing language, but did not help overcome an aversion to constructionist analysis in mainstream environmental science – an aversion expressed in the "omission strategy" that involved ignoring and quietly devaluing and discouraging constructionist analyses of mainstream environmental science.

Apprehension about deconstructions of environmental science is apparent in peer-reviewed literature, but commonly kept to more private conversations. Environmental sociologists have been especially critical of constructionist STS studies of climate science, albeit rarely in the open. For example, in informal comments to me, two top environmental sociologists expressed, separately, mindfulness about anti-environmental actors' thirst for ammunition. In that context, they shared their opinions that (1) those who produce constructionist analyses of mainstream climate science are "naïve" and, in the words of another, (2) that little of value has emerged from STS perspectives on climate change. In reviewer comments I received as editor with a journal, one of them suggested that the limited value of such scholarship was reinforced by "LaTour's

[sic] admission that it is hard to tell the difference between a 'strong' constructivist position in STS and climate change denial," so little can be expected from expanding this line of work. This reviewer was referring to Bruno Latour, a leading figure in STS scholarship. This reviewer defined "the problem of a social constructionist view of science" as consisting in its usefulness to climate change denial: "If science is a social construction strongly influenced by power relations and professional and organizational ambitions, then we can assert that climate change is just a construction by climate scientists seeking to increase research funding and/or promote their socio-political agenda." The phrasing reveals a set of interlined assumptions that rarely are explicit and tested: social constructionist analyses of environmental science (1) are negative for environmental policy, (2) yield findings about power relations and incentive structures in science that contrast with common (idealized) understandings of science, and thereby (3) weaken the authority of science for environmental policy. Similarly, a 2008 report (Nagel, Dietz, and Broadbent, 2010) from a sociology workshop funded by the US National Science Foundation shows how misgivings about constructionist studies of mainstream science have shaped the climate research agenda in environmental sociology. The two-day workshop convened 40 sociology faculty, graduate students, and policy experts to define how sociological research can contribute to global efforts to understand the human dimensions of climate change and support and design strategies for mitigation and adaptation. Summarizing existent research and defining research needs going forward, the report presented one of sociologists' tasks as consisting in mapping and analysis of "social and cultural processes that shape attitudes, discourses, and ideological dimensions of climate change in public debates and policy processes" (Nagel, Dietz, and Broadbent, 2010, p. 16). It did not mention mainstream climate science as meriting sociological attention, however. While it devoted a chapter to the backlash coalition and associated scientists, it made only a single, brief reference to the possibility of studying the scientific mainstream. This reference was in a sentence that called for analysis of the consequences of contrarians' use of non-scientific outlets for their work versus the scientific mainstream's reliance on traditional, refereed journals (p. 69). In other words, when it came to mainstream climate science, Nagel, Dietz, and Broadbent (2010) called for sociological analyses only when it was a positive foil serving to highlight the negatives in the case of contrarians.

Anthropology shows similar tendencies to environmental sociology. For example, few anthropological studies exist on the socio-political dimensions of integrated assessment models (IAMs), even though these are deeply social, political, and central in climate policy, where they help justify delay of aggressive mitigation (Dyke, Knorr, and Watson, 2021). Socio-political dimensions are built into how they model the interactions of climatic and economic factors. Both present trends and supposedly desirable and possible alternative future pathways.

Apprehension about constructionist theory and symmetrical analysis of climate change science is found beyond the fields of environmental sociology and anthropology. One of the most influential books on climate science over the past two decades also steers clear of symmetry. Written by historians, *Merchant of Doubt* (Oreskes and Conway, 2010) received special attention and endorsement by *Science*, which expressed a desire to make it required reading for "all those engaged in the business of conveying scientific information to the general public."[1] A hard-hitting exposition of the network and campaigns of some high-level scientists and science advisers working to undermine public faith in scientific knowledge, *Merchants of Doubt* documents four decades of efforts to cast doubt on the science showing that global warming, smoking, acid rain, and the ozone hole are real and dangerous phenomena. The book performs a thoroughly researched historical and sociological analysis of these actors' efforts to undermine mainstream science. Although one of the authors is a trained STS scholar, the book does not present a balanced account of scientists involved in the knowledge controversies that one would expect from a constructionist perspective. It does not present or discuss the complexities of the range of positions on climate change and related issues found among scientists (Lahsen, 2005a, 2013a). Rather, as STS scholar Reiner Grundmann observed in a review of the book, "It provides a partisan account of scientific controversies embedded in policy disputes. It is written as an attack on the [contrarian scientists], justified by the importance of the issues at stake and the alleged influence of the protagonists" (Grundmann, 2013, p. 370). The book exposes cultural bias, but only for one side of the debate. It carefully avoids recognizing that cultural bias, interests, and politics also shape respected mainstream science in ways that can weaken environmental policy (Lahsen and Turnhout, 2021).

Key Questions and Premises

STS scholars criticize the lopsided, asymmetrical approach to social study of climate knowledge as politicized. They are not all equally explicit about what is at stake, or of countervailing considerations. Those who perform such asymmetrical studies likely defend them as a lamentable but necessary precaution, with the aim of achieving much-needed protective environmental policies. This stance rests on a series of interlinked, mostly implicit and untested premises: (1) rigorous examination of extra-scientific influences on mainstream climate science will expose things that publics do not currently know about science; (2) knowledge of such extra-scientific influences will weaken public faith in science and undermine environmental policy that depends on it; (3) science needs to be perceived as an unassailable authority if it is to convince publics and decision-makers of its value, including its merit as a basis for policy; and (4) exposing extra-scientific influences on the scientific mainstream will cause a fall into total "anything goes" relativism, where all knowledge, false or not, is granted the same epistemological value based only on subjective preferences.

These premises need to be questioned. How much do publics in fact perceive science along the lines of the traditional idealized image? To what extent is public trust in science indeed the defining factor in whether effective environmental policy emerges in democratic societies? How much do anti-environmental forces need constructionist peer-reviewed studies and facts to make their impact? The high-profile attacks that this coalition has launched to date show that they do not need constructionist social scientific analyses of the mainstream to be effective. They very capably *produce* false facts and appearances whenever needed. While they indeed would be inclined to cite peer-reviewed social scientific analyses as sources to back their political activism, such analyses are unnecessary for their purposes.

Since the omission strategy tends to be made in private, its core premises are not subjected to debate and analysis. Important questions are rarely asked and discussed, including higher level questions such as: to avoid giving ammunition to the anti-environmental forces, is it right, necessary, and worth it in utilitarian terms to sacrifice social scientific rigor in the form of unsparing constructionist analysis across the board following the principle of symmetry? Leaving ethics aside, how can the policy value of this strategy be estimated and weighed against the risks? A risk that should be considered is the more wholesale rejection of science and descent into ideology. However, this risk is more likely to occur when the idealized image of science is sustained but the real face of science publicly shows itself. This occurred in 2009 with the famous Climategate controversy that emerged with leaked emails by IPCC lead authors that allegedly revealed fake data (Lahsen, 2013b). Climategate showed how serious a threat such leaks can be to the authority of science – and, thus, to the credibility of environmental policy processes – when the legitimacy of science depends on a cultivated idealized image of science. Analyzing the episode, I have argued for the necessity of helping publics to a more nuanced and empirically based understanding of science, both its strengths and its weaknesses, to minimize the risk of its wholesale rejection. Such understanding can emerge from constructionist studies of environmental science.

Addressing Common Assumptions

From the premise that stronger climate policy is needed, the danger of constructionist analyses of mainstream climate science depends on the extent to which exposure to more complex and realistic understandings of science in fact weakens its perceived value as a guide and stimulus for environmental policy. In what follows, I discuss this issue.

Relativity can be Relativized

Whether or not awareness of extra-scientific influences on science will weaken public faith in science and undermine environmental policy depends on what is revealed and on preexistent assumptions about the correct, possible, and ideal nature and roles of science. Barring major manipulation of public opinion as well as irregularities in how science is conducted, it is quite possible that publics can be helped to more realistic expectations of science without undermining their support. Some studies even suggest that it is counterproductive and a disservice to (at least) some segments of the public to shun – rather than directly address – politics bearing on climate change science (Corner et al., 2015). More realistic expectations may help sustain public trust in science, not least when social and political dimensions become too evident to deny. Publics can perceive and accept that climate science has socio-political influences and limitations revealed through constructionist analysis, and that it nevertheless is a particularly rigorous means of producing and adjudicating knowledge on which environmental decisions are made.

The omission strategy rests in part on a fear of a fall into total relativism once science is taken from its (undeserved) pedestal. The basis for this fear needs examination. At the least, accepting that value- and perspective-free objectivity is an impossibility and, often, a power tool of oppression does not mean that all knowledge claims become equally valid (ibid.). It is possible to relativize relativity (Harding, 1992). Descent into complete relativism is not at all a necessary result of constructionist analyses of mainstream climate science. More importantly, a descent into relativism might be *best avoided* where constructionist analyses exist to provide transparency, because transparency can reveal *differences*. The capacity to relativize relativism depends on a symmetrical analytical approach because it is only with the benefit of analyses of all relevant actors in any given scientific controversy that differences (asymmetries) between them can be made apparent, such as degree of bias and politicization. In what follows, I back these claims up with my experiences publishing a critical analysis of uncertainty distribution around global climate models.

Even-Handed ("Symmetrical") Approaches are of Limited Use to Anti-environmental Forces

Through my constructionist 1990s doctoral fieldwork among US-based atmospheric scientists and their critics, I became privy to how modelers and their scientific colleagues spoke, thought, and felt about climate models (numerical General Circulation Models, GCMs). GCMs are a cornerstone in the atmospheric sciences and provide key evidence supporting claims about human-induced climate change. In 2005, I published ethnographic evidence of historical and sociological dimensions and dynamics shaping when and how climate models were accepted, questioned, or rejected within the atmospheric sciences. Although model developers and defenders acknowledged many uncertainties, I presented interview data from both close observers and modelers themselves that illustrated how socio-cultural and psychological factors at times reduced their ability to retain critical distance from their own creations. I stressed that climate models are impressive scientific accomplishments with importance for science and policy making, but I brought into focus that they also have important limitations and that a combination of psychological and socio-political factors encouraged perceptions and representations of the models as greater "truth-machines" than warranted. I judged that this phenomenon was significant in the sciences and that it was important knowledge by which to calibrate GCM-based assertions brought to bear on policy decisions.

Climate change modelers were not my only research focus. I also focused on US climate science politics where anti-environmental actors were as important as mainstream climate scientists. Including both in my analyses yielded evidence of some similarities between the two sides. However, this "symmetrical" analytical approach also revealed important differences in the degree of bias and politicization between them. In line with my training as a cultural anthropologist, I approached all scientific subgroups from an empathetic, cultural perspective. But I also depicted honestly the values and attitudes that set them apart from important norms and trends in science and society. I showed how especially politicized anti-environmental scientists – often led by the conservative think tank-based physicist S. Fred Singer – systematically distorted actual events and orchestrated attacks based on ill-founded, unfair, and conspiratorial claims and motive attributions. I also traced his connections to a wider network of conservative groups and fossil fuel interests that also funded his think tank. The key point here is that a symmetrical analysis of mainstream climate modelers and anti-environmental scientists not only found similarities in how their science was affected by social, psychological, and political factors, but also key differences that implied more legitimacy of the former and less for the latter. I suspect that this level of understanding would increase trust in climate change science rather than reduce it. This calls into question the implicit assumption that it is necessary to protect mainstream science from constructionist analysis. Furthermore, as I discuss below, to the extent that constructionist analysis reveals problematic aspects in mainstream science, it is better to know and address these than to allow them to persist and even grow. Constructionist analysis offers opportunities for corrections that can help keep science *closer* to the ideals, and thus *sustain* positive but *realistic* public perceptions of science.

Optimal Trust is not Perfect Trust

Lacey et al. (2018) note the emergence of a body of research dedicated to understanding how to increase the uptake and use of climate science by decision-makers at all levels, public and private. This research shares the assumption that trust between producers and potential users of scientific knowledge largely determines its adoption and use. However, it is mostly uncritical of and leaves under-examined *how* trust is produced and operates at science-society-policy interfaces. It ignores that optimal trust is not total trust, because total trust can be betrayed and even exploited (ibid.). For example, scientists can be "captured" by politics or policy makers by pursuing specific streams of research at the exclusion of others. They persistently approach science and policy through their own exclusionary lenses. Total trust of these scientists may induce blindness to other research streams that might be good and even more socially beneficial. Optimal trust is where persons are positively disposed but retain a healthy level of distrust and, therefore, an inclination to see evidence, to test, and to question factual statements passing as best available knowledge and research priorities (Lahsen, 2009). There is therefore a level of "optimal trust" beyond which additional trust can undermine rather than enhance the benefits of a trusting relationship (Stevens, MacDuffie, and Helper, 2015). Optimal trust is not complete trust. Rather, optimal trust retains some critical distance, and this distance can help correct improper and otherwise problematic practices, reducing their prevalence. Constructionist studies of mainstream science can offer the transparency required for building optimal trust in climate change science.

The Dangers of the Omission Strategy

The suppression of constructionist inquiry creates institutionalized blind spots. This suppression is dangerous considering the importance of science for environmental policy. Constructionist analysis brings needed questioning, debate, and depth of policy reforms (Lahsen and Turnhout, 2021; Swyngedouw, 2010). Constructionist knowledge of GCMs, for example, might have seemed inconvenient when anti-environmentalists focused their attacks on evidence of temperature changes and their anthropogenic origin. But without this knowledge policy makers become overly reliant on climate models (and economic models that collectively form what is known as **integrated assessment models**) without an understanding of the assumptions, social forces, and politics that underpin them. Even scientific leaders, including the former Chair of the United Nations Intergovernmental Panel on Biodiversity Ecosystem Services (IPBES), now observe that integrated assessment models "severely restrict which policy options are considered" and "ignore complex social and political realities, or even the impacts of climate change itself" (Dyke, Knorr, and Watson, 2021, p. 42). Rare works – for example, Dykes, Knorr, and Watson (2021), Pielke Jr and Ritchie (2021), and Lahsen and Turnhout (2021) – reveal and analyze how and why mainstream climate scientists can also be forces against deeper-cutting climate policy changes.

The omission strategy is supposedly necessary to protect the public image of science and science-based policy, but discouragement of constructionist analysis can obstruct rather than facilitate more rigorous climate policy. It can abet rather than help challenge the socio-political and economic status quo that is obstructing needed environmental reform. Constructionist studies of mainstream climate science could help publics become aware of this phenomenon and take measures to reshape institutions and funding priorities to encourage deep-cutting, system-critical research, and better protect research against political influences. Protection of current institutions is counterproductive to needed changes. The IPCC has existed for over 30 years. Its conclusions have consolidated but remained largely similar throughout. Yet this scientific assessment process has not been sufficient to force the more needed, fundamental overhaul of policy arrangements. But (how much) can this be attributed to the weakness of science, in the eyes of decision-makers and the public, and to contrarian activity? This problem construction justifies the omission strategy. By contrast, deeper-cutting analyses from a more constructionist perspective suggest that the very institutions entrusted to define climate policy are ineffective *and that this often is by design* (Dimitrov, 2020). Deeper analysis reveals the extent to which science often is subservient to power. Science has mostly served the status quo of the slow unfolding of climate policy without altering the world order rather than prompting the deeper changes toward sustainability and greater well-being.

Conclusion

Tendencies toward deliberately asymmetrical study of climate science and politics were flagged more than a decade earlier by leading environmental sociologists who noted "dramatic interrelations" of science and politics in the case of global environmental change (Buttel, Hawkins, and Power, 1990, p. 66). They suggested that global change was one of a growing number of instances of "scientized policy" and "politicized science" in need of detached, critical, and cautionary social science inquiry. Their suggestion has not been heeded, however; scholarship is tending in the opposite direction, as illustrated above. Constructionist analyses of science and technology ask probing questions about the limits and implications of "truths," and about the basis of their authors' authority and claims. However, this line of questioning has been quietly discredited and widely avoided in a context of contested environmental issues in which science mediates understanding and action. Examples presented above yield insight into the largely implicit and under-discussed underpinning logic and values that help explain why few social scientific studies exist which reveal the extent to which mainstream climate science is neutralized and politicized, how this happens, and the purposes and interests that are thereby served, intentionally or not.

The omission *tendency* does not always result from deliberate political calculation at the individual level. It has become cultural and implicit, embedded in researchers' intuitive feelings about what feels comfortable, right, and worthy as lines of research to be pursued and funded. Considerations that underpin the omission *strategy* rest on a series of interlinked, mostly implicit and under-examined premises: rigorous examination of extra-scientific influences on mainstream climate science will expose things that publics do not currently know about science, weakening public faith in science and undermining environmental policy. An "anything goes" relativism is assumed to result from such exposure, with all knowledge, false or not, being granted the same epistemological value and left open to subjective preferences and power politics.

I call each of these premises into question, at least as significant bases for foreclosing constructionist scrutiny of mainstream climate science. I suggest that the omission tendency is dangerous. To the extent that this tendency is a function of deliberate strategy, it is unnecessary. Though minor gains may be had in the short-term sparring with climate denialists, it is also, ultimately, counterproductive if the goal is more rigorous environmental policy. Climate models serve to illustrate this danger. It seems logical for climate policy promoters to protect climate models from constructionist scrutiny, since anti-environmental forces made them central targets in their contestation of the need for climate policy. Indeed, revealing extra-scientific influences on models might raise doubt about their veracity, especially in a context of idealized understandings of science. Ironically, the protection of climate models from constructionist scrutiny only makes sense if one accepts the (factually solid) premise that models, like all science, *also* have socio-political aspects. Admitting this and embracing examination of these aspects is in line with scientific ideals. Ultimately, embracing examination of the socio-political aspects of climate science is therefore the best way to protect the integrity – and, thus, the authority – of science.

Embracing examination of the socio-political aspects of climate science is *necessary*, not *counter*, to achieving the needed deep-cutting environmental policy innovation. The omission strategy creates blind spots that can be deeply consequential. Protecting climate models from scrutiny has, for example, reduced scrutiny and questioning of the extent to which societies should rely on climate-economic integrated assessment models. The latter have offered a scientific justification for the delay in urgently needed aggressive mitigation on the hope that largely hypothetical negative emissions technologies will be invented in time to meet the temperature targets under the United Nations 2015 Paris Convention.

Moreover, why should mainstream science be protected if it is the product of earnest, rigorous (but still social and fallible) scientific pursuit? If publics are taught realistic understandings of science, its merits should be as clear as its limitations. In a context of realistic understandings of science, which constructionist studies help foster, honest constructionist symmetrical analysis of climate change science may encourage more, not less, public support of climate science and policy, because such analyses are likely to reveal the value of mainstream science. Environmental sociologists (Nagel, Dietz, and Broadbent, 2010) perceive that potential, but they should also embrace a more universal commitment to transparency in the form of constructionist studies of the mainstream. Transparency encourages positive feedback in favor of positive reform, where needed, by limiting space for illicit and undemocratic interests; problematic aspects must be known and visible to be addressed. Ultimately, the risks of performing constructionist analyses of mainstream climate science outweigh the dangers of not performing them.

Note

1 See www.merchantsofdoubt.org/home/praise/, accessed July 7, 2023.

References

Barnes, B. and Bloor, D. (1982) "Relativism, rationalism, and the sociology of knowledge." In Hollis, M. and Lukes, S. (Eds.), *Rationality and Relativism*, pp. 21–47, Oxford: Basil Blackwell.

Buttel, F. H., Hawkins, A. P., and Power, A. G. (1990) "From limits to growth to global change: Constraints and contradictions in the evolution of environmental science and ideology," *Global Environmental Change*, *1*(1), pp. 57–66.

Corner, A., Roberts, O., Chiari, S., Völler, S., Mayrhuber, E. S., Mandl, S., and Monson, K. (2015) "How do young people engage with climate change? The role of knowledge, values, message framing, and trusted communicators," *WIREs Climate Change*, *6*(5), pp. 523–534. doi:10.1002/wcc.353

Dimitrov, R. S. (2020) "Empty institutions in global environmental politics," *International Studies Review*, *22*(3), pp. 626–650.

Dyke, J. G., Knorr, W., and Watson, R. (2021) "Why net zero policies do more harm than good." In Böhm, S. and Sullivan, S. (Eds.), *Negotiating Climate Change in Crisis*, Cambridge, UK: Open Book Publishers.

Fuchs, S. and Ward, S. (1994) "What is deconstruction, and where and when does it take place? Making facts in science, building cases in law," *American Sociological Review*, 59, pp. 481–500.

Grundmann, R. (2013) "Debunking sceptical propaganda," *BioSocieties*, *8*(3), pp. 370–374.

Harding, S. (1992) "After the neutrality ideal: Science, politics, and 'strong objectivity'," *Social Research*, *59*, pp. 567–587.

Jasanoff, S. 1996. "Beyond epistemology: Relativism and engagement in the politics of science, " *Social Studies of Science*, 26, pp. 393–418.

Lacey, J., Howden, M., Cvitanovic, C., and Colvin, R. (2018) "Understanding and managing trust at the climate science–policy interface," *Nature Climate Change*, *8*(1), pp. 22–28.

Lahsen, M. (2005a) "Seductive simulations? Uncertainty distribution around climate models," *Social Studies of Science*, *35*(6), pp. 895–922.

Lahsen, M. (2005b) "Technocracy, democracy, and US climate politics: The need for demarcations," *Science, Technology, & Human Values*, *30*(1), pp. 137–169.

Lahsen, M. (2009) "A science-policy interface in the global south: The politics of carbon sinks and science in Brazil," *Climatic Change*, *97*(3–4), pp. 339–372. doi:10.1007/s10584-009-9610-6

Lahsen, M. (2013a) "Anatomy of dissent: A cultural analysis of climate skepticism," *American Behavioral Scientist*, *57*(6), pp. 732–753. doi:10.1177/0002764212469799

Lahsen, M. (2013b) "Climategate: The role of the social sciences," *Climatic Change*, *119*(3–4), pp. 547–558. doi:10.1007/s10584-013-0711-x

Lahsen, M. and Turnhout, E. (2021) "How norms, needs, and power in science obstruct transformations towards sustainability," *Environmental Research Letters*, *16*(2), 025008. Retrieved from https://iopscience.iop.org/article/10.1088/1748-9326/abdcf0/pdf

Nagel, J., Dietz, T., and Broadbent, J. (2010) "Workshop on sociological perspectives on global climate change." In. www.asanet.org/research/NSFClimateChangeWorkshop_120109.pdf. National Science Foundation.

Oreskes, N. and Conway, E. M. (2010) *Merchants of Doubt: How a Handful of Scientists Obscured the Truth on Issues from Tobacco Smoke to Global Warming.* New York: Bloomsbury Press.

Pielke Jr, R. and Ritchie, J. 2021. "Distorting the view of our climate future: The misuse and abuse of climate pathways and scenarios, " *Energy Research & Social Science*, 72, 101890.

Stevens, M., MacDuffie, J. P., and Helper, S. (2015) "Reorienting and recalibrating inter-organizational relationships: Strategies for achieving optimal trust," *Organization Studies*, *36*(9), pp. 1237–1264.

Swyngedouw, E. (2010) "Apocalypse forever? Post-political populism and the spectre of climate change," *Theory, Culture & Society*, *27*(2–3), pp. 213–232.

5 Political Economies of Climate Science

Beyond Technological Villains and Scientific Saviors

Zeke Baker

Introduction: Some Partial Truths

Let's begin with a simple narrative—a story neatly structured with characters, a grim setting, and an opportunity for redemptive action. It goes something like this. First, there enters a set of people who invent technologies fueled by burning extracted and refined stocks of fossil energy. Let's call them industrial technologists. The impacts are marvelous: a new world of machines that accelerates the human capacity to transport themselves and goods (think cars and container ships). A world marked by the globally transformative factory system of commodity production. And for many—but by no means all—of us, an electrified world with temperature-controlled environments and gadgets in our pockets. And yet, of course, the story has a twist. Next, the production, consumption, and life cycles of these very technologies generate climate change that threatens to deeply disrupt, if not wholly upset, the social and economic systems they helped to create. What's more, those people with the most money and power, whose interests in commodity production, economic growth, and profits helped push the fossil economy into so many aspects of life, apparently cannot risk an alternative path. Still further, elected government officials and those nominally in charge of fulfilling the will of the people have yet to develop solutions that have meaningfully mitigated global warming by bringing down global greenhouse gas emissions. Many in turn cry foul: greed, injustice, evil! Perhaps we are slaves to technologies that bring benefits, but which may spell doom? But alas, the sounding calls for another way can be heard, growing louder. Scientists, working worldwide, gain consensus on the nature of the problem, the issues societies likely face in the future, and ways to transition energy and economic systems away from fossil fuels and economic models based on compound growth to successfully mitigate major climate disasters. And so, an inflection point, a climax, is reached: will people come together, listen to scientists, and save themselves whilst casting down the old guard fossil-industrial technologists—those damned wolves in sheep's clothing?

If it is not yet apparent, the preceding story is a sketch, a gross caricature of some four centuries of modern history and a whole range of social groups in the space of a single paragraph. Even so, I believe it contains some provocative partial truths hidden in the grander myth. Stepping out of the story, we may take away the following questions and consider them more seriously as issues for intellectual and public debate: First, what role did technological forces play in global warming—if not as villains now unmasked, then what? How do scientific and technological developments actually relate to the economic and political forces that have come to dominate modern fossil-industrial, capitalist societies? Second, can scientific practices and products—coming from climate science, environmental science, green design, engineering, and so on—form the critical solutions to climate crisis? Can technological innovation and scientific ingenuity, if not "saviors," go so far as to "decouple" our economy from our current modes of

DOI: 10.4324/9781003409748-7

over-exploitation, waste, ecological destruction, and pollution? How can we situate technical ideas and products that are, on the one hand, central to our complex, global, and energy-hungry society and, on the other hand, invested in radically innovative ways of resolving environmental problems? If we move beyond villainizing technology and holding science up as savior, what perspectives and possibilities might open up for exploring, questioning, even rebuilding, the relationship between science and technology, politics, and the economy? In the remainder of this chapter, I draw from my own and others' research, primarily historical in nature, to take up these questions. This chapter will thus help to introduce some basic ways of thinking about the economic and political aspects of climate change and science, while going beyond the partial truths—the villains' and saviors' narrative—sketched out above. The concepts introduced can then serve as a guiding theoretical framework, called the political economy of climate change, that may be usefully applied or challenged in the other chapters in this book, and indeed, when engaging other studies and media regarding climate change.

Did Technology Get Us Into This Mess?

Technology, Economy, and Engines of Modernity

Fossil fuel extraction and energy consumption most rapidly began to increase globally around 1950—when scholars roughly date the initiation of **the Great Acceleration** (Steffen et al., 2015). At this time, global integration of financial and commodity markets (and in a more limited sense, integrated governance and culture), correlated with a wide range of socially and ecologically impactful patterns: increasing rates of GDP (albeit with clear ups, downs, and inequalities), rapid human population increase, increasing rates of deforestation, increase in greenhouse gas emissions, decline in fisheries stocks worldwide, steep increases in automobile use, air miles flown, cement production, and more. It is tempting to focus on this time period as the most relevant context in which to explore the political and economic causes of global warming. Indeed, it is clear that the technologies of **globalization** (ranging from supply chain logistics and freight shipping to corporate conglomeration and automated production, to the internet and innovations in financial instruments, to bioengineered agriculture and synthetic soil inputs) have bound together and accelerated economic production, exchange, and consumption. This has, in turn, led to environmental disruptions that are global in scope, with climate change among them.

Yet, I would argue that it was the time of the Industrial Revolution that is centrally important to the social, economic, and technological context in which fossil fuels took hold, and that this context shows how fossil fuels were inseparable from industrial capitalism as an emergent economic and political system. So, let's talk about coal. Coal and the steam engine, when wed together in the late eighteenth century, provided a new means of transforming through combustion the energy held in England's rich coal stocks into motion that could power machinery. This wedding is a critical moment in the history of technology and of capitalist economic processes. Machines, particularly standardized ones with interchangeable parts produced on assembly lines, could be manufactured and put to use in empowering other mechanical processes, for example, English textile looms and mills. James Watt, often credited with the invention of the transformative Watt engine, initially did so under the logic of efficiency, thus his famous 1769 patent, titled "A New Invented Method of Lessening the Consumption of Steam and Fuel in Fire Engines." Previous models, especially the Newcomen steam engine, were remarkably wasteful in terms of energy use. As it took shape as a technology, the steam engine can reasonably be called a revolutionary engine of modernity. It connected diverse parts of the world through resource extraction (e.g. raw cotton dependent upon slave labor in the U.S. South), factory production (in a rapidly urbanizing England that entailed the growth of a working class), and the early development of industrial-capitalist economies that globally linked people, markets, states, and colonies with natural resources and manufactured goods.

So, fossil fuels and the technologies for converting energy stocks into mechanical energy helped power the Industrial Revolution and, ultimately, globalizing economic markets. Importantly, the technologies (including the steam engine, innovations in coal extraction and burning, the factory system of production, etc.) are only poorly understood in isolation from the economic interests and social investments in the systems that have produced them, put them to use, and made them meaningful. Marxist geographer Andreas Malm's (2016) treatment of coal and steam power in his historical account of the rise of capitalist English manufacturing helps demonstrate a leading approach in STS, namely **the social construction of technology**. This approach treats technologies as artefacts that only make sense when analyzed and situated within their contexts of production, use, and meaning. For his part, Malm draws upon archival evidence to show that coal's utility to steam-powered manufacturing industries was not just about its material qualities. Indeed, water-powered wheels were widely in use in powering cottage industries and some larger-scale manufacturing. Waterpower was quite efficient at the job—not to mention renewable. Rather, the wedding of coal and steam power had more to do with the capacity for industrialists and factory owners to *centralize* production in economically advantageous places. Like landless peasants who could populate urban slums and factory floors, and machines that could be installed and made to run, coal could be dug up in the countryside, stored, and transported for use whenever it was needed. Thus, coal, wage labor, machines, large factories, shipping facilities, and the capital to finance manufacturing operations could *all* be physically brought together into a new social and ecological system, namely modern industrial cities populated by a growing class of urban workers and powered by fossil energy and steam-powered machinery.

The development and use of technology and the birth of **fossil capital** (the accumulation of capital through the exploitation of human labor and fossil fuels) fit hand in glove. This basic principle can be extended to other commodities that link together fossil fuel, natural resources, technological development, and economic interests: automobiles and highways, airplanes, plastics, oil rigs, liquified natural gas terminals—all of it. Perhaps these and many other technologies that characterize a fossil-fueled world can be understood best when situated as economic artefacts, rather than as stand-alone inventions that move, run, and perform on their own.

The social and economic construction of technologies can also help us think anew about science. Drawing from studies of early modern England and its colonization of Ireland, STS scholar and sociologist Patrick Carroll (2006) argues that modern science was buttressed by a mechanical philosophy and the use of experiments. Pioneered by the likes of Robert Boyle, William Petty (and, later, yes, James Watt), modern science emerged most powerfully as what Carroll calls **engine science**. Engine science is less characterized by the refinement of ideas, the testing of theories, and the making of knowledge through standard methods, and more about the development, engineering, tinkering, and use of engines, diversely understood to include meters, scopes, graphs, and chambers. James Watt, in his work with the steam engine and economic investment in factory production, specifically Birmingham's Soho Foundry that mass produced steam engines, was a quintessential disciple of engine science. Engineering, mechanical philosophy, and economic investment in the capitalist production of commodities mark the initial expression of a fossil-based, incipiently global economic system. If this is the case, then it stands to reason that technology and science are fundamental causes of global warming, even as technological artefacts are hardly villains in their own right.

The Treadmill of Production: Economic Growth and the Role of "Production Science"

A leading political-economic theory that helps explain the modern industrial economy in ecological terms—including the science and technology that help comprise it—is **treadmill of production** theory. In this theoretical model, the treadmill represents the cyclical, but expanding, use of resources to produce commodities, often with increasing energy intensity and complexity over time.

In this model, business owners, who exist in competition with one another, are interested in expanding their production to maintain market share in a given sector of the economy. For example, when U.S. corn producers introduce new farming methods to increase yields (say, through GMO technologies and new forms of chemical pest control), Mexican corn farmers are structurally required to do likewise if they want to stay in business—regardless of whether the new methods are environmentally costly or toxic. (Free trade agreements mean, in this case, that U.S. and Mexican farmers are not necessarily protected through tariffs or price controls.) Likewise, governments are invested in the growth of the treadmill, because making and selling more commodities leads to economic growth, in turn bringing jobs, wealth, and opportunities to levy taxes on incomes and economic exchange. If production (that is to say, the treadmill) slows down, as in a recession or depression or through regulation on economic activities, governments risk losing legitimacy from their citizens and investors will likely try to pack up, shift their capital, and move production and business elsewhere. Workers and consumers are often also invested in the expansion of the treadmill because they want jobs and opportunities for higher levels of consumption. The more successful that working people are in advancing their vested economic interests (e.g. by winning high wages in a time of labor demand, or through advancing the political power of labor unions), the more businesses will try to innovate technologically to displace workers. This in turn causes problems of unemployment and an increasingly technical division of labor. But more important for our concerns, it also leads to situations of highly complex systems that may boost economic productivity but will tend to involve environmental risk, for example, health and biodiversity costs associated with pesticide use in agriculture, or higher energy costs of production that has turned to machines or robotics instead of human labor to perform tasks.

Technology is important here. Treadmill of production theory explains technological innovation primarily insofar as it helps business owners and states boost economic capacity, often at the expense of the environment and resulting in a more complex ecological system. How does *science* fit into this process? Treadmill theorists have a useful account of science as a social institution. Allan Schnaiberg, in his 1980 book *The Environment: From Surplus to Scarcity*, initially proposed the treadmill of production theory to explain aspects of what was discussed above as the post-WWII Great Acceleration. Schnaiberg situated science with reference to the post-war treadmill, namely differentiating production science from impact science. **Production science** is that which is institutionally, practically, and intellectually connected to the growth of the treadmill. Although post-WWII science, in the U.S. and the Soviet Union most prominently, related also to geopolitical rivalry—what Stuart Leslie (1993) and others have termed the "military-industrial-academic complex" of Cold War science—the development and use of science to generate economic growth has been a basic tenet of science policy and government research funding. On the other hand, Schnaiberg argues that **impact science**, especially in the environmental sciences but also in public health and some social sciences, fundamentally sheds light on the impacts of the treadmill of production to human life and ecologies, thus making legible the relative (un)sustainability of current social and economic systems.

The treadmill approach to science/technology is important when considering the topic of global warming because it calls into question the relevance of economic interests, values, and goals that play a hand in directing scientific and technological developments. By extension, this approach would have us analyze not only "climate science" to understand the scientific and technical aspects of global warming, but also the technologies and sciences that have played a role in climate change, whether in centuries past or on the other end of university campuses from the halls of "climate and environmental science." The geology of hydraulic fracturing, the chemistry of oil refinement, the aerodynamics of wind turbines, the invention of financial instruments that move money around the world and finance economic production—these are all relevant aspects of modern technoscience implicated in global warming. To compare, I would argue that **social constructionist** approaches (see Lahsen, Chapter 4, this volume) that center around the technological, scientific/intellectual, cultural, and political construction of "climate" may miss important dynamics linking science/technology to climate, namely the dynamics of *economic production* that link technology, science, and climate. Climate change, from a political-economic view, is the outcome of structural processes, and focus should thus remain on the basic underlying economic mechanisms.

It is worthwhile to note that distinguishing science in an either/or fashion as "production"/"impact" science is overly simplistic. Interestingly, in my own historical research on three centuries of climate science, I do not find that climate (and related) sciences ever neatly fit into this binary model (Baker, 2021). It would seem reasonable that contemporary climate science is centrally about registering among the largest *impacts* of the Great Acceleration. As an impact science, climate science has fundamentally helped to make legible, explain, and predict how the fossil economy is impacting human and ecological systems. Yet, climate researchers who began to build the first mathematical models of the global climate in the 1950s and 60s—just as the treadmill and its impacts were accelerating globally—were more invested in the use of their science to inform economic gains (e.g. through improved forecasting or even climate engineering—see Schubert, Chapter 31, this volume) than to call attention to environmental problems. And today, the burgeoning field of "climate services" remains as much about protecting economic investments from climate impact shocks as it is in informing systemic change with respect to the economic drivers of climate change. Recently, meteorologists and climate scientists themselves have brought attention to the vast inequalities in access to climate impact science: In other words, those that may be most vulnerable to climate impacts are systematically less likely to have scientific and related resources to forecast and anticipate those impacts (Otto et al., 2021). Therefore, neat delineations between climate-relevant sciences along the lines of production/impact have their limits. Even so, the lesson here is that to understand the contours, emphases, and questions of a given scientific field, it is worthwhile to consider how its institutions, organizations, and individuals relate to prevailing economic interests and the political institutions that support them. Such is the guiding hypothesis in the field called the **political economy of science** (Tyfield et al., 2017).

Are Innovation and Science our Salvation?

Ecological Modernization and the Promise of Innovation

If connecting change in science and technology to the dynamics of economic interests and political institutions makes sense, then it stands to reason that we would expect science/technology to change if economic actors became deeply concerned about, or otherwise pressured to deal with, climate change and related environmental problems. For example, if the investment risk of constructing a new coal-fired power plant was too high, or concern for climate impacts too salient, or renewable energy systems more easily exploited, then governments or corporate utilities would not invest in coal plants. Instead, they would take them offline and invest more heavily in renewables. In turn, the available technology in this sector would be scaled up by orders of magnitude, and research and development would proceed apace in line with calls for energy transitions (see Part VII, this volume).

Ecological modernization theory (EMT) is an approach in the environmental social sciences (and to some extent engineering and design fields) that explains how environmental values, issues, and processes are now independent of, but increasingly influential in existing economic, industrial, and governance processes. EMT argues that **modernization** centrally entails reflexivity about society and nature. Modernity, in other words, has historically entailed a recognition, embrace, and active pursuit of a society that shapes, if not controls, its own destiny. The modern world makes *itself*, its history, and its future, rather than being subject to divine or natural forces. Industrialism (and engine science) had harnessed the power of science and technology to bolster productive capacity. Capitalism likewise harnessed the power of markets to exchange goods and meet human needs. Of course, these historical forces caused problems too, namely exhaustion of some natural resources, disruption of ecological processes, and pollution. Yet, just as industrialists (think, James Watt and his engine) sought to innovate their way out of the economic problem of scarcity through machine production, rather than relying on God to ease their suffering, so too have major social institutions sought to innovate their way out of environmental problems. They are institutions of **ecological modernization**. The environmental movement, environmental agencies in most national governments around the world, economic pricing of pollution, international organizations like the IPCC (see Part VI, this volume), and the rise of environmental sciences—all represent institutional manifestations of environmental values and interests in sustainability.

Ecological modernization and related programs for sustainable development rely heavily upon the promise of technological solutions to problems of unsustainability and pollution as well as poverty and scarcity. Only through technological innovation can the current chains linking energy use and economic growth be broken, a controversial prospect often called the **decoupling** of energy and economy. As *An Eco-Modernist Manifesto* (2015) has put it, the goal is to innovate technical, economic, and social systems to "liberate the economy from nature." In this view, the problem is not so much that human societies are alienated from their natural environments, but rather that they are *too reliant* upon their exploitations of nature. Cutting down Amazonian rainforests to plant soybeans to fatten cattle to feed beef-loving people halfway across the world represents an extreme reliance on an exploitative use of nature to meet human needs and wants. It's technically possible, but absolutely stupid. The same could be said with structures that require collecting wood to cook food over polluting, indoor ovens in rural villages around the world. Returning to nature—as traditional environmentalism would have it—is not the answer. Rather, the powers of technological innovation should invest in methods of deepening our ability to reorganize, reinvent, and manipulate nature so that people, businesses, and governments can increase efficiency, enhance ecosystems' functions, and escape the whims and harms of nature. Examples include product development in the field of "industrial ecology," which uses materials and systems sciences to create more cyclical production-consumption-waste systems, often termed "cradle to cradle" (as opposed to cradle to grave) production. Industrial ecologies cannot escape the rules of thermodynamics, but they can indeed use design and cutting-edge science to create products that involve minuscule waste of energy in resources—say, compared to the original steam engines or today's "fast fashion" that is halfway to the landfill by the time it is first worn.

Questioning Innovation in Sociotechnical and Economic Systems

EMT acknowledges that "innovation" is hardly sufficient to explain or predict the uptake of technologies. This is because technological change or stasis necessarily happens within **sociotechnical systems**, including economic pressures and incentives (see Part VII, this volume). For example, it is widely accepted that airplane travel is a significant contributor to global warming. Yet, air miles flown per capita are exploding. Furthermore, in a recent study of commercial aviation, Bruce and Spinardi (2018) show that although more efficient airplanes, fuel systems, and air travel systems exist, current forms of air travel are entrenched, or "locked in," because large corporate players like Boeing face economic risk in innovation and face significant barriers in seeing new, available technologies brought into the mainstream. This example shows that ecological modernization may be a partial or segmented process, rather than a general theory useful for explaining widespread economic trends. Even so, EMT helps to center the possibilities of design and innovation that incorporate "natural capital" and ecological values. Science and technology in themselves may not have a salvation function, but they can help confront environmental problems caused by the "aberrant" industrialism of old that failed to effectively value nature and relied too heavily upon unsustainable resource use, including non-renewable fossil fuel stocks.

Although EMT may help to explain growing concern in civil society about environmental sustainability, and growing state and business interest in dealing with environmental problems, the theory has been criticized for being idealistic. Let's return to coal. As an energy source, coal is decreasingly used in some places, like the U.S. and Europe. Ecomodernists may see this as a sign of ecological modernization. Globally, in the case of coal and renewable energy technologies, there is a different and perhaps more treadmill-like story: there has yet to be a year since 2000 when *retirement* of coal-based energy productive capacity outpaced growth in global energy output from coal (I encourage you to explore the Global Energy Monitor to observe global and national trends in energy production and use, specifically the Coal Plant Tracker: https://globalenergymonitor.org/projects/global-coal-plant-tracker/dashboard/). In other words, despite coal plants being retired in many countries, more energy is being drawn from coal right now than last year, and so on. In the important case of China, the share of coal-based power dropped from 53 percent to 44 percent in less than five years (from 2018 to 2022). Even so, the total amount of coal-based power in China rose over that same period (from 1010 to 1121 GW). The International Energy Agency found over a 1 percent increase in total worldwide coal consumption in 2022, and the Agency has projected around the same level of total annual consumption through 2025. This would suggest that the sociotechnical system that involves coal is relatively robust. It is a treadmill with significant business and policy momentum despite general recognition or reflexivity about environmental costs.

Conclusion

This chapter situated climate change and science with reference to economic and political structures. In the process, I argued that the technologies and sciences we think of as related to global warming should be expanded. The steam engine installed in a new textile mill in 1840s Manchester and the petrochemical fertilizer being sprayed on a wheat field in Alberta, Canada, each have their place: these, and many others besides, are products of technoscience; they form moments in the growth and functioning of a capitalist economy, manifest in the last 75 years as the Great Acceleration and the economic drivers of the treadmill of production. To consider climate, science, and society together, central attention must remain on the simultaneous economic and technological basis of a fossil-based global society.

Even so, the process of ecological modernization is undeniable: climate change and environmental sustainability are increasingly woven into nearly all scientific disciplines. And the chances your hometown, state, or province's government has a "Climate Action Plan" are pretty high. Environmental values are a feature of the modern world, as is the will to incorporate them into policy and economic decision-making. STS scholars, including many in this volume, see this as an opportunity to study, evaluate, and even help design technological systems and ways of knowing that are democratically accountable, transparent about their underlying values and assumptions, and when necessary, modest or critical about their "salvation" potential.

Technology as villain. Science as savior. These are simplistic and problematic propositions. They have little basis in the history of science and the history of fossil capital. Worse, they obscure the variegated, and indeed contradictory, ways that science/technology relate to global warming, environmental problems and attempts to deal with them. But the partial truths that opened this chapter remain salient: *follow the economic interests*, and the relationship between climate, science, and society will likely come into better focus.

Further Reading

Alkhalili, N., Dajani, M., and Mahmoud, Y. (2023) "The enduring coloniality of ecological modernization: Wind energy development in occupied Western Sahara and the occupied Syrian Golan Heights," *Political Geography* 103, 102871.https://doi.org/10.1016/j.polgeo.2023.102871

Bonneuil, C. and Fressoz, J. (2021) *The Shock of the Anthropocene: The Earth, History and Us*. New York: Verso.

Ellul, J. (1990) *The Technological Bluff*. Grand Rapids, MI: Eerdmans.

Jorgenson, A. K. and Clark, B. (2012) "Are the economy and the environment decoupling? A comparative international study, 1960–2005," *American Journal of Sociology* 118(1), pp. 1–44. https://doi.org/10.1086/665990

Hawken, P., Lovins, H., and Lovins, A. (1999) *Natural Capitalism: Creating the Next Industrial Revolution*. New York: Little, Brown & Company.

References

Baker, Z. (2021) "Agricultural capitalism, climatology and the 'stabilization' of climate in the United States, 1850–1920," *British Journal of Sociology* 72, pp. 379–396. https://doi.org/10.1111/1468-4446.12762

Breakthrough Institute. "An ecomodernist manifesto," (June 2015) DOI:10.13140/RG.2.1.1974.0646

Bruce, A. and Spinardi, G. (2018) "On a wing and hot air: Eco-modernisation, epistemic lock-in, and the barriers to greening aviation and ruminant farming," *Energy Research & Social Science* 40, pp. 36–44. https://doi.org/10.1016/j.erss.2017.11.032

Carroll, P. (2006) *Science, Culture, and Modern State Formation*. Oxford: Oxford University Press.

Leslie, S. W. (1993) *The Cold War and American Science: The Military-Industrial-Academic Complex at MIT and Stanford*. New York: Columbia University Press.

Malm, A. (2016) *Fossil Capital: The Rise of Steam Power and the Roots of Global Warming*. New York: Verso.

Otto, F. E. L., Harrington, L., Schmitt, K., Philip, S., Kew, S., van Oldenborgh, G. J., Singh, R., Kimutai, J., and Wolski, P. (2020) "Challenges to understanding extreme weather changes in lower income countries," *Bulletin of the American Meteorological Society* 101(10), pp. E1851–60. https://doi.org/10.1175/BAMS-D-19-0317.1

Schnaiberg, A. (1980) *The Environment: From Surplus to Scarcity*. New York: Oxford University Press.

Steffen, W., Broadgate, W., Deutsch, L., Gaffney, O. and Ludwig, C. (2015) "The trajectory of the Anthropocene: The Great Acceleration," *The Anthropocene Review* 2(1), pp. 81–98. https://doi.org/10.1177/2053019614564785

Tyfield, D., Lave, R., Randalls, S., and Thorpe, C. (eds.) (2017) *The Routledge Handbook of the Political Economy of Science*. London: Routledge.

Part III

Media and Public Communication about Climate Change

Introduction

Stephen Zehr

As climate change was constructed as an environmental problem in the 1980s and 1990s, one of the first topics Science and Technology Studies (STS) researchers addressed was the movement of scientific knowledge and information into political deliberations and the public realm. These topics were not unique to STS. Other social science disciplines (especially sociology, political science, and communications studies) also contributed research, but as we see with the chapters in this Part, STS asks a mostly unique set of questions and takes a different approach. While Part VI addresses the policy context for climate change, this Part addresses the media's role in knowledge translation to publics.

It is important to understand both "media" and "publics" as pluralities. Media is plural in its form (television, social media, newspapers, etc.), ideological orientation (politically conservative, liberal, etc.), source of revenue (advertising, public, or government support), and targeted audience (international, national, regional, or local). Publics are diverse along the typical lines studied by social scientists – social class, gender, race, ethnicity, urban/rural, religion, political leaning, and so on. Consequently, the same knowledge or information may be interpreted, represented, and understood differently by diverse media and publics. That doesn't necessarily mean that some media are irresponsible, and some publics uninformed or unintelligent. Rather, they bring with them different interests, mental models, values, and so on, through which climate change knowledge is processed.

A dominant approach in the social sciences is to study the types of frames, storylines, topics, and so on that would be of interest to diverse publics and persuade them to take climate change seriously and change their behavior. This approach is sometimes referred to as the "science of science communication". Social scientists also research how different media are influenced by the fossil fuel industry and other social actors who oppose transitions to renewable energy in what they claim as the disruption of economical and dependable fossil fuels in the energy supply sector. These actors work through subtle or overt political pressure, or by directly financing media outlets that reproduce their claims.

While these topics also are raised in STS research, the primary foci are media translations of scientific knowledge on climate change, sociotechnologies proposed as solutions, and the interaction of knowledge and values amongst diverse publics. STS critiques both the "deficit model" of public understanding of climate change and the "linear path model" between scientific knowledge production, its uptake in the media, and public reception. Working outside these models, STS research emphasizes the interactions among scientific knowledge, media attention, and public knowledge, values, and local concerns as climate change in all its complexities reaches them as an environmental problem. Attention is placed on diverse framings of climate

DOI: 10.4324/9781003409748-8

change and their distribution across the media and the sorts of mental models and concerns that influence publics' reception of media accounts.

In the opening chapter, Zehr outlines the STS critique of the deficit model and its implications for media representations of climate change. Zehr then discusses how the heterogeneity and "wicked problem" nature of climate change is handled in media accounts, introducing the notion of diverse news frames and what he terms "hybrid framing". An empirical study of climate change in newspapers across five nations from 2000–2015 is presented, the results of which highlight how climate change has been represented disproportionately as a political problem rather than a problem of the economic system or everyday lifeworlds.

In their chapter, Loose and Carvalho turn our attention to media representations of climate change and public perceptions in Brazil. As the authors emphasize, Brazil is an important location for research given its position as one of the world's largest greenhouse gas emitters and the crucial role it performs in preserving (or not preserving) the climate-regulating Amazon rainforest. Loose and Carvalho discuss several types of Brazilian media, focusing on the degree to which they construct climate change as a global, universal, and scientific issue or as a locally relevant issue to people's everyday lives. While there are some differences between mainstream, regional, and alternative media in their coverage of climate change, Loose and Carvalho note that they tend to emphasize globalizing and universal aspects of climate change (e.g., IPCC reports, universal climate change models) rather than issues more directly relevant to, especially, rural Brazilians, North/South differences, and other forms of social inequality.

In the final chapter of this Part, Schäfer and Yan focus on climate change imagery in news and social media. The chapter first reviews STS research on climate imagery and visualization. Among other foci, Schäfer and Yan emphasize how visual images of climate change can form boundary objects that bridge different climate change actors (e.g., scientists and politicians), become the centerpiece of science-based controversy (e.g., the famous "hockey-stick" graph in the so-called "Climategate" controversy), and serve as key parts of imaginaries of climate change futures. Schäfer and Yan then review some key differences between climate change imagery in mainstream and social media and outline areas where further research, potentially by readers of this primer, is still needed.

Across these three chapters one sees differences in media accounts across nations and types of media, but importantly similarities as well. Thus, a perceived universality of climate change, discussed critically in other chapters in this volume, is reproduced in many media accounts. One might question how cross-national media accounts might look very different were it the case that they more heavily focused on local experiences of climate change rather than international policy proceedings and scientific reports. One might also think about the dominant framings of climate change reported in these chapters and what different media are leaving out.

6 Climate Change Communication

Simple, Right?

Stephen Zehr

Introduction: STS Approach Versus the Deficit Model

A Science & Technology Studies (STS) approach to climate change communication might, at first glance, muddy the waters for many people concerned about the effects of climate change. As discussed below, it does not unduly emphasize simplifying climate change, excoriating the fossil fuel industry, criticizing politicians, or accentuating climatic changes and their impacts. Along with other social sciences, STS scholars research the framing of climate change in the media and its effects on how audiences perceive climate change. But STS goes further with its emphasis on the complexities, heterogeneity, and "wicked nature" of climate change and whether they are present or absent in media accounts. STS opens up climate change, recognizing that it really is about the combination of many things including politics, social inequality, public knowledge, technological solutions, and economics as well as scientific knowledge and environmental consequences. What is emphasized and what is not and how these different aspects of climate change are framed together in the media are important questions for STS researchers. This questioning is addressed below where I focus on the presence of diverse framing in media accounts and introduce the concept of hybrid framing.

The STS approach can be contrasted with a viewpoint commonly held about climate change communication, especially among people concerned about the problem but without the vantage point of STS and other social science researchers. In this viewpoint, several facts are clear. Climate change is occurring and has major environmental consequences. Scientists have studied it, agree about the nature and severity of the problem, and have communicated that agreement to politicians and the public. Means for reducing greenhouse gas emissions exist such as renewable energy technologies. Consequently, the media simply need to tap into this knowledge base, describing the severity of the environmental problem and solutions in simple, direct language for public consumption. They also might use visual imagery of climate change such as parched ground, flooding, severe weather events, heat waves, and so on to magnify the problem (see Schäfer and Yan, Chapter 8, this volume).

Thus, when social scientists report that the (American) public holds minimal and sometimes incorrect knowledge about climate change, mixes it up with other environmental problems, or is apathetic about altering lifestyles, it might be assumed that the media have done a poor job or are unduly affected by the fossil fuel industry and conservative politicians. Within this view, what the public needs is more, and more accurate, information. Greater quality control is needed, perhaps through scientifically trained journalists, better communication from and more oversight by climate change scientists, or science media centers (like those in Australia, New Zealand, and England) that facilitate the transfer of information from the scientific community to journalists. For example, a New Zealand Science Media Centre staff person indicated the

DOI: 10.4324/9781003409748-9

following in a personal interview several years ago. "[I]n New Zealand we just see lots of inept media, and issues that complex just getting brushed to the side because no one understands how to cover them properly .... There needs to be that climategate hook or some personality conflict, drama ... to make a climate change story work in New Zealand" (Personal Interview, New Zealand Science Media Centre staff).

STS scholars have a concept for this reasoning called the **deficit model of public understanding**. This model assumes a linear scientific knowledge to public uptake pathway with the media serving as intermediaries. Thus, gaps between clear scientific understanding of a phenomenon and public knowledge are assumed to result from failures somewhere along this pathway. Perhaps scientists have not been forceful enough at getting the message out. Perhaps the media have not been receptive or have misunderstood scientists' messages. Perhaps the media have not given enough space and time to the issue as it competes with other issues of public importance. Perhaps the public has been uninterested, grown tired of hearing about it, and started ignoring media accounts, instead focusing on other issues that are more personally relevant or interesting. Or perhaps all these points along the linear path have been negligent. The solution? More, more accurate, and more dramatic information to reduce the deficit.

Critique of the Deficit Model

STS researchers have critiqued the deficit model since the early 1990s, a critique especially pertinent to climate change communication. The critique focuses on several flaws in the deficit model, some more significant and relevant to climate change than others. First, STS differentiates among *diverse publics* rather than a single or homogeneous public, each bringing different values, vested interests, local knowledge, and conceptual models to climate change. As Mike Hulme in his book *Why We Disagree About Climate Change* (2009) and other researchers (e.g., Corner, Markowitz, and Pidgeon, 2014) point out, people hold different meanings about climate, values relevant to the environment, beliefs about human/environment relationships, fears about environmental catastrophe, social positions that increase or decrease risk of climate change, and levels of trust in science and government. Consequently, climate change looks very different depending upon the compilation of life experiences that a public brings to media accounts. Whereas one public may interpret an account as indicating that climate change effects will be dire in their lifetime, another may be dismissive based on their distrust of science, high risk tolerance, or understanding of climate as fickle, constantly changing, and independent of human activity.

Second, STS recognizes the presence of *diverse media* that bring different ideological orientations to climate change. Different media in the form of newspaper articles, television news networks, browser images, and social media orient themselves to different clientele often appealing to their ideological orientations, language, and dominant values. Also, media themselves are often owned by corporations or form a distinct corporation and are dependent upon advertising and other profit-oriented concerns. Distinct messaging on climate change becomes targeted to receptive and self-chosen audiences, which is easily transferable to corporate profits through advertising or subscription sales. Some corporations may not wish to advertise with media sources that highlight climate change damage, while other corporations may welcome that messaging. Publics holding diverse beliefs and values may choose among media sources most amenable to their views.

These two critical points are not unique to STS research. They are found in other social science research as well. Where STS provides a more distinct critique of the deficit model (the third point) is in its conceptualization of climate change (and other environmental problems) as a *complex, heterogeneous, and "wicked" problem* that cannot be reduced to a linear scientific knowledge—policy—solutions pathway. There are several dimensions to climate change that make it this way, some briefly discussed here.

(a) Climate change involves much uncertainty and indeterminacy regarding future impacts, greenhouse gas emissions levels, and societal, governmental, and economic responses to both the problem and its many proposed solutions. Thus, while many vehemently claim that we need to act immediately, it is unclear exactly what some measures will accomplish and how people will react to their effects. For example, using renewable sources for electricity production may lower the price of fossil fuels to a level where it encourages their (over)use in other activities or in other nations.

(b) It is complex in its diverse cross-national relevance. Actions by myself or readers to reduce greenhouse gas emissions, while important and necessary, will not accomplish much if developing nations continue their (rightful?) development along at least partial fossil fuel-driven western pathways of modernizing industry, agriculture, transportation, consumption, and so on. Climate change is also unequal cross-nationally due to unequal greenhouse gas contributions and consequences (see Part V, this volume). While it is clearly necessary to emphasize these inequalities in climate change actions, it is often less clear how to do so. For example, does each nation deserve equal access to the atmosphere based on population size (i.e., equal per capita emissions) or on some other distributive formula? Are reparations owed by rich to poor nations based on unequal historical emissions levels or on some other ethical/political/practical grounds?

(c) Climate change is heterogeneous and complex when considering the types of sociotechnologies that need development and implementation. Unfortunately, there is no quick technological solution as was mostly the case with CFC substitutes in the ozone depletion problem. We can readily name numerous technologies that reduce greenhouse gas emissions. However, STS researchers emphasize how these technologies are inherently integrated with social arrangements and economic and political consequences. Each new technology has winners and losers. Thus, grid-tied, rooftop solar systems, while seemingly an important development, are unequally accessible due to their upfront costs and alter, within many nations, an entrenched system for centralized (in the US monopolized) provision of electricity. Rooftop solar is seamlessly connected to societal institutions and norms and the distribution of resources, among other things, in what Bruno Latour called an "actor-network" where non-human actors hold social agency to act upon other actors in the network.

These points reflect only part of the complexity but, hopefully, the point is made. Climate change and transitions to renewable energy are complex, wicked problems. For these and other reasons, STS scholars know that media attention to climate change is not a simple matter of getting the (scientific) information right. It involves more than journalists listening to scientists and then reporting, accurately, what scientists said to open-minded, attentive audiences. Instead, effective climate change communication requires representation of the complexities and heterogeneities of climate change to diverse audiences. That is not a simple task. Most climate change journalists agree.

The STS Focus

A few summary points capture key aspects of STS research on media representations of climate change. One focus, shared with other disciplines, is how the media **frame** climate change. The general idea of framing is that journalists cannot depict all potential issues relevant to climate change, or any other issue, at one time. Thus, they focus attention on specific, limited features considered important to the story or of interest to the audience. They produce a "frame" around specific parts of the story while leaving out others, much the same way that photographers frame their subject through a viewfinder to construct desirable photographs. The content of these frames varies. STS researchers tend to focus on science- or technology-related frames and the constructed view of nature that emerges from them. Quite typical are frames that depict recent or forthcoming scientific research, often purported as answering or more often *potentially* answering key climate change questions. Less often used are frames that depict scientists as active and embodied actors in the research process and even less often are frames focused on scientific infrastructure for climate change such as the funding, tools, or organizational arrangements that make research and knowledge possible.

Frames may depict scientific certainty, but STS researchers have noticed that scientific uncertainty is quite common. The media constructs scientific uncertainty in different ways. Journalists and reporters may interview a scientific expert who says one thing and another scientific expert who says the opposite. This action is often referred to as **journalistic balancing**, where it appears that journalists and reporters are being objective by not favoring one position over the other. Journalistic balancing of climate change science was common in the 1990s and early 2000s but receded once environmental journalists recognized its polarizing effects on publics. However, scientific uncertainty can be constructed in other ways by emphasizing forthcoming climate change research (i.e., answers will be forthcoming but we're uncertain now) or through interviewed scientists' qualifying or guarded language about their research results, implications, or policy relevance. For scientists, these may be deeply socialized, routine ways of communicating to prevent embellishment or undisputable interpretations of results before research is complete and definitive. But publics may receive these frames as indications of scientific uncertainty in climate change knowledge.

In addition, deeper **indeterminacy** (we don't know enough about this to place parameters around it) and **ignorance** (we don't know what we don't know) may be depicted as **uncertainty** (we don't know something exactly, but we have parameters and will know the answer soon) in media accounts. For example, scientists are indeterminate and to some extent ignorant about the range and severity of future climate impacts of different degrees of global warming. But it may be represented as uncertainty in the media when claims that limiting global temperature increases to 2°C or 1.5°C are necessary to remain safe. In this case, parameters are placed around "safe" levels of global warming and its consequences, albeit with uncertainty about the exact outcome, rather than more realistically depicting indeterminacy or ignorance about what is safe. This framing makes the problem seem more manageable. When an environmental problem is represented as manageable, it appears as less of a threat. Social control is more easily maintained within current institutional arrangements and power structures. However, when deeper indeterminacies and ignorance become salient, the problem may seem out of control and potentially stimulate questioning and challenging of existing social structures and institutions. Hence, the absence of indeterminacy and ignorance in media accounts of climate change may make publics more complacent, continuing their everyday practices, while assuming that larger institutional and authoritative structures will address the problem before it gets out of hand.

In addition to the handling of uncertainty, STS researchers are interested in how the media represents climate change's *heterogeneity*. By this we mean that climate change constitutes more than greenhouse gas buildup and environmental consequences, though obviously an important part of the story. Climate change is also about politics and governance; causes such as fossil fuel burning, consumption habits, and land use change; economic impacts of climate change and efforts to reduce emissions; scientific research and scientists; social inequalities; public understanding, knowledge, perceptions, and values; and mitigation/adaptation technologies, designs, and practices. We can think of these as different potential **news frames** for representing climate change and would expect, over time, that each is extensively covered and combined with each other in interesting and informative ways. We might research whether this heterogeneity is built into media accounts by following one or another media source over time to see how well reporting moves across these frames, while recognizing that any one reportage allows insufficient time and space to address them all.

STS researchers might also be interested in how these news frames are combined within reportage episodes (e.g., a newspaper article or tv news report) to truly reflect climate change's heterogeneity. In other words, climate change is not distinctly about each separate frame, but seamlessly about all of them at once. Environmental impacts of climate change are also inherently about scientific research, social inequality, politics, economic impacts, and so on. These seamless connections reflect the heterogeneous, wicked nature of the problem. We can research whether and how the media makes seamless connections among these frames such that they are inseparable, for example, enabling a reader or viewer to understand an environmental impact in Bangladesh as also inherently about social inequality and politics. We might term this **hybrid framing** – the seamless connection of multiple climate change news frames such that they capture the heterogeneity of climate change. Of course, climate change might also be about other things as well (vacations, food practices, and norms, etc.) but addressing the above news frames is a good starting point.

News Frames and Hybrid Framing: An Empirical Study

I conducted a study that researched this last issue discussed above. The study involved climate change news frames and hybrid framing in major national newspapers of five nations from 2000–2015. Two major newspapers were analyzed from each of these nations: Australia, India, New Zealand, United Kingdom, and the United States (see Table 6.1). Readers might rightfully question the methodological focus on newspapers with their decline in importance as sources of news. However, I would defend their use during this time interval because they were still widely read, especially early in the 2000s, and maintaining a focus on one media outlet over

Table 6.1 Newspapers analyzed in the project

Australia	*Sydney Morning Herald*
	The Australian
India	*Hindustan Times*
	India Express
New Zealand	*New Zealand Herald*
	Dominion Post
United Kingdom	*The Guardian*
	Financial Times
United States	*The New York Times*
	Washington Post

the 16 years provided methodological continuity for the longitudinal study. Using *LexisNexis* (now *Nexis Uni*) articles were identified with search terms "climate change" or "global warming" appearing in the headline. Approximately 3500 articles were read and coded for the presence of major news frames or "subframes" (i.e., a specific subcategory of a news frame such as "drought" as a subframe of environmental impact). Table 6.2 lists frames and subframes that were identified and coded in the analysis. They also were coded for presence of hybrid frames – when two or more news frames were seamlessly merged such that the elimination of one would largely negate the meaning of the text. A hybrid frame addressed a heterogeneous climate change frame, rather than two or more separate frames.

Table 6.2 Climate change news frames and subframes

Frames	Subframes
Economic	Costs of corporate actions
	Economic impacts of GCC
	Economic opportunities of mitigation or adaptation
	Policy impacts on economy
Causal factors	Fossil fuel extraction & use
	Capitalism/consumption
	Land use change/deforestation
	Natural causes
Environmental impact	Adaptation (non-human adaptations)
	Biodiversity/species loss
	Health
	Sea level rise/flooding
	Desertification/drought/fire
	Extreme weather events
	Environmental imposed security threat
	Agriculture/fishing
	Fresh water
Political	Policies
	Policy actors
	Political deliberation
Science	Discoveries, new studies, new reports
	Science funding or infrastructure
	Scientists
Public	GCC education
	Public norms, values, behaviors
	Understanding, knowledge, beliefs
	Environmentalist/civil society action
Social inequality	Unequal contributions to GCC
	Unequal mitigation-adaptation obligations
	Unequal vulnerability to GCC
Technological/Design developments	Adaptation
	Mitigation
	Geoengineering

News Frames

The quantitative results indicate that political, science, environmental impacts, and economic frames and subframes dominated the time interval across newspapers. Of these four, political framing was the most dominant. Receiving much less attention were frames around causal factors; public knowledge, education, understanding, and environmentalist/civil society action; social inequality; and technological or design developments. There was some variability across newspapers and over time (e.g., the *Financial Times* gave more attention to economic issues, and across the newspapers and nations attention to public and social inequality framing slightly increased over time), but the dominance of political framing stood out across both dimensions. This finding suggests that climate change was primarily represented as a political problem rather than a problem endemic to the capitalist economic system, consumer practices, lack of public knowledge, or national or global inequalities.

How might we interpret the significance of these results? My preferred option draws upon the social theory of Jürgen Habermas. Habermas distinguished between the "system," within which he included the economic and political institutions, and the "lifeworld" which consisted of the everyday world of social and community relations. His argument was that these have become differentiated in capitalist societies. Furthermore, crises occurring in capitalist societies are often transferred from economic to political institutions where the government may be blamed for flaws within the capitalist system (e.g., a president, prime minister, congress, or parliament blamed for high inflation or high unemployment). The political system may, in turn, transfer the crisis to the lifeworld where people blame each other or themselves (e.g., a crisis around a pandemic is transferred to governments which in turn may transfer it to the lifeworld where people blame each other for not following rules, not getting vaccinated, etc.). Turning to climate change, we understand that at its roots the problem is generated through economic production and consumption activities. However, over time it could easily be transferred to governments whom we hold responsible for the problem and for generating solutions. It may also be transferred to the lifeworld where we blame ourselves and others' wasteful lifestyles.

Without addressing Habermas' theory in more detail, the empirical results mentioned above indicate that climate change was primarily framed as a political problem in major newspapers from 2000–2015, rather than as a problem of capitalist production and consumption or everyday life activity. Newspaper reading publics were predominantly pushed to the view that governments and politicians were responsible for finding solutions, rather than economic leaders or communities and individuals. Because of its global and complex nature – its "wickedness" – governments were ill-equipped at developing treaties or passing legislation to reduce greenhouse gas emissions. Meanwhile, economic organizations such as corporations or people in their everyday lives may have felt exonerated from engaging in profit-reducing or life-changing behavior. The crisis was not represented as inherently part of their world.

This imbalance in newspaper framing of climate change may have facilitated inaction. Rather than pulling together because we are all at fault and part of the solution, newspaper frames pushed members of these nations in the direction of holding governments and politicians accountable. Through an STS approach that recognizes climate change is about many heterogeneous things (represented here simply as news frames), we then empirically see how newspaper coverage from 2000–2015 did not represent this diversity in proportionate amounts.

Hybrid Frames

What about hybrid frames? As noted above, concern has been expressed that the media "get the science right" when writing about climate change. Within the deficit model there is the expressed concern that members of the media may not sufficiently understand the science of climate change and consequently misrepresent or ignore important features. An STS perspective deconstructs this way of thinking by challenging the meaning of "getting the science right." Is getting the science right simply describing scientific knowledge accurately as scientists communicate it in scientific papers, professional conferences, or personal interviews, perhaps simplifying it a bit for public consumption? Or is getting the science right a process of combining scientific knowledge with other climate change concerns as they relate to public knowledge and values, social inequality, economic matters, and other frames described above? This is where hybrid framing is relevant. We might consider hybrid frames as more accurate descriptions of climate change because they combine different frames together in ways that describe deeper, relational, and more complex aspects of climate change. That is, they tap into the wicked nature of the problem.

The 2000–2015 study indicates that hybrid framing was common in these newspaper accounts. In general, journalists did an effective job of seamlessly combining frames to tap into the problem's complexity. However, one notable shortcoming involved situations where science frames were part of hybrid frames. As expected, there were many instances of hybrid frames involving science and environmental effects. In these articles, journalists seamlessly integrated environmental effects with the scientific research and researchers who teased them out. However, there were far fewer hybrid frames involving science and other news frames such as politics, public knowledge or values, or social inequality. The combination of science and environmental effects made up around 50% of all instances where science was involved in a hybrid frame, with some variability across newspapers. From an STS perspective, this finding suggests that journalists were reticent about making strong associations between the science of climate change and other dimensions. To say it differently, climate change science was not often incorporated into climate change's complexity or wickedness.

Why might this be the case? Some clues came from personal interviews with journalists. For example, a New Zealand journalist mentioned that he was afraid to get something scientifically wrong due to the angry pushback he'd receive. "[T]he sensitivity is something I'm very aware of. People jump at the chance to take us to the Press Council, which is our newspaper standards council here. If you make one slight slip up there are a lot of eager eyes .... I just have to be careful to be very accurate about what I'm saying" (Personal Interview, New Zealand Journalist). Other journalists expressed difficulties interviewing academic scientists who considered themselves "burned by the press" in the past. An acceptance of the deficit model and related **linear model** (i.e., scientific knowledge comes first followed in a linear path to political action) by scientists and some journalists may also be a reason. Within these models, an assumption (false according to STS research) is that objective scientific research is conducted first without regard to political, public, or economic implications. Only later is that knowledge interpreted for its political, public, and economic relevance. Translating this to journalist activity, then, perhaps some journalists felt pressure to precisely describe scientific knowledge without further interpretation, elaboration, or combination with other news frames to make certain they "got the science right." Opportunities to bring science into the heterogeneous, wicked nature of climate change were consequently lost.

Conclusion

It is important for the media to give good balance across climate change issues, but not necessarily in a form that gives equal voice to climate change believers and naysayers. Balance, instead, emerges with representations of diverse dimensions of climate change, as depicted in the news frames and subframes mentioned above, and in representing its heterogeneous, wicked character. Success in doing so gives publics fuller information about climate change – its effects on them, public knowledge about it, its integration with global inequality, policy possibilities, and so on.

We can ask how well the media has performed in providing balance. The above empirical study provides a partial answer by examining news frames and hybrid frames in national newspapers from 2000–2015. The results are mixed. While a range of news frames appear, there was also an imbalance toward political framing and away from closely tying climate change to the lifeworld and to production and consumption activities that cause the problem. As for hybrid framing, one detected limitation is a gap in representing climate change science as seamlessly associated with other frames, with the exception of environmental impacts. What this study does not tell us is how the media has been doing over the past several years, especially as social media has become a more salient mode for obtaining news. Perhaps readers of this chapter can design a study to address that gap in knowledge. They might also consider other types of climate change frames not covered in this study.

More generally, an STS approach to media coverage of climate change steers clear from an overly simplistic deficit model, that all people need is more information in more striking language and images to change their minds and importantly their behaviors. Climate change and most other environmental problems are far too complex. My research above only looks in one direction – at diversity in news frames and hybrid framing in newspapers. STS research also looks at many other research problems such as different patterns between national and local media (see Loose and Carvalho, Chapter 7, this volume), the use of images in climate change reporting (see Schäfer and Yan, Chapter 8, this volume), representations of uncertainty and ignorance, distinctive features of social media representations of climate change, and so on. Readers of this chapter might, as a matter of course, reflect upon how climate change and its different dimensions are constructed in the media they attend to. What is included? What is excluded? What are the implications of those media decisions?

Further Reading

Boykoff, M. (2019) *Creative (Climate) Communications: Productive Pathways for Science, Policy and Society*. Cambridge, UK: Cambridge University Press.

References

Corner, A. J., Markowitz, E., and Pidgeon, N. F. (2014) "Public engagement with climate change: The role of human values," *Wiley Interdisciplinary Reviews: Climate Change* 5 (3), pp. 411–422.
Hulme, M. (2009) *Why We Disagree About Climate Change*. Cambridge, UK: Cambridge University Press.

7 Public Communication and Perceptions of Climate Change in Brazil

Eloisa Beling Loose and Anabela Carvalho

Climate Change Communication in Brazil Matters

How is China's growing appetite for pork and the USA's massive consumption of beef linked to Brazil's contribution to climate change? How does political populism exacerbate climate change? What is the social distribution of gains and harms involved in greenhouse gas emitting activities? These questions illustrate some of the complexities of human-environment relations in the current world. They also point to connections between different scales and spaces involved in climate change, as well as to important social justice issues. Climate change communication plays a critical role in shaping public perceptions. The way in which the mentioned connections are represented and debated in public spaces, including mainstream and alternative media, can reinforce current trajectories of climate (in)action or stimulate transformation. Therefore, it is essential to consider whether different media outlets make the relationships between economic, political, and social systems visible or opaque, whose voices and perspectives are given prominence, and what needs and values are privileged in media discourses on climate change.

Before we delve into media and communication, let us look at some important climate-related aspects of Brazil. It is widely recognized that the Amazon, a vast portion of its territory, possesses exceptional biodiversity and plays a vital role in the natural regulation of the climate system. Yet, both the Amazon and other critical biomes, like the lesser-known Cerrado, have been disappearing at alarming rates over the last few decades. Most of the deforestation, deliberate burning, and other forms of environmental destruction are carried out to "free" land for cattle raising and food production, which are tied to national and international economic interests. In recent times, vast swaths of Brazil's territory have been planted with soy, which is exported to China to feed a rapidly growing pig farming industry (meeting the demand for meat of a fast-expanding middle class). Meat and agricultural exports to the USA and to Europe have been contributing to changes in Brazil for longer. All of this helps explain why Brazil is one of the top ten countries in terms of greenhouse gas (GHG) emissions. Land use changes (especially deforestation), together with agriculture and animal farming, produce about three quarters of the country's greenhouse gas emissions (SEEG, 2021).

The Economic Commission for Latin America and the Caribbean (ECLAC) has highlighted in a report on the economics of climate change that the low environmental sustainability development model adopted in the region is obstructing the fight against climate change (CEPAL, 2015). Like other Latin American nations, Brazil heavily relies on environmental resources and is highly susceptible to the consequences of climate change due to its demographic and socioeconomic status. Environmental governance in Brazil faces several challenges. For years, public policies and protections have been weakened, and under the government of Jair Bolsonaro (2019–2022), state supervision of destructive practices was largely halted. Arguing that

DOI: 10.4324/9781003409748-10

the country's economic growth should be prioritized even if that meant taking down a "few trees," populist Bolsonaro stimulated the plundering of Brazil's natural spaces and resources. During his tenure, deforestation increased by 60% compared to the previous four years, and illegal mining activities expanded significantly in the world's largest rainforest, resulting in severe environmental and humanitarian consequences for indigenous populations such as the Yanomami. In a very unequal society like Brazil's, a powerful agricultural and industrial elite tends to accumulate wealth at the cost of all others. A host of international corporations also lie on the earning side. On the other side lies a large part of the population that struggles to make a living and that is highly vulnerable to extreme weather events due to poor quality housing or other forms of insecurity.

Given the above, Brazil represents a crucial case to explore public communication and public perceptions of climate change. To address climate change, it is important to understand its causes and perceive it as a risk that already impacts the present and will be even more challenging in the near future. Quality journalism and engaged citizens may contribute to public debate and exert pressure toward public policies and other actions to mitigate climate change. This public engagement can also help reduce the risks of climate disasters and prepare for a reality with higher temperatures, more irregular rainfall patterns, and more intense and frequent extreme events. In 2020, 78% of Brazilians rated climate change as a very important issue, and this figure increased to 81% in 2021 (based on surveys by ITS-Rio and Yale University, 2022). Despite high levels of concern, knowledge about climate change remained insufficient, with only 21% of respondents indicating that they had extensive knowledge of the topic; 46% felt that they had a moderate understanding of climate change; 24% said they knew little; and 8% said that they knew nothing (1% did not choose any answer).

Public perceptions of climate change are largely influenced by the media. While public understanding of climate change is also influenced by psychological, social, cultural, and political factors, news media's perceived credibility and reach make them a vital area to study social representations. However, not all media outlets are the same and their coverage of climate change may differ significantly. The remainder of this chapter looks at trends in media coverage of climate change in Brazil and examines differences and similarities between various types of journalism, namely national mainstream media, a regional media outlet with hegemonic characteristics, and two digital alternative media platforms (see the next section for explanations on mainstream and alternative media).

The Media Landscape of Brazil

The media play a crucial role in informing the public about the causes, consequences, and stakes involved in climate change, as well as in discussing response pathways and proposals. They have the ability to influence public opinion, and to legitimize or contest public policies. Despite being commonly referred to as a singular entity, the media are made up of multiple and often distinct institutions, ranging from public to privately-owned, with different communication styles and methods. Media diversity is an important indicator of social and political health, as well as a crucial factor in the quality of democracies. In Brazil, control of the media has historically been in the hands of a few families. According to the Media Ownership Monitor (https://brazil.mom-gmr.org/en/), the Brazilian media system exhibits significant concentration in terms of ownership, audiences, and geography.

Mainstream media can be distinguished from smaller, alternative, media. The former are typically owned and/or have strong connections with large corporations, and prioritize profit as their goal. Given their economic and cultural dominance, mainstream media may be termed

hegemonic. Their approach to journalism is "top-down," with a prevalence of official sources and a relatively rigid and hierarchical organization of news production. **Alternative media** adopt a "bottom-up" perspective in their journalistic practice, giving more space to citizen sources and to those without a voice in the news of major media groups. Alternative media are financially sustained by public contributions or temporary subsidies, which makes the production process more flexible and freer from the constraints of a single model. In principle, this can allow for more critique of the established political and economic systems.

In Brazil, access to journalistic information remains a significant challenge for many small towns located far from major urban areas, particularly those situated outside of the South and Southeast regions. The rise of messaging applications like WhatsApp and Telegram has contributed to the spread of disinformation, partly filling the information void left by traditional media outlets. Television still dominates in terms of advertising expenditure within the media industry, although it is gradually losing ground to digital alternatives (Reuters Institute, 2022). Although Brazil boasts more smart devices than people, internet access is still uneven across regions and social groups.

While in power as President of Brazil, Jair Bolsonaro often tried to discredit journalism. Numerous attacks on the press and journalists were carried out, and public information became more difficult to obtain. This led some news media to repeatedly question Bolsonaro's policies and, as the disruption of environmental protection was flagrant, a large part of the Brazilian press began to cover environmental issues more frequently, including climate change.

Trends in Coverage of Climate Change by Brazilian Mainstream Media

According to Climate Radar's monitoring of climate coverage in Latin American newspapers (https://conexioncop.com/radarclimatico/), there has been an increase in climate-related news in recent years. However, such news items, which may discuss the climate crisis or simply mention it, only constitute around 2% of the total news analyzed. Over time, there has been significant fluctuation in media coverage of climate change, primarily influenced by international factors. In *Folha de São Paulo*, a highly influential Brazilian newspaper, the number of news stories quadrupled between the second half of 2006 and the first half of 2007 (Fioravanti, 2007). This surge in attention was attributed to significant events during that period, such as the release of multiple Working Group reports by the IPCC and a Group of Eight (G8 – group of eight leading industrial nations) meeting. Similar trends were observed in an analysis of 50 newspapers between 2005 and 2008 (Vivarta, 2010), confirming the impact of international events on news agendas. Coverage of climate-related topics is often driven by major scientific reports like those from the IPCC, United Nations Framework Convention on Climate Change (UNFCCC) conferences, or disasters that may be connected to climate change.

What do we know about mainstream media coverage of climate change in Brazil? Various studies have demonstrated that there is a strong dependence on news agencies, both domestic and international, resulting in limited coverage of climate change at the regional or local level. News reports primarily rely on sources associated with the government. Representatives of the scientific field also feature frequently. In contrast, mainstream media do not give much visibility to environmental activists, indigenous peoples, traditional communities, and other citizen sources. Political-economic frames have dominated news coverage of climate change. For instance, *Veja*, *Isto É*, *Época*, and *Carta Capital*, Brazil's largest newsmagazines, have emphasized the costs involved in addressing climate change (Girardi et al., 2013) and *Folha de São Paulo* has focused on political decisions and disputes in international summits, giving prominence to governmental sources (Rodas and Di Giulio, 2017). There is a notable research gap in the coverage of climate change on television, which is particularly concerning as television remains a leading source of information and opinion-shaping in Brazil. Of particular interest are major broadcast networks like TV Globo, both in their regular news coverage of climate change and in their special programs and documentaries that explore the topic, such as the 2019 series "Planet Extremo" ("Extreme Planet").

What about media coverage of the Amazon? Given its biophysical and symbolic significance, it is worth considering how the Amazon is covered in Brazilian media. Despite their clear connection, the climate crisis and the destruction of the Amazon are often addressed separately. The Amazon has historically been depicted as a pristine and untouched landscape, with the human presence often erased (except for indigenous populations), rendering the relationship between local communities and the forest invisible. For many years, the focus was on the forest's exuberance and what was considered exotic, while only a few reports of environmental crimes were produced, and not systematically.

In recent years, the media have increasingly covered issues related to the Amazon, particularly illegal logging, the expansion of the agricultural frontier (including cattle ranching and soybean cultivation), and deforestation. This coverage is due to national public policies and the intensification of climate change consequences. During the period of the Bolsonaro government, there was an increase in media coverage on issues such as illegal mining (supported by the President), as well as failures of or opposition to the preservation of indigenous lands. Additionally, the growth of the global climate debate and the need to reduce greenhouse gas emissions and preserve the forest's ability to capture carbon dioxide have contributed to the Amazon receiving more attention.

National coverage of the Amazon is typically produced by newsrooms located in the distant Rio de Janeiro-São Paulo axis, which adds to the challenge of adequately reporting on the region. Although there have been some improvements in media coverage about the Amazon, significant problems persist. The historical metropolis-colony relationship, stemming from the colonial era, still persists in what is referred to as **internal colonialism**. Local communities continue to be excluded from decision-making processes that determine the future of the region, and nature is often viewed solely as an economic resource, benefitting groups that do not reside there. The notion of the Amazon as an empty and backward region, waiting for development, has been prevalent for a long time and was reinforced by Bolsonaro's rhetoric in favor of exploiting it for economic gain.

Overall, the IPCC's perspective – that climate change is happening and is a result of anthropogenic action – has been prevalent in the Brazilian press. However, with the anti-environmental government of Bolsonaro, some political authorities and media outlets have aligned with policies propagating climate change denial. The widespread dissemination of misinformation through social media and messaging apps, the former government's attempt to discredit the press, and an increase in political polarization are factors contributing to the strengthening of views opposed to scientific evidence. An opinion poll conducted in July 2019 by DataFolha (https://datafolha.folha.uol.com.br/opiniaopublica/2019/07/1988289-para-85-dos-brasileiros-planeta-esta-ficando-mais-quente.shtml) found that 15% of the Brazilian population did not believe in global warming – almost double the number from a survey conducted about a decade earlier (8%).

Regional Media Focusing on National and Global Scales

How is climate change represented in media that operate at the local or regional scales? Given that climate change is a global issue, do they matter at all? Local and regional journalism is an important arena for connecting climate change with citizens' daily lives. It can link climate change to specific social sites and experiences and can offer qualified information for exercising environmental and political citizenship. Research has shown that the local scale is more accessible to citizens and offers more opportunities for engagement with climate change. Local journalism is also better suited to expose local vulnerabilities to climate change, examine socioeconomic inequities, and scrutinize official policies to tackle the sources and consequences of climate change in a given area. After all, all greenhouse gas emissions occur in particular spaces (although very unequally around the world) and climate change impacts also materialize in specific locations.

In Curitiba, the capital of the state of Paraná and the most populous city in southern Brazil, *Gazeta do Povo* is one of the main news media. Although Curitiba is often cited as a model city in environmental terms, it faces climate change-related risks such as floods, landslides, and storms, especially hail. Precarious dwellings and underprivileged social groups are particularly exposed to these impacts, and there are also forecasts of a higher incidence of climate change-derived diseases.

A study of news pieces published in *Gazeta do Povo* in 2013 (Loose, 2020) revealed interesting hints on climate change journalism at the local scale. The analysis showed that the general message was that the climate was changing and that humans contribute significantly to the acceleration of this phenomenon. This message was consistent with propositions of the IPCC. Scientific knowledge was the dominant macro-frame in the newspaper's coverage.

Despite the infrequent use of the term "risk," its meaning was conveyed frequently, particularly in news headlines, subtitles, and leads covering scientific reports and projections regarding the impacts of climate change. The coverage, however, emphasized climate hazards that were distant from the local community, such as the melting of polar ice caps. These findings provide insight into the decisions made by journalists when reporting on scientific studies. Although risks were utilized to capture readers' attention, the detachment of climate-related risks from people's daily lives reduced readers' awareness of the proximity of these risks.

The constant use and reproduction of materials from news agencies created a distance between the reports of *Gazeta do Povo* and the reality of the state of Paraná or the city of Curitiba. While the global scale is relevant and necessary to discuss climate change, its disconnectedness with citizens' daily lives tends to obscure nearby responsibilities over a problem that affects everyone, albeit in different ways. The newspaper's reports did not address the governance of climate change beyond international politics nor encourage public debate on such matters. Possible forms of action to avoid or reduce GHG emissions or to develop adaptation and resilience at the local or regional levels were hardly explored.

Similar to its coverage of climate change policies, *Gazeta do Povo* drew extensively on materials from national and international news agencies when reporting scientific knowledge on climate change. It predominantly contributed to the dissemination of prevailing scientific views on climate change, especially those of the IPCC and Brazilian Panel on Climate Change (PBMC), with little room for other perspectives, such as those of Brazilian and Latin American scientists interested in local aspects.

Different, But Not So Much: Climate Change Coverage by Alternative Media

To explore journalistic discourses beyond those constructed by national and regional mainstream media, this section focuses on alternative media outlets. While both types of media share a set of values and techniques, they tend to differ in their coverage. Alternative media outlets, being non-commercial, have greater freedom to report on the causes and responses to the climate crisis, highlight diverse voices and actions from the Global South, and value local perspectives and knowledge. Historically, alternative journalism has taken a stance against injustices and inequalities, rejecting the notion of neutrality, impartiality, or objectivity that dominates mainstream media. The question then arises: how does alternative climate journalism fare in Brazil?

Conexão Planeta (https://conexaoplaneta.com.br/) and Colabora (https://projetocolabora.com.br/) are two digital news projects specializing in environmental issues that are part of the alternative media landscape. Their editorial lines are dedicated to promoting a more sustainable society, and they frequently cover climate change-related issues. In a study of the 2019–2020 period, Loose (2022) analyzed how and when these media outlets covered climate change. Three peaks in coverage were observed: during the COP-25 period, the student climate strikes in September 2019, and in April–June 2020 in connection with the COVID-19 pandemic. Content related to climate change was categorized based on its focus on causes, effects, solutions, pro-climate actions (such as demonstrations and campaigns), and criticism of inaction. Discursive markers were also examined to reveal silences, representations, and the most recurring players in the news.

A significant portion of news stories focused on global impacts rather than on scales and realities closer to the audiences of these alternative media outlets. The stance was predominantly characteristic of the Global North. Most coverage presented science as the sole authoritative source of knowledge, similar to the approach of mainstream media. Scientists held considerable influence in the climate debate within these alternative media and were often cited in support of the discourses of activists and promoters of climate solutions. While scientists are crucial actors in combating denialism and misinformation, their prominence reinforces a single perspective (the scientific one), which can hinder the pluralization of voices in climate coverage. Traditional knowledge based on a closer relationship with nature, for example of indigenous peoples, riverside communities, and *quilombolas* communities (descendants of enslaved Brazilian-African people), was underrepresented in news reports about climate change in alternative media. The need for economic growth was left unquestioned, with only occasional criticism of the **neoliberal discourse** (i.e., discourse that emphasizes free market economics) when compared to the promotion of a green economy during the post-pandemic recovery. Managerial solutions, such as increased energy efficiency, less polluting transportation, and carbon pricing, were often mentioned. These observations suggest that the dominant values, routines, and criteria of newsworthiness in the journalistic field strongly influence climate change coverage even in alternative media.

Despite this, several distinctive traits were identified in the coverage. Regular reporting was dedicated to pro-climate actions, such as street demonstrations, celebrity statements and political pledges, exhibitions, fundraising and campaigning, as well as other events aimed at raising awareness and promoting climate action. This reflects an engaged journalism that also emphasized blunt criticism, a rare occurrence in mainstream media due to commercial interests. Civil society sources were given more coverage than official sources, particularly political sources. However, the activists that were most frequently consulted or quoted were only a few individuals, namely celebrity-activists that were already well-known figures such as Greta Thunberg. Local activists, including indigenous people fighting for the protection of Brazilian forests, did not receive as much attention as might be expected. This coverage limited the diversity of voices and consequently the range of perspectives.

One distinctive aspect of the alternative media that were analyzed concerns the positioning of journalists alongside activists. In their editorial self-presentation, these media outlets openly declared their commitment to expanding environmental awareness by providing qualified information, and did not try to hide subjectivity behind discursive strategies that pretended to be objective. Instead, they openly exposed their perspective by applauding those who are willing to confront the climate crisis, contrasting them to the government or market depicted as failing to respond adequately to its urgency and seriousness. This approach to journalism driven by journalists who believe in the transformative role of environmental journalism differed from mainstream journalism in Brazil. They used a range of rhetorical devices, including irony, questioning, praise, and disapproval, to express their opinions and critique those deemed responsible for environmental problems.

Although the alternative media differed from the mainstream media in various aspects, they encountered some of the same issues as conventional journalism. Their focus on the global level often overshadowed national, regional, and local perspectives on the climate crisis, leading to the marginalization of discourses from the Global South. While the alternative media offered a platform for activists who are frequently disregarded by mainstream media, their coverage still fell short of representing the expected diversity of perspectives. Climate change was portrayed as an urgent and inevitable threat, yet the underlying causes of the climate catastrophe, specifically the **capitalist-colonialist system**, were seldom discussed. Capitalism and colonialism are phenomena that operate together: both rely on the asymmetrical distribution of power and the limitless extraction of resources from colonized territories (see Callison, Chapter 3, this volume). Up to the present day, Brazil's economy remains largely dependent on the global market's demand for agricultural and mining products, which, as mentioned above, contribute heavily to GHG emissions and bring little benefit to its peoples. In a different example, international policy mechanisms created in the last few decades like carbon markets and carbon-offsetting schemes enable large emitters of GHG to continue profiting from burning fossil fuels while paying a small price to communities in the Amazon. The fact that these systemic issues do not gain more visibility in alternative media indicates that structural blindness to some forms of power and exploitation (especially from the Global North) exist even in media outlets that are not considered part of the mainstream.

Concluding Remarks

Journalistic representations have significant influence on the formation of worldviews and subsequent actions. The way in which journalism frames climate change can bring the issue closer to or further from people's everyday reality, and can present solutions or portray it as inevitable. Focusing solely on the negative impacts of climate change can create fear and lead to inaction. It is important to have a discussion about the root causes of the current situation and what measures can be taken to mitigate it. In addition to individual actions, it is essential to examine the roles of companies and governments in addressing climate change. The social inequalities inherent in the issue must be addressed and solutions must be inclusive and equitable.

Communication studies on climate change in Latin American countries have tended to focus on traditional or hegemonic media, leading to similar findings: a dependence on international media sources, a lack of specialization among journalists, an emphasis on international agreements, and a disconnect from local realities. Despite Brazil's significant responsibilities and vulnerabilities with regard to climate change, media coverage has not adequately addressed the issue and suffers from various limitations. Local and regional actions to respond to climate change are rarely featured in media coverage, which tends to prioritize international politics (i.e. summits and official events, rather than structural inequalities and dependencies). Even within alternative media, there is a tendency to privilege sources with existing institutional or media importance, thus reproducing a hegemonic logic. Although their positions are closer to activists and they are more committed to positive social change than mainstream media, they still fall short of appropriately debating the systemic, multi-scalar political and economic relations that are implicated in climate change.

Although media coverage of climate change has increased in recent years, it remains a challenge to connect this complex global crisis to people's everyday lives. In a recent study, Brazilian activists maintained that mainstream media often overlook sources with direct experience of climate impacts, failing to consider local factors that are essential for engaging audiences (Loose et al., 2022). These findings underscore the challenges faced by journalists in covering climate change from a multicultural and multi-scale perspective that promotes fairness and justice. To gain a broader understanding, it would be interesting to compare these conclusions about Brazilian media with research on media discourses about climate change in other countries, including their portrayal of Brazil and its role in the climate crisis.

Further Reading

Broadbent, J., Sonnett, J., Botetzagias, I., Carson, M., Carvalho, A., Chien, Y. J. … Zhengyi, S. (2016) "Conflicting climate change frames in a global field of media discourse," *Socius: Sociological Research for a Dynamic World*, January–December, 2.

Carvalho, A. and Loose, L. (2018) "Climate change in Brazilian media," in Brevini, B. and Lewis, J. (eds.). *Climate Change and the Media*, 2nd ed. New York: Peter Lang, pp. 79–94.

Deutsch, S. and Fletcher, R. (2022) "The 'Bolsonaro bridge': Violence, visibility, and the 2019 Amazon fires," *Environmental Science & Policy*, 132, pp. 60–68.

References

CEPAL (2015) *La Economía del Cambio Climático en América Latina y el Caribe Paradojas y Desafíos del Desarrollo Sostenible*. Santiago de Chile: Naciones Unidas.

Fioravanti, C. (2007) "Climate change reporting in Brazil," presentation at the workshop *Carbonundrums: Making sense of climate change reporting around the world*, University of Oxford.

Girardi, I. M. T., Massierer, C., de Moraes, C. H., Loose, E. B., Neuls, G., Schwaab, R., Camana, Â., and Gertz, L. (2013) "Discursos e vozes na cobertura jornalística das COP-15 e 16," *Em Questão*, 19, pp. 176–194.

ITS-Rio and Yale University (2022) *Mudanças Climáticas na Percepção dos Brasileiros: Relatório de Análise*, Rio de Janeiro: ITS-Rio.

Loose, E. B. (2020) *Jornalismo e riscos climáticos: Percepções e entendimentos de jornalistas, fontes e leitores*. Curitiba: Editora UFPR.

Loose, E. B. (2022) "Cobertura climática desde o Sul: análise crítica de discursos jornalísticos não hegemônicos," *Estudos de Jornalismo e Mídia*, 19, pp. 219–232.

Loose, E. B., Fante, E. M., Jacobi, C. M., and Thiesen, L. J. (2022) "A cobertura climática pode levar à ação?," *Revista Ciências Humanas*, 15, pp. 8–21.

Reuters Institute (2022) *Digital News Report 2022*. United Kingdom: University of Oxford.

Rodas, C. and Di Giulio, G. (2017) "Mídia brasileira e mudanças climáticas: Uma análise sobre tendências da cobertura jornalística, abordagens e critérios de noticiabilidade," *Desenvolvimento e Meio Ambiente*, 40, pp. 101–124.

SEEG (2021) *Análise Das Emissões Brasileiras de Gases de Efeito Estufa e Suas Implicações Para as Metas Climáticas do Brasil 1970–2020*, São Paulo: Observatório do Clima.

Vivarta, V. (2010) *Mudanças Climáticas na Imprensa Brasileira: Uma Análise Comparativa de 50 Jornais nos Períodos de Julho de 2005 a Junho de 2007–Julho de 2007 a Dezembro de 2008*. Brasília; Agência de Notícias dos Direitos da Infância (Andi).

8 News and Social Media Imagery of Climate Change

Analyzing the Role and Impact of Visuals in Public Communication

Mike S. Schäfer and Xiaoyue Yan

Why Media Imagery of Climate Change is Relevant

Public and media communication is crucial when it comes to climate change. Journalistic media – in print, radio, television, or online – are still important sources of news about climate change for many people. In addition, social media from social networks sites like Facebook, Instagram, and Twitter to video-sharing platforms like YouTube and TikTok have strongly risen in importance, particularly (but by no means exclusively) for younger generations. Even though a considerable proportion of the public say they perceive the impacts of climate change in their everyday life and immediate surroundings already, media sources still play an important role in shaping public perceptions and attitudes toward climate change.

In recent years, visualizations such as photos and infographics, but also memes, gifs, and short videos have become more important in news and social media. This applies to many topics, including communication about climate change. Imagery is used more widely in climate-related news. On social media, the importance of visuals has been heightened by the fact that several platforms are centered around visuals (like Instagram) or audio-visual content (like YouTube or TikTok). Visuals can serve as eye-catchers for audiences and users, generate and focus attention and engagement, and can be distributed easily across national, cultural, and linguistic boundaries (even though they have been shown to be interpreted differently in different sociocultural contexts). Ed Hawkins' "warming stripes" and "climate spirals" visuals, indicating average global temperatures and their changes over time (Figure 8.1), are well-known examples. They have gained prominence in news and social media and are used across the globe.

Researchers have responded. Studies of climate-related media imagery have gained importance in recent years. They have focused on stakeholders' use of climate change-related visuals (Wozniak, 2020), visual representations of climate change (Schäfer, 2020), and their uses and effects (Metag, 2020).

This chapter describes, first, how researchers have approached these questions, emphasizing the contributions of STS. It then summarizes the studies conducted on news media imagery of climate change, and social media imagery, respectively. Finally, it outlines future pathways for the field.

Analyses of Climate Change Imagery in STS and Beyond

Research into climate change communication started to grow in the 1990s and particularly in the mid-2000s. Soon afterward, studies on visual climate change communication appeared (see O'Neill and Smith, 2014 for an overview). Scholars analyzing climate-related visuals in news and social media come from different disciplines (such as sociology, geography,

DOI: 10.4324/9781003409748-11

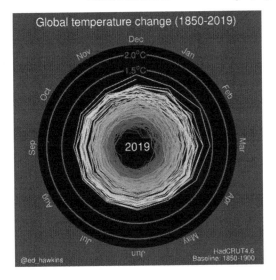

Figure 8.1 The "climate spiral" and the "warming stripes".

Source: Both developed by climate scientist Ed Hawkins (https://ed-hawkins.github.io/climate-visuals/).

communications, political science, or computer linguistics) and interdisciplinary fields like environmental studies. They mostly apply qualitative or quantitative variants of visual content analysis, visual framing analysis, or discourse analysis. Notably, they have mostly focused on print media imagery in English-speaking countries.

STS approaches play a role in this field, but not a prominent one yet. This may be due to STS's strong focus on communication *within* science rather than public communication, and its perspective on news and social media as socially shaped technologies rather than as places of public debate. The same is true for imagery: While STS scholars have turned to visual aspects more strongly in recent years, they have concentrated on images and the surrounding social practices within science. As a result, STS and studies on public communication have intersected and collaborated only selectively so far.

Nonetheless, STS has a lot to contribute to the analysis of climate-related visuals, starting with some of its fundamental tenets. Its basic assumption that science and scientific developments are socially embedded and constructed and must be studied accordingly also applies to climate science. STS's emphasis on news and social media's sociotechnological characteristics and their impact on communicators and audiences as relevant means for enabling societal debate and creating social meaning is important to keep in mind when analyzing any form of mediated communication.

STS also approaches imagery from a fundamentally constructivist point of view. It underlines that "seeing and recognition are historically and culturally shaped" (Burri and Dumit, 2008, p. 299), and that "scientific images and visualizations are exceptionally persuasive because they partake in the objective authority of science and technology" because they may "appear universal and neutral while selectively privileging certain points of view and overlooking others" (Burri and Dumit, 2008, p. 299). STS emphasizes that this persuasiveness also applies to mediated, public communication, for example when scientific images are used and distributed beyond science among stakeholders or the broader public.

In addition to these general points, specific concepts from STS lend themselves well to analyses of climate-related visuals and have partly been employed in analyses of mediated communication already. Examples are:

- **Boundary work** (Gieryn, 1999), which refers to practices through which different fields of knowledge and the boundaries between them are negotiated, often repeatedly and continuously. This can be the fundamental boundaries between science and its surroundings, or negotiations about expertise and about the question of who is an expert in a given field. Related concepts are **boundary objects** (Star and Griesemer, 1989) or **boundary organizations** (Guston, 2001), which bridge different knowledge communities. The concept of boundary organizations has been employed to analyze public communication of scientific institutions and Science Media Centers, for example, and images have repeatedly been interpreted as important boundary objects.
- **Science-related controversies**, in which knowledge communities exchange and negotiate different perceptions of scientific knowledge or technological applications, are another focus of STS that has been taken up by scholars analyzing mediated communication. The latter have analyzed, for example, how advocates and opponents of stem cell research or human cloning hold clashing viewpoints, how different communities debated fracking, or how discussions about applications of artificial intelligence develop. Sometimes using methods developed in STS such as "controversy analysis" (e.g., Marres and Moats, 2015), they have aimed to reconstruct how stakeholders engage in public, how they try to make their views and positions prominent in news and social media, and how successful they are in doing so.
- A third focus of STS studies that lends itself well to analyses of mediated communication is the concept of **imaginaries**, understood as "collectively held, institutionally stabilized, and publicly performed visions of desirable futures" (Jasanoff and Kim, 2015, p. 4) that can manifest themselves in various ways and influence how people think and act toward issues such as climate change. Imaginaries can manifest in imagery as well. Such imaginaries, and their visual representations in news media or social media, can influence how members of the public envision the future development of scientific fields and certain technologies.

Prevalence and Characteristics of Climate Change Imagery: Findings from Research

Analyzing News Media Visuals of Climate Change

In recent years, more scholars have analyzed climate change visuals in news media. They have mostly focused on print media outlets and their online equivalents, including tabloids and broadsheet publications, and occasionally also television news. Most of them analyzed Western countries, particularly Anglophone countries such as Australia, the UK, and the US, but recently research has diversified to include some countries from the Global South.

Scholars found, first, that the *use of imagery in news media has been increasing.* News media utilize more images and position them more prominently – which is true for many topics and also applies to climate change.

Second, *news outlets throughout the world adopt relatively similar visual representations of climate change.* Images depicting polar bears on (too) small ice shelves, national landmarks submerged under water due to rising sea levels, or the joyous pose of world leaders after signing the Paris Agreement at the Conference of the Parties (COP) climate summit in 2015 have been used in news coverage around the world. Quantitative analyses have shown that in many countries, the most widely used images are those showing large-scale, harmful effects of climate change such as melting polar ice, floods, droughts, or mudslides. The second most prominent category is visualizations of politicians, celebrities, or stakeholders, which are often shown speaking at international summits or conversing with each other. Less common but still prevalent in news coverage are infographics that show temperature curves and other climate developments, images reflecting the causes of climate change such as greenhouse gas emissions, visualizations of impacts on citizens' everyday lives, pictures of unspoiled nature, visualizations of activists and their actions, and images of solutions.

Third, studies on the effects of news media imagery have shown that *this selection of visualizations may be problematic.* They have shown, for example, that the most widely used images of large-scale, harmful effects of climate change draw audience attention, but also give them a feeling of powerlessness and hopelessness, diminishing people's perceived self-efficacy and their intentions to act. In turn, visuals that may be more potent in motivating action are utilized less frequently in news media, such as pictures of solutions or the everyday consequences of climate change.

Fourth, research has shown that the *visual portrayal of climate change in news media is not only influenced by journalists and news producers, but also by stakeholders from politics, civil society, business, and science.* Throughout the past few decades, a wide range of visual arguments and assertions regarding climate change have been generated strategically by environmental groups and activists. For instance, the environmental non-governmental organization (NGO) Greenpeace first used visuals to reflect the potential dangers and destructions coming from the rising temperature in the 1980s and 1990s, then shifted its visual messaging to identifying the causes, responsible parties, and potential solutions to the issue in the late 1990s and 2000s, and finally, it turned to pressuring policymakers. These climate visuals were widely adopted by journalists to explain the rather complex issue and engage the public. Another active group of stakeholders are climate scientists, who often employ infographics and photographs to visualize research findings for decision-makers and the public. A typical, influential example is the Intergovernmental Panel on Climate Change (IPCC) assessment reports that widely, and influentially, use imagery as heuristic cues to convey scientific messages. In recent years, stakeholders' use of visuals has evolved, with many turning to more audience- and impact-focused visualizations that sometimes move beyond the realms of environmental activism and climate research and into spheres like arts and advertising.

Fifth, *visual content has been shown to have considerable influence on viewers,* potentially more so than textual information. Scholars have analyzed the effects of climate visuals on audiences' awareness, emotions, and behaviors toward climate change. Regarding awareness, findings show that images depicting the consequences of climate change are better comprehended by audiences and lead to a feeling that climate change is significant. It has been found that climate visuals – including audio-visual content like climate-related films and documentaries – can heighten audiences' concerns about climate change, elicit positive or negative emotions, and influence audiences' behavioral intentions. Studies exploring the affective and emotional

Table 8.1 Core findings from the research field

	Analyses of News Media Imagery	Analyses of Social Media Imagery
Typical research approach	Mostly quantitative or qualitative content analyses of print media imagery from Western countries.	Mostly manual variants of content analysis of small samples, usually from one social media platform.
Core findings	The use of imagery in news media has increased.	Visuals on social media differ from news imagery.
	News outlets around the world adopt relatively similar imagery of climate change.	Number of videos supporting consensus views about climate change and skeptical videos differ considerably from platform to platform.
	The selection of news imagery can be problematic.	
	News imagery of climate change is influenced by journalists, news producers, but also many stakeholders.	Activists and activist groups have been shown to use social media visuals extensively.
	Visual content has considerable influence on audiences.	Different types of visuals receive different levels of engagements on social media.
Limitations	Lack of multimodal analysis that is taking imagery as well as written text, sound, or video into account.	Mostly single platform studies, often analyzing Twitter.
	Focus often on case studies and lack of large-scale (e.g. computational) analyses.	Lack of quantitative and computational analyses.
	Focus largely on English-language data and Western countries.	Focused largely on English materials and Western countries.

impacts of visuals have shown that fear appeals (messages aiming to elicit fear by highlighting dangers and suffering), which often appear in climate change communication, can elicit negative emotions and stifle action – while other, less commonly used visuals (like depictions of solutions) evoke positive emotions and can improve people's belief that they can take action to address climate change.

Analyzing Social Media Visualizations of Climate Change

The emergence and rise of social media have changed how issues like climate change are publicly communicated, perceived, and engaged with. Social media have become important sources of climate-related content for many. They have changed the fundamental logic of public communication by allowing for many-to-many communication, bypassing established gatekeepers like journalists, and enabling members of the public to author, distribute, share, or comment upon content and take an active part in the public discussion of climate change.

So far, however, only very few studies have analyzed climate change imagery on social media. They mostly employ discourse analysis, qualitative or quantitative content analysis that rely on manual coding and modest sample sizes (from one to 200 visuals). Often, they are case studies focusing on single well-known organizations or persons, like Greenpeace and Greta Thunberg, or prominent events like the COP summits. They frequently employ multimodal approaches, taking visual and textual attributes of posts into account. Several findings can be distilled from the field:

First, even though few studies have compared this systematically, *visuals on social media seem to differ from news media imagery*. On Twitter, memes, motivational quotes, and screenshots have been shown to be the most common visualizations of climate change, followed by portrayals of individuals like politicians and celebrities. Visualizations of climate change consequences, the most popular visual category in the news, are less important on social media. When focusing only on "top tweets" (defined by Twitter as "the most relevant" tweets for a search query based on the platform's "popularity" measure that contains interactions, shares, and other factors), the use of visuals changes. While the imagery in top tweets often portrays individuals, these are more often citizens rather than politicians and celebrities who are prominent in news media. Images of climate consequences and solutions are the second most prevalent category among top tweets, followed by images depicting protests and scientific imagery. Memes on Instagram and Facebook have been shown to call mostly for awareness and action against climate change, followed by memes attacking liberal and conservative political views and politicians.

Second, studies analyzing audio-visual content have shown that the *number of videos supporting consensus views about climate change versus climate-skeptical videos differs considerably from platform to platform*. A YouTube analysis in 2018 showed that among 200 randomly selected videos on climate change, the majority opposed the scientific consensus (Allgaier, 2019). On TikTok, the science of climate change is rarely a topic, but a large majority of videos on the platform support the scientific consensus, often coupling sincere appeals with humorous text or visuals when mentioning the issue. It is notable, however, that some widely viewed videos exist on TikTok as well that refute anthropogenic climate change, and that a substantial proportion of TikTok videos with climate-related hashtags are irrelevant to the issue. They just "hijack" the buzz generated by the hashtags to draw attention to themselves.

Third, *activists and activist groups use social media visuals extensively* to broadcast their views, mobilize their audiences, and illuminate the absence of news media coverage. This was especially true for young activists, who are more accustomed to social media logics. The prime example is activist Greta Thunberg who communicated her weekly climate strikes on social media and has become a global icon and an important communicator of the issue. She has been shown to frame climate change as a moral and ethical issue on Instagram and to use visuals to motivate collective action, for example, by depicting protest signs and smiling peers while engaging in activism. Activists sometimes come together and protest during important climate events like COP meetings. In their protest videos on YouTube during COP15, activists visually portrayed themselves as soldiers or freedom fighters. The visual discourse focuses on the nodes of war, injustice, and resistance. Among activist groups, Greenpeace uses three main visual themes on Instagram to communicate a "climate crisis" in Indonesia: climate crisis threats, an urgent need to switch to renewable energy, and calls for more environmentally friendly political regulations. Environmental NGOs also post advertisements about climate change on Facebook, in which they frequently combine texts on pollution and efficacy with visuals of climate impacts and texts about protest with visuals of collective action.

Fourth, *different visuals receive different levels of engagement on social media*. Social media allow individuals to become both content producers and reproducers by sharing, liking, and commenting on specific content. Therefore, understanding the drivers of social media engagement is important for analyses of (visual) communication of climate change. Several scholars have focused on this question. They have shown that despite their infrequent occurrence, protest visuals regularly received the highest engagement on Twitter, followed by people-related visuals. In contrast, memes, motivational quotes, and screenshots generated limited amounts of engagement despite their prevalence. On YouTube, videos in favor of the scientific mainstream perspective barely outnumbered those against it in terms of views. On TikTok, videos of natural disasters and the environmental effects of climate change typically get more views, likes, and comments than other videos. Although making up only a small portion of climate videos on TikTok, videos that spread misinformation about climate change earned many views as well.

Limitations and the Way Forward

Generally, researchers have paid less attention to climate change images than merited by their importance. Consequently, several of the findings above are less definitive than they should be. Since the amount of climate change visuals is clearly rising in news and social media and their effects on audiences are significant, more research is urgently needed in this field. In addition, this research needs to be more diverse – in the cases it analyses, the contexts and countries it draws them from, and the news and social media it focuses on. With their unique conceptual and methodological approaches, more STS research would be particularly helpful in providing an additional conceptual grounding.

In addition, current research on climate change communication in news and social media has clear gaps and limitations. Most research analyses print media and often focuses on single news media or social media (mostly Twitter) within each study. Even though visuals are disseminated across borders and socio-political contexts, and can travel across language barriers, research continues to concentrate almost entirely on English-language materials and Western nations. Often media within cultures and nations more vulnerable to climate change are ignored such as coastal countries in south and southeast Asia. Despite the importance of understanding how climate visuals affect audiences, existing research frequently does not investigate impacts.

A significant gap in news media studies is their lack of multimodality. Studies usually concentrate only on visuals such as still images and photographs. But visuals function in concert with other modalities, such as texts, and can have varying effects on audiences accordingly. More studies examining news visuals in conjunction with other modalities should be conducted.

Social media analyses should go beyond Twitter and expand their view to include social media platforms with large user bases like Facebook and centered around visuals like Instagram. In addition, studies across platforms are required since different social media platforms have unique logic and affordances. Scholars should also pay attention to specific online visuals like memes, screenshots, and gifs and consider their potential for use by climate change skeptics. In general, the role of visuals for the dissemination of dis- and misinformation and climate-related conspiracy theories needs questioning.

Finally, the role of generative artificial intelligence, which provides original responses to user prompts based on supervised and reinforcement machine learning techniques, for the visualization of climate change should be analyzed (Schäfer, 2023). Tools like DALL.E, Midjourney, or Stable Diffusion can already produce photo-like visualizations that could change visual communication about climate change (among other issues) considerably.

Further Reading

Burri, R. V. and Dumit, J. (2008) "Social studies of scientific imaging and visualization." In Hackett, E. J., Amsterdamska, O., Lynch, M., and Wajcman, J. (Eds.). *The Handbook of Science and Technology Studies.* Cambridge, MA: MIT Press, pp. 297–317.

Metag, J. (2020) "Climate change visuals: A review of their effects on cognition, emotion and behaviour." In Holmes, D. and Richardson, L. (Eds.), *Research Handbook on Communicating Climate Change.* Cheltenham, UK: Edward Elgar Publishing, pp. 153–160.

O'Neill, S. J. and Smith, N. (2014) "Climate change and visual imagery," *Wiley Interdisciplinary Reviews: Climate Change*, 5(1), pp. 73–87.

Schäfer, M. S. (2020) "Introduction to visualizing climate change." In Holmes, D. and Richardson, L. (Eds.), *Research Handbook on Communicating Climate Change*. Cheltenham, UK: Edward Elgar Publishing, pp. 127–130.

Wozniak, A. (2020) "Stakeholders' visual representations of climate change." In Holmes, D. and Richardson, L. (Eds.), *Research Handbook on Communicating Climate Change*. Cheltenham, UK: Edward Elgar Publishing, pp. 131–142.

References

Allgaier, J. (2019) "Science and environmental communication on YouTube: Strategically distorted communications in online videos on climate change and climate engineering," *Frontiers in Communication*, 4: 36, pp. 1–15.

Gieryn, T. F. (1999) *Cultural Boundaries of Science*. Chicago: University of Chicago Press.

Guston, D. H. (2001) "Boundary organizations in environmental policy and science: An introduction," *Science, Technology, & Human Values*, 26(4), pp. 399–408.

Jasanoff, S. and Kim, S. H. (Eds.) (2015) *Dreamscapes of Modernity: Sociotechnical Imaginaries and the Fabrication of Power*. Chicago: University of Chicago Press.

Marres, N. and Moats, D. (2015) "Mapping controversies with social media: The case for symmetry," *Social Media+ Society*, 1(2), 2056305115604176.

Schäfer, M. S. (2023) "The Notorious GPT. Science communication in the age of artificial intelligence," *JCOM – Journal of Science Communication*, 22(2), Y02.

Star, S. L. and Griesemer, J. R. (1989) "Institutional ecology, 'translations' and boundary objects: Amateurs and professionals in Berkeley's Museum of Vertebrate Zoology, 1907–39," *Social Studies of Science*, 19(3), pp. 387–420.

Part IV

NGOs, Civil Society, and Social Movements

Introduction

Mark Vardy

Building on the work of Part III, the chapters in this Part continue to explore how knowledge about climate change is framed and mediated. But instead of looking at the media, the chapters in Part IV turn their attention to social and cultural movements. These movements, and the non-governmental organizations (NGOs) who are often referred to as social movement organizations, have a profound impact on society. Critical sociologists used to think about social movements in terms that were largely informed by Marxist analysis; the working class was considered as the locus of social change. But this changed during the 1960s and 1970s. Rather than the working class, it was a different set of social movement actors who emerged to profoundly impact the shape of society in the Global North, including the Civil Rights, Women's, Gay Liberation, Peace, and Environmental Movements. Sociologists and other social movement scholars spend considerable effort to study these social movements, examining topics such as how, when, and where social movements emerge, grow, mobilize, and dissipate, as well how they shape society.

Science and Technology Studies (STS) has important areas of affinity with social movement scholarship. Social movements are an integral part of democratic societies, giving voice to positions of dissent or opposition. For this reason, a large part of STS is concerned with how ordinary people can have greater participation in the decisions that affect them (Chilvers and Kearns, 2016). In addition, many STS scholars argue that STS should help open up issues of public importance that are dominated by the presumption that scientific expertise has a taken-for-granted authority to determine how social issues should be understood and what should be done about them (Wynne, 2003). Here we can see the constructivist roots of STS showing through. That is, STS is well suited to analyzing how environmental issues are debated in public because it refuses to grant science the ultimate say in how problems should be defined.

Sociologists and historians working in areas adjacent to STS also research the climate denialist movement as a "countermovement." Tracing connections between the fossil fuel lobby, conservative think tanks, public relations firms, politicians, and individual scientists, these scholars have argued that climate denialism is motivated by the belief that governments should not regulate the market and should not impede the ability of individuals to pursue private profit (Brulle, 2021; Dunlap and McCright, 2015). In other words, just as there are social movements urging governments and businesses to act on climate change by reducing emissions, so too are there social movement actors who urge the opposite. STS scholars point out how a key area of similarity between these two seemingly contradictory movements is an image of science that is unified and capable of producing certainty through a singular scientific method (Latour, 2015). For environmentalists, climate science grounds their claims that we should reduce fossil fuel use; for climate denialists, any supposed deviation or misstep from a highly idealized version of science is reason to discredit the entirety of climate science.

DOI: 10.4324/9781003409748-12

As Steven Yearley details in Chapter 9, there is a plurality of ways that environmental NGOs intervene in the public framing of climate change. Considering the School Strikes and Fridays for the Future movement, Yearley asks if the climate justice movement might have found a new historical actor. That is, the School Strikers can be considered on the same scale and scope as the new social movements identified earlier, which emerged in the 1960s and 1970s. But new climate movements (NCMs), such as the Extinction Rebellion, face similar challenges as the earlier generations of environmental NGOs, which pertain to their reliance on the social authority of science. In considering these issues, Yearley raises important points to consider about how the plurality that STS sees in science can be extended into the climate justice movement.

In Chapter 10, Adam Fleischmann draws from his ethnographic research with NGOs to engage with questions of how to make climate change a tractable social problem. Through his in-depth and long-term research with Climate Interactive, Fleischmann shows how the framing of science and policy can be reworked by actively engaging ordinary people in simulations of international climate negotiations. In these sessions, individuals adopt the roles of various nation-states and enter into negotiations as if they were acting on that country's behalf. In this way, Fleischmann shows us how climate change is not just a matter of communicating the correct science but also engaging with difficult, contentious, and all-too-human politics.

While the first two chapters in Part IV expand our understanding of the plurality of ways NGOs are acting to make climate change a social and political problem, the third chapter considers a conservative cultural movement associated with climate denialism. More specifically, Allison Ford draws from her ethnographic research to show how people who identify themselves as "preppers" understand and deal with disaster. Preppers actively prepare for future disasters by making themselves as self-sufficient as possible. The preppers who Ford discusses in Chapter 11 envision a future in which disasters have rendered many of the infrastructures, goods, and services that are part of life in industrialized nations unavailable or unworkable. Ford shows us how preppers share a common reality with environmentalists, namely a realist understanding of the potential for disaster, such as those caused by floods, fires, and droughts. But while they are aware of the potential for disaster, preppers are skeptical of climate change precisely because of their conservative and libertarian beliefs, which are against governmental regulation. Ford shows how their commitments to ideological beliefs lead them to "skirt the frame" of anthropogenic climate change that is promoted by the very NGOs that Yearley and Fleischmann consider in their chapters.

As the chapters in Part IV remind us, climate change is not just a scientific issue. In order for us to understand it as a social and political problem, we need to pay attention to the myriad ways it becomes known, communicated, and understood by people, including the NGOs and other cultural and social movement actors.

References

Brulle, R.J. (2021) "Networks of Opposition: A Structural Analysis of US Climate Change Countermovement Coalitions 1989–2015," *Sociological Inquiry*, 91(3), pp. 603–624.

Chilvers, J. and Kearnes, M. (2016) *Remaking Participation: Science, Environment and Emerging Publics*. Abingdon: Routledge.

Dunlap, R.E. and McCright, A.M. (2015) "Challenging Climate Change," in Dunlap, R.E. and Brulle, R.J. (eds.) *Climate Change and Society: Sociological Perspectives*, New York: Oxford University Press, pp. 300–332.

Latour, B. (2015) "Telling Friends from Foes in the Time of the Anthropocene," in Hamilton, C., Bonneuil, C., and Gemenne, F. (eds.) *The Anthropocene and the Global Environmental Crisis*. London: Routledge, pp. 145–155.

Wynne, B. (2003) "Seasick on the Third Wave? Subverting the Hegemony of Propositionalism: Response to Collins & Evans (2002)," *Social Studies of Science*, 33(3), pp. 401–417.

9 Non-Governmental Organizations and the Environmental Movement

Challenges in Climate Change Framing

Steven Yearley

Introduction

Environmental non-governmental organizations (NGOs) and advocacy groups have long been renowned for their stunts and campaigning, not least in relation to issues around climate change. They mount noisy protests in the face of airport construction projects; at international climate negotiations they stage marches and "shadow events" to exert pressure on the delegates; and they have adapted earlier anti-nuclear mobilizations to oppose new coal-fired power stations, fracking, and those carbon capture and storage facilities that depend on links to fossil fuel industries. In the last few years, new groups have emerged that apply direct action in novel ways, such as the Sunrise Movement in the USA. In the USA and Britain, and in numerous other countries, one prominent group is the Extinction Rebellion, known as XR, which was founded in the UK in 2018. XR focuses on direct action events to draw attention to the climate crisis and threats to biodiversity. Its methods resemble those of the Occupy movement that protested banks and capital in the context of economic austerity after the financial crisis of 2007–2008. In one stand-out XR protest, in 2019 a former Paralympic athlete managed to super-glue himself to the top of a British Airways plane at a London airport favored by business travelers, disrupting many flights for the day.

Urgency is the key theme of XR, as memorably communicated by its logo, which uses a stylized "X" to evoke an hourglass or glass timer, highlighting that our time is running out. Some members of XR in Britain then reshaped themselves into the more precisely focused Insulate Britain (2021), a campaign organization demanding that government intervene so that new social housing and the existing stock of dwellings be adequately insulated (it is widely agreed that the UK wastes a lot of natural gas in heating poorly designed domestic spaces). Insulate Britain supporters caused consternation and delays by blocking roads, thus highlighting society's addiction to fossil fuels. A later spin-off, Just Stop Oil (dating from 2022), built on this tradition of non-violent protest and began by "occupying" trucks transporting petroleum products and by using various means to block traffic on freeways and major bridges. In October 2022, two Just Stop Oil activists threw canned tomato soup at a celebrated Van Gogh picture of sunflowers in London's National Gallery as part of a move toward protests in the cultural sectors. Safe behind protective glass, the Van Gogh painting was undamaged, but the incident, which was intended to stimulate discussion about what society values and why, sparked outrage.

The stance of these groups is radical. Protestors are often arrested for not compromising their commitments to their goals. They mobilize through social media and to some extent online but have very little administration or overhead, making them distinct from groups such as Greenpeace and Friends of the Earth, who – these days – are professionalized groups with offices and rents to pay, and who need to solicit donations from foundations and middle-class supporters. But this makes it all the more notable that on Just Stop Oil's website they choose to have a

DOI: 10.4324/9781003409748-13

prominent quote in very large letters attributed to Sir David King, the former Chief Scientific Advisor to the British Government: "What we do over the next three to four years, I believe, is going to determine the future of humanity" (Just Stop Oil, 2023). King made this comment in a speech he gave at the 2021 Climate Emergency Summit in Australia where he went on to say, "We are in a very, very desperate situation." King's remark has been cited by XR also; their website invokes him, stating "This stuff is real. The science is clear. Our future is not" (Extinction Rebellion, 2023). The key point is that, despite their activism and spontaneity, their super-gluing and soup-based protests, these climate pressure groups are keen to show that their claims are ratified or endorsed by senior scientific figures. This indicates something important about environmentalism and climate change.

Climate Framing and the Role of Science

This conspicuous role for scientific authority arises precisely because the convincingness of these groups' message depends on the notion that their claims have a basis in factual accuracy – that they are not simply matters of opinion or ideology, but can withstand expert, scientific scrutiny. Environmentalists, more than any other type of campaigner, need to persuade the public that things are *in fact* the way they say things are, even when some of the claims they are making seem – at first glance at least – to be counter-intuitive or implausible: that methane-heavy "burps" from cows and sheep can warm the atmosphere significantly, that minute plastic spheres in cosmetic products can end up accumulating in ocean creatures, or that burning coal, gas, and oil can unsettle the entire global climate. Most other social movement claims are based around justice, fairness, or rights, as with the Civil Rights movement, the Women's Movement, and activism around LGBTQ+ identities. In the case of climate change, a big challenge has been to express the strength of evidence for the reality of climate effects and to combat those who have set out to sow doubt. The difficulty for environmentalists arises from two sources. In part, there is the fact that climate change has generally been a gradual process so that ordinary people have mostly not been able to detect it or distinguish it from general weather variability on a casual basis. This means that environmentalists have had to rely on the social authority of science to argue that the climate is indeed changing and that particular instances of observed changes are attributable to anthropogenic causes. Second, since climate change has arisen primarily from fossil fuel consumption (and is thus tied to all sorts of economic activity), attempts to take steps to combat global warming have been opposed or questioned by many in industry and intensive agriculture, by lots of vehicle manufacturers, by right-leaning politicians and policy makers (who are often inclined to view it as a left-wing attempt to regulate the market), many bankers, and most directly by fossil fuel industries and producers themselves. Even some established labor unions have voiced skepticism, based on perceived threats to workers' livelihoods. All of these groups, motivated by ideological, economic, or political concerns, have questioned the scientific basis of climate change, which can make the environmentalists all the more insistent that science be granted authority.

Since their formation in the 1970s, celebrated environmental movement organizations in the Global North have often protested against the establishment, including establishment scientists, over issues such as nuclear power, agricultural chemicals, and the desirability of genetically modified crops and foods. In the climate case, environmentalists have thus found themselves in an unusual situation. What they see as the world's leading environmental problem is fully endorsed by the mainstream scientific community and, in principle at least, by most world governments whose representatives have now signed off on six sets of Intergovernmental Panel on Climate Change (IPCC) reports and, overwhelmingly, signed up to the 2015 UN

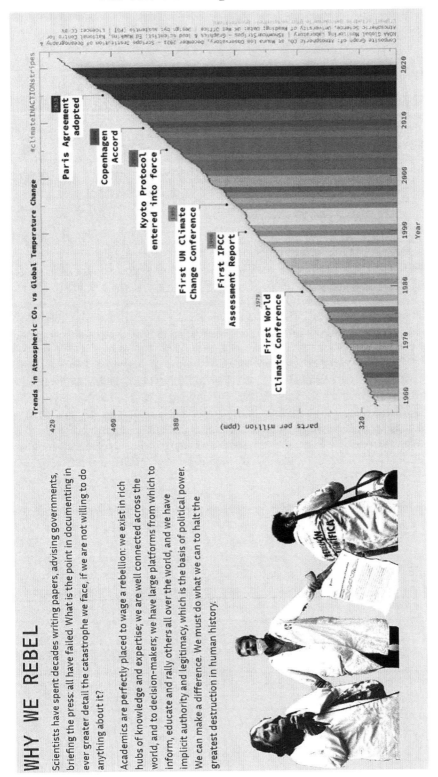

Figure 9.1 Screenshot of homepage of Scientist Rebellion.

Source: See https://scientistrebellion.org/.

Paris Agreement to combat climate change. A large part of environmental NGOs' efforts – even those with scientists on their staff – have accordingly been directed at restating and emphasizing mainstream findings, identifying novel ways to publicize the message, and countering the claims of global-warming skeptics.

It is in this context that new climate movement (NCM) groups, such as Just Stop Oil, have come to focus on the need for urgent action and have picked tactics that are designed to shock societies into change. They accept governments' decarbonization ambitions, but they want the action to be greatly accelerated. In the face of attempts by some political figures and some industry lobbyists to cast doubt on the growing international scientific consensus on climate change, NGOs of all sorts have tended to highlight the central scientific findings and to emphasize that scientific findings are rigorous and methodical. Famously, at the "Camp for Climate Action" at London's Heathrow Airport in 2007, environmentalists protesting plans for further airport expansion carried a huge banner declaring "we are armed only with peer-reviewed science" (Bowman, 2010, p. 177). They were expressing – they insisted – not their own views but the message of science, and they were stressing that scientific claims should be trusted because they are based on robust quality-control mechanisms such as peer review. This close relationship between NGOs and scientific warnings about climate change was further intensified in 2019 in the leading scientific journal *Science* where a short article (a letter) by Gregor Hagedorn of Berlin's Natural History Museum and several co-authors sought to demonstrate that the climate concerns of young protesters are scientifically justified. Hagedorn was also one of the founders of Scientists for Future (S4F), an intended scientific counterpart to the Fridays for Future movement, which is discussed below (Scientists for Future, 2023). A similar group, Scientist Rebellion, is composed of scientists who are "uniting against climate failure" and who support "civil disobedience to demand emergency decarbonisation and degrowth, facilitated by wealth redistribution" (Scientist Rebellion, 2023). The group is well known for having leaked an advanced draft of the latest IPCC report (IPCC AR6 WGIII) for fear that it would be watered down through horse-trading between national delegations and governments.

The force and clarity of NCM campaigns are buoyed by avoiding uncertainty and discord about scientific views, an observation that is supported by Rödder and Pavenstädt's (2023) work on NCM groups such as XR and the Sunrise Movement. Drawing on recent NCM publications and on fieldwork with protestors, these authors note that "A striking feature, again shared across the NCMs under study, is that they picture science as a unified actor, which communicates with one voice" (2023, p. 35; for XR specifically, see also Hinks and Rödder, 2023). It is worth noting here that the NCM's vision of a unified science speaking with a single voice is rather different from that found in STS, which tends to regard science as, typically, more plural.

Although this alignment between a unified voice of science and the invoking of top establishment scientific figures such as Sir David King would seem to confer appreciable benefits onto environmental NGOs, there is also a sense in which it places these NGOs in a dilemma. One aspect of this dilemma relates to the aforementioned urgency. However, many scientific investigations take a lot of time. The cycle for producing the reports of the IPCC, for example, typically lasts five to seven years. It is hard to be urgent and as thorough as possible. The occasional errors that have crept into IPCC reports have been relentlessly exploited by those who wish to discredit the IPCC, so there are clear incentives to be careful and painstaking. Second, NGOs' statements in favor of the objectivity of the scientific establishment's views mean that it is hard to distance themselves from scientists' conclusions on other occasions without appearing arbitrary or inconsistent. The case of genetically modified organisms (GMOs) used in agriculture is revealing in this light since it is difficult for NGOs to insist on the simple correctness of scientific views over climate change but to disagree with the apparent establishment position on GMOs. Relatedly, a reluctance to criticize mainstream science threatens to make NGOs too accepting of establishment policy positions, for example, about the scope for future technologies to remove carbon dioxide from the atmosphere. As a consequence, this leaves NGOs with only a derivative stance on all policy matters since – lacking supercomputers or polar research stations – they cannot easily generate new, fundamental knowledge about the climate themselves. Environmental activists know that support from the scientific community is a key asset in their campaigning. But it does not always deliver the boost that they hope for since scientific processes may lack the urgency that activists seek and there is a risk that activists become the "junior" partner in their relationship with scientific experts.

Alternative Framings

Although NGOs' link to science is very close, even in the case of the most recent organizations such as Just Stop Oil, that does not exhaust the strategies open to environmental groups. There are, for example, things that they can act on within the parameters of accepted science and already-agreed policies. One of the most straightforward options is to focus on the extent to which existing greenhouse gas pledges are being carried through into practice. International climate agreements call on countries to meet specific emissions targets. NGOs can accordingly involve themselves in monitoring and in publicizing countries' successes (or failures) in meeting commitments. Of course, greenhouse gas monitoring tends to be a technical or scientific exercise, but the logic of this strategy is not about science. Rather, it is more a question of making sure people do what they have promised. Indeed, one of the distinctive things about the 2015 Paris Agreement was that the countries that signed up set their own detailed targets, so there is a lot of monitoring work to do. Research and monitoring groups have emerged to play key roles here – Climate Action Tracker seeks to monitor pledged and actual emissions, while the Global Carbon Project offers an independent check on the overall carbon budget.

In a related way, climate NGOs have sought to get governments and sub-national political entities (cities and regions, for example) to declare a "climate emergency." No specific policy measures are tied to such a declaration, but it has been a popular campaigning objective precisely because it obliges governments to acknowledge how serious the climate issue is, and makes them answerable for taking action. In parliamentary democracies, the idea of declaring a climate emergency has also been popular with parties in opposition or out of government because it allows them to position themselves as more inclined to act than the party currently in office.

On a similar basis, environmental pressure groups can target actors and institutions who are responsible for or are investing in activities with large associated greenhouse emissions. Activists can address institutions or holders of capital who have – or wish to be seen to have – high ethical principles. US-based lobby group 350.org (named for the target CO_2 concentration of 350 parts per million) called upon universities and other institutions to divest from companies tied to fossil fuels. Student bodies have been effective in putting pressure on universities, particularly US and other private universities (for example, the Colleges at Oxford and Cambridge in the UK), that may hold large investment portfolios, to get them to move their money out of carbon-intensive investments. Subsequently, 350.org linked up with the left-liberal UK newspaper, *The Guardian*, to run a joint campaign targeting large-scale private funding bodies, including the UK's Wellcome Trust, to persuade them to divest from fossil fuel shares. The gigantic Norwegian sovereign wealth fund has also been keen to be seen as an environmentally sensitive and sustainable form of investment. Universities and research-funding bodies, it should be noted, have no specific reason to hold energy shares, other than the idea that such investments are likely to be of long-term value. Promoting divestment becomes symbolically important and may also act to put downward pressure on the value of this type of asset.

University and high-school students have pioneered other initiatives too, including promoting vegetarian or vegan dining facilities for students and calling on their institutions to use catering budgets in climate-friendly ways. The distinctive thing with this kind of approach is that, aside from continuing to affirm the reality and urgency of the problem of climate change, there is no significant science communication challenge involved. Campaigners no longer have to argue about the adequacy of emissions targets; they focus instead on creating a moral concern not to invest in certain kinds of stocks or to avoid serving students methane-producing meat products.

Over the last 15 years, environmental NGOs have taken prominent roles in another kind of approach with practical policy relevance: re-conceptualizing the issue as about turning off the supply or leaving carbon unburnt. NGOs have also adopted a pioneering role in taking forward such arguments. Thus, Oilwatch – a network NGO set up in Quito, Ecuador, in 1996 with members from Latin America, Africa and Asia – was established to oppose the expansion of hydrocarbon extraction especially in tropical, biodiverse regions. During the negotiations over the United Nations treaty that preceded the Paris agreement, Oilwatch proposed a moratorium on new oil activities instead of emissions targets. It developed this idea in a report a decade later (2007), arguing that the UN agreement had failed to stop the expansion of the oil industry, that tropical forests were under threat from hydrocarbon prospecting, and that the only successful strategy would be to agree to leave large quantities of oil in the ground (this became known as LINGO, leave it in the ground). This was adopted as the strategy of the then-president of Ecuador to avoid the development of oil extraction in the forests of the Yasuní National Park (the specific zone is known as Yasuní-ITT, for the Ishpingo-Tambococha-Tiputini prospecting block, and is a celebrated biodiversity hotspot). He sought to raise international funds equivalent to half the projected value of the reserves in order to compensate his nation for keeping the oil underground and to allow Ecuador the resources to keep habitats intact. The scheme attracted high-level international support and the money was to be administered by a Trust Fund of the United Nations Development Programme (set up in 2010). The project was further endorsed by environmental NGOs and by celebrity backers including Leonardo DiCaprio. In the end, insufficient funds were offered in the initial years and President Correa reversed the policy in 2013 blaming a lack of international support. Oil extraction later began.

This approach, focusing on the role of suppliers and concentrating on ways to keep hydrocarbons in the ground rather than on reducing emissions, was presented in an adapted form four years after Oilwatch in a report by the Carbon Tracker Initiative (Leaton, 2011), based in London. Carbon Tracker's analysis highlighted that there is only so much carbon that can be emitted before the targeted 2°C rise will be exceeded. Anyone whose wealth relates to fossil fuel reserves after that point will find that the reserves may be unrealizable and therefore of much-diminished value. Carbon Tracker directed this message to investors and institutional shareholders rather than to oil-rich states, warning that their long-term assets could become devalued. As Jacobs (2016) expresses it:

> If governments acted on their own commitments, it would leave many of the world's fossil fuel companies with "stranded assets," unable to continue planned production and with heavily devalued share prices. The world's stock markets and pension funds were effectively sitting on a "carbon bubble."
>
> (2016, no pagination)

Seen in this way, therefore, holding oil or gas investments beyond a certain level is financially very risky, and current investment portfolios underwritten by the presumed future value of fossil fuel reserves may be drastically overvalued. Carbon Tracker presented this idea very cogently and in a manner far more tailored to institutions than Oilwatch, and achieved recognition for their idea from significant market actors.

Environmental NGOs have discovered various ways of acting on climate change that manage to side-step the challenges created by sticking very close to science. They have focused on policy promises rather than on scientific targets and they have campaigned by following the money rather than the scientific results. With these alternative strategies in mind, let us now turn to the School Strike protests and their identification of young people as a distinctive kind of political actor.

Generational and Related Framings

As befitted *Time* magazine's person of the year for 2019, Greta Thunberg was hardly out of the news from the start of her School Strike for Climate through to the abrupt temporary end of politics-as-normal caused by Europe's COVID-19 virus outbreak in early 2020. She gave speeches to crowds of activists and young people across the world, and addressed politicians and UN bodies. Her personality and guileless approach spoke to her authenticity and conviction while, through the development of her School Strike campaign (generalized as Fridays for Future), she achieved what many analysts of social movements identify as a critical success. That is, she identified in the school-age activists a new and potentially coherent political actor that could press for action on climate change without the compromises and split loyalties of existing political leaders.

Part of the triumph of the School Strike idea was that, unlike in most regular strikes, participants had little to lose by striking: they lost neither wages nor pension benefits and their actions were often endorsed by educational authorities. Nor was it much of a hardship for the strikers; many school students may well have preferred to join in strikes once a week than to attend classes. But the essential point was that school strikes represented and crystallized a segment of society that felt the urgency of the climate issue without having really been complicit in causing the problem. Here was a cohort, mostly born after the signing of the first major international climate agreement at the end of the 1990s (the UN Kyoto Protocol), who felt that their adult lives were going to be overshadowed by a vast problem that they had not caused and from whose causing they felt they had derived no great benefit.

In her numerous speeches, Greta Thunberg focused on two kinds of claims. One was essentially the argument reviewed at the start of this chapter: that politicians the world over are not listening to the grave warnings coming from the scientific community. She implied that there was a single scientific view and preferred to point political leaders to scientific reports than to repeat the claims herself. Her demand was that leaders listen to what was already known. Her second claim was moral: today's adults have known about the problem for decades but have failed to do enough about it. Though they speak about a sustainable future for coming generations, they are not delivering on that ideal. On the contrary, they are leaving a problem for their children which they (the adults) had been too weak to address.

In the sociological literature on social movements, one principal line of thought argues that the truly historic social movements are the ones that develop a new collective actor with newly identified goals and ambitions. The Women's Movement, for example, articulated the interests of women, diagnosed the sources of their oppression, and devised strategies for their emancipation; the movement has developed through several "waves" in which these steps have been broadened and enlarged. The School Strike movement and Fridays for Future started to achieve the same in relation to climate. They have defined an actor on a generational basis, an actor who faces a very uncertain future through no fault of their own. This outlook is well expressed in some movement organization names: XR clearly makes possible the thought of extinction. In Italy, a high-profile group in 2023 has adopted the name Ultima Generazione (Last Generation) (The Guardian, 2023).

This sense of a distinctive orientation toward climate change is potentially powerful. It could lead to a willingness to adopt more rapid or far-reaching measures than have been seen to date, where official targets expect net zero even in the pioneering nations only by around 2050. But the sense that younger citizens might literally be members of the last generation appears to be correlated with detectable levels of anxiety. In one large-scale international study, with a thousand respondents in 10 countries, a majority of respondents answered that they believed "the things I most value will be destroyed" and that "humanity is doomed" (Hickman et al., 2021).

Environmentalism's close connection with scientific claims and with the authority of science offers many benefits to the movement. But, as STS studies have shown, holding too close to the authority of science can bring disadvantages too. Scientific results are often provisional and may take a long time to produce. In some cases, scientific views are based on scientists' judgment. In these ways, science does not have the unified character ascribed to it by NCMs. There is also a great deal of difference between STS understandings of science and how it is likely taught as a singular scientific method in high schools, including the high schools of the student strikers. As this chapter has indicated, there are many ways that climate activists can effect change and many of them involve side-stepping scientific claims, focusing instead on governments' pledges or the investment decisions of schools and colleges. In this sense, the climate anxieties of a new generational actor can perhaps beneficially be channeled into the abundance or plurality of ways citizens can engage productively with climate change.

Further Reading

Bocking, S. (2004) *Nature's Experts: Science, Politics, and the Environment*. New Brunswick, NJ: Rutgers University Press.

Dunlap, R. E. and Brulle, R. J. (eds.) (2015) *Climate Change and Society: Sociological Perspectives*. New York: Oxford University Press.

Rödder, S. and Pavenstädt, C. N. (2023) "'Unite behind the science!' Climate movements' use of scientific evidence in narratives on socio-ecological futures." *Science and Public Policy*, 50(1), pp. 30–41.

Thunberg, G. (2019) *No One is Too Small to Make a Difference*. Harmondsworth: Penguin.

Yearley, S. (1992) "Green ambivalence about science: Legal-rational authority and the scientific legitimation of a social movement." *British Journal of Sociology*, 43(4), pp. 511–532.

References

Bowman, A. (2010) "Are we armed only with peer-reviewed science? The scientization of politics in the radical environmental movement." In Skrimshire, S. (ed.), *Future Ethics: Climate Change and Apocalyptic Imagination*. London: Continuum, pp. 173–196.

Extinction Rebellion (2023) "Extinction Rebellion." Available at: https://rebellion.global/ (Accessed May 22, 2023).

Hagedorn, G., Hagedorn, G., Kalmus, P., Mann, M., Vicca, S., Van den Berge, J., van Ypersele, J., Bourg, D., Rotmans, J., Kaaronen, R., Rahmstorf, S., Kromp-Kolb, H., Kirchengast, G., Knutti, R., Seneviratne, S.I., Thalmann, P., Cretney, R., Green, A., Anderson, K., Hedberg, M., Nilsson, D., Kuttner, A., and Hayhoe, K. (2019) "Concerns of young protesters are justified." *Science* 364(6436), pp. 139–140.

Hickman, C., Marks, E., Pihkala, P., Clayton, S., Lewandowski, R. E., Mayall, E. E., Wray, B., Mellor, C., and van Susteren, L. (2021) "Climate anxiety in children and young people and their beliefs about government responses to climate change: A global survey." *The Lancet Planetary Health*, 5(12), pp. e863–e873.

Hinks, E. K. and Rödder, S. (2023) "The role of scientific knowledge in Extinction Rebellion's communication of climate futures." *Frontiers in Communication* (Published online ahead of print March 27, 2023). DOI: 10.3389/fcomm.2023.1007543.

Jacobs, M. (2016) "High pressure for low emissions: How civil society created the Paris climate agreement." *Juncture*, 22(4).

Just Stop Oil (2023) "Just Stop Oil." Available at: https://juststopoil.org/ (Accessed May 22, 2023).

Kendon, M., McCarthy, M., Jevrejeva, S., Matthews, A., Sparks, T., and Garforth, J. (2021) "State of the UK Climate 2020." *International Journal of Climatology*, 41(2), pp. 1–76.

Leaton, J. (2011) "Unburnable carbon – Are the world's financial markets carrying a carbon bubble?" *Carbon Tracker Initiative*. Available at: https://carbontracker.org/reports/carbon-bubble/ (Accessed May 22, 2023).

Oilwatch (2007) *Keep Oil Underground: The Only Way to Fight Climate Change*. Bali: Oilwatch.

Rödder, S. and Pavenstädt, C. N. (2023) "'Unite behind the science!' Climate movements' use of scientific evidence in narratives on socio-ecological futures." *Science and Public Policy* 50(1), pp. 30–41.

Scientist Rebellion (2023) "Scientist rebellion: Our positions and demands." Available at: https://scientist-rebellion.org/about-us/our-positions-and-demands/ (Accessed May 22, 2023).

Scientists for Future (2023) "Scientists for future." Available at: https://de.scientists4future.org/ (Accessed May 22, 2023).

The Guardian (2023) "Rome climate protesters turn Trevi fountain water black." Available at: www.theguardian.com/world/2023/may/21/rome-climate-protesters-turn-trevi-fountain-water-black (Accessed June 13, 2023).

10 Expert Activists and NGOs

Understanding and Acting on Global Climate Change

Adam Fleischmann

Let's begin with a fundamental insight in Science and Technology Studies (STS) and related fields like anthropology: the incredible but simple fact that things have been different in the past, they are different in some places on Earth right now, and they can and will be different in the future. This idea, in one way or another, also motivates the actors and organizations whose work is the focus of the research I will discuss in brief in this chapter. For nearly a decade, I have conducted a series of studies of a network of non-governmental organizations (NGOs) operating at the intersection of expertise in climate change politics and climate science in North America. Through their work, these expert activists recognize that things can be different than the present state of things, however entrenched, including accelerating climate change. The people and organizations who make up the network address the challenges of global anthropogenic climate change in different ways: through science communication and education, policy and data analysis, coordination and convening, computer modeling, and grassroots organizing. However, they all have this in common: they are all experts in shaping the space where political and scientific knowledge meet. In my research, I use tools and perspectives from STS and related fields like anthropology and social movement studies to study how these groups help people understand and act on global climate change.

This study is part of a larger research agenda about the challenges and possibilities of addressing global anthropogenic climate change both politically and epistemologically (that is, regarding knowledge about climate change, its history, how we know what we know about it, and, even, how climate change can be known in the first place). In doing this work, I regularly contend with the high-stakes questions that trouble researchers, expert activists, and, I'm sure, many students. How do we understand and act on the scale of the global climate? How can humans even grapple with the vast consequences of our collective action? My research shows how, in my interlocutor's spaces of expertise, both climate change science and politics are negotiated, problematized, and made intervenable. In other words, it shows how these groups make climate change a vivid and concrete issue for their publics.

In this chapter, I first explore some STS insights about global anthropogenic climate change, which help us understand both the challenges couched in the above questions and the work of expert activists to shape the space of climate science and politics. I then turn to this work more directly, focusing on an example of ethnographic field research in climate movement NGOs, providing ethnographic details and a narrative anecdote, before concluding by considering the insights STS and NGOs offer for rethinking climate change, its challenges, and possibilities.

STS Insights: Understanding Global Climate Change

Climate change as it is understood by climate science is global. That is, it is a phenomenon that encompasses the whole planet and all five spheres of the Earth's planetary system—the

DOI: 10.4324/9781003409748-14

atmosphere (gases), the lithosphere (rocks), the hydrosphere (water), the cryosphere (ice) and the biosphere (living things). Global climate change is also **anthropogenic**, meaning it is caused by humans (some more than others). Global anthropogenic climate change, as a way of describing *human-caused* changes in the *global* climate system, is an issue that is known through the concepts and institutions of Western science. This global thinking has an ongoing history, often enmeshed with the histories of colonialism and imperialism. As some STS scholars have put it, "Conceptions of the world as a globally connected system ordered by physical, chemical, and biological laws have a long history, animated not just by abstract theoretical advances but by processes of European expansion and the imperial thirst for both facts and resources" (Beck et al., 2016, pp. 1060–1061). This means that the scientific concept of a global climate is inherently tied in with—cannot be understood disconnected from—social, cultural, and political histories. It is in this sense that it is "constructed" as a concept. And this construction has a history and is actively reproduced. While now understood with increasing degrees of specificity, that specificity is made up of a complex constellation or assemblage of phenomena, people, things, and events.

The idea of climate as a single, global unit had already been proposed by the mid-19th century. However, early theoretical models of atmospheric circulation—the movement of energy and air through the atmosphere and oceans—had serious limitations in tackling the staggering complexity of the problem of a global system. This place-based, at most regional, conception of the climate remained until technological advances following the Second World War. It was the advent of computer models, during and after the war, that could begin to handle the practical task of confirming a theory of general atmospheric circulation. This allowed for the realization of the concept of a *global* climate, previously imagined in terms of physical laws but not practically calculable. This breakthrough led the way to systems theories of general circulation, which connected the oceans, land, geology, living things, and ice (all five spheres of the Earth's system) in the latter half of the twentieth century (Fleischmann and Yip, 2019). The notion of a global climate thus emerged, produced by the complex set of relations *through which* we have come to know and understand it (Edwards, 2010): a constellation of scientists and meteorological phenomena, discourses and institutions, national meteorological services and massive computer models, satellites and weather stations, and archives—physical things and actual events in time.

Yet as STS scholar Paul Edwards put it in his influential history of climate science and modeling, "No one lives in a 'global' climate. Without scientific guidance, not even the most cosmopolitan traveler could perceive a global average temperature change of about [1.2°C], the amount we have seen so far" (Edwards, 2010, p. 4). We can't *see* changes in climatic averages over 30 years and we can't perceive the global climate itself, in all its globality, per se. We *can*, of course, see its meteorological impacts and cumulative effects, the "natural" elements that form part of its constructed nature. We *can* witness how its accumulating, interacting systems change how we experience the day-to-day variation of the state of the atmosphere with respect to its effects on human life (otherwise known as the weather). We *can* measure the global climate, and model it, projecting it into the future and the past. But in its very globality, it is a sum greater than its knowledge-production-system parts. No one lives in a global climate. Global climate change is both constructed by and beyond humans.

The dynamic and perhaps counterintuitive challenges of understanding and acting on climate change described above are due in large part to the global nature of climate change. But, of course, this does not mean the global climate is not "real." It is grounded in observations and other empirical data, a global knowledge infrastructure, and requires active reproduction throughout this system—in other words, its realness is reproduced in all the elements that make it up. The labor and maintenance of this knowledge-production system is the very reason why we can even think of a planetary climate as something to be observed, understood, affected by human activities, cared about by the general public, and managed through the political regulation of the composition of the atmosphere. Knowing this concept and its history can help us understand and address many of the challenges surrounding climate change that can feel overwhelming and inevitable.

In sum, STS perspectives have taught us that perhaps more than other complex and challenging social and natural problems of a certain intercontinental magnitude, the changing global climate is always a combination of phenomena and knowledge. Yes, climate change is partially made up of massive and changing phenomena like weather patterns. And STS teaches us that we can only know these changing phenomena on a global scale because of a global scientific knowledge system and the history of imperialism and expansion that proceeds it. It is both constructed by humans and exists beyond us. This tension is inherent to the issue of climate change and must be continually taken up anew, including in research. Next, I'll share some of what I've learned from one group within the network I've studied.

STS Research: Fieldwork in Climate Movement NGOs

Climate Interactive is a small US-based nonprofit climate change NGO and one organization whose staff and participants formed a key group of interlocutors for my study. Through my research with the Director and Co-founder of Climate Interactive, Drew Jones, I have learned that the dynamics of the global climate system are not intuitive. We drive cars, producing carbon emissions in the US or Canada, yet the effects are seen, much sooner and more intensely, faraway in Bangladesh or Greenland, for example. People produce emissions *today*, but it is our children, grandchildren, and great-grandchildren who will deal with growing consequences like drought, sea-level rise, and increasingly extreme weather. To put it another way, climate change is complex and heterogeneous, with dynamic characteristics across time and space. All of this makes effective action on climate change difficult to inspire or enact, or even to imagine, for most people. Yet just telling people this information doesn't do much. In fact, Drew Jones integrated this insight into the organization's very foundation.

STS insights on the dilemmas of the complex, global problem of climate change help explain why simply telling people about the challenging global characteristics of climate change is not enough to help galvanize action to solve the issue. As MIT Sloan School of Management professor and Climate Interactive senior adviser John Sterman liked to say during my study, "Research shows that showing people research doesn't work." Through his pithy and clever turn of phrase, Sterman is summarizing the insufficiency or weakness of the **deficit model**, which is sometimes referred to as the information deficit model or science deficit model. The deficit model suggests that if only the public had the correct information, they would understand the problem, take appropriate action, and create political or social change. Under the assumptions of this model, political action on climate change plays out in particular ways: climate scientists bestow knowledge about climate change upon diverse publics, who are then rationally incited to take action in the form of lobbying, petitioning, protesting, and other environmental work; this, in turn, influences expert leaders to act through legal and policy engagement.

While this model is prevalent in public discourse, it is criticized extensively by STS scholars, as well as by Climate Interactive. As my interlocutors at Climate Interactive explained, just telling people what "the science says," that is, explaining what scientific research claims, does not work. For issues like climate change, telling people what to think or how to act doesn't have an impact on convincing people of the importance of the issue or inspiring them to act. As interdisciplinary STS, anthropology and journalism scholar—and author of Chapter 3 in this book—Candis Callison, writes, climate change "enables questions beyond what the realm of what science offers" (Callison, 2014, p. 23). That is, simply knowing the complexity of global climate change does not help people form answers to the high-stakes emotional and existential questions that climate change, and inaction on it, provokes.

Ethnographic data from my study on climate change NGOs builds upon these insights. Today, following decades of work from scholars in STS, anthropology, sociology, and related fields, researchers conduct ethnography in organizations like NGOs, among scientists and activists, as well as remotely in "spaces" without a specific geographic location (Fleischmann, 2022; Knox, 2020). Researching climate change ethnographically in-person and remotely, I have been able to study and learn from expert activists who teach people the non-intuitive dynamics of global climate change and—moving beyond the deficit model—inspire them to take action in their own lives.

Climate Interactive was founded on the idea that values and experiences, not information, are what really shape people's perceptions and actions. The organization's simple computer models, tools, and games create opportunities for people to *learn for themselves* about the climatic, economic, and geopolitical systems that shape our world. Climate Interactive creates interactive simulations, timely analysis, decision-support tools, and experience-based educational games and workshops that endeavor to empower people, from school children to the US President's climate team, to reach their goals in addressing climate change. Their work combines innovations in climate modeling and education. More specifically, they use simple, interactive computer models that can run for free on a laptop, simulating changes in global temperature, carbon emissions, or energy policy in literally one second. They use these models in experience-based games and workshops that offer participants the opportunity to role-play national and international efforts to reduce global greenhouse gas emissions. With roots in systems dynamics modeling and open-access, experience-based design, and education, Climate Interactive contributes a unique intervention on climate change that engages people socially, emotionally, and physically. In doing so, they provide insights into how a social scientist and climate action practitioners alike work to grasp global anthropogenic climate change as an emergent object of study and political action.

Employing ethnographic methods like participant observation in my study of Climate Interactive, I was able to participate in and observe training for and renditions of the organization's model-based educational role-playing games. Here, I provide a very brief ethnographic anecdote, in combination with insights on the global nature of climate change, in order to demonstrate an STS approach to studying climate change and climate activists.

Box 10.1

The World Climate role-playing game is being held in an intimate room in a building off the main chapel at Grace Cathedral in San Francisco. The group of us, about 20 people, range in age from late 20s to 60s or 70s are moved into groups of two to five, with each group representing a country or grouping of countries. We huddle together and prepare our negotiating approaches based on the provided printed position briefing. My group, the United States of America, is made up of the three youngest people in the room, myself, Brent, and Elena, plus a late-comer, a white-haired man named Abe.

For each negotiating round, we move across the room, gather in groups. We make our demands and concessions then gleefully scuttle, whispering, back to our huddle of teammates. After each round, back in our groups, we record what we've negotiated: 1) our intended reductions in greenhouse gas emissions (peak year, reductions start year, and percentage of reduction per year), 2) our monetary contribution to the Green Climate Fund, and 3) how much we'll reduce deforestation and increase afforestation (planting trees). After short speeches are made and proposals announced, our facilitator, a Reverend playing the UN Secretary General, quickly enters the numbers into the instant climate model, C-ROADS. Changes appear in global temperatures, CO_2 levels, sea-level rise, and more. Our goal is under 2°C warming by 2100, and preferably 1.5°C.

At first, the negotiations are engrossing, but polite, not too urgent, playing into the stereotypes I'd constructed in my head about soft-spoken older religious folks. Teams China and "Other Developing" advocate for their right to develop, India emphasizes needing help from richer countries. The European Union (EU) is playing polite hardball, though. A middle-aged woman with short, graying hair and sharp glasses, she's uncompromising in her steely insistence that the US and "Other Developed" countries must match the EU's leadership in the fight against climate change. We on Team US, for one, do not give in, maintaining the recalcitrant position of a Trump Administration-era US that has pulled out of the Paris Agreement.

After several rounds of negotiations, the Secretary General sternly warns us of the consequences to come should we not negotiate stronger emissions-reduction commitments. He shows us Shanghai underwater, London submerged by the Thames. As temperatures increase, disaster looms. Participants soon realize how little their countries' modest contributions are changing the results in the computer model. Negotiations get nastier, more urgent. As the timer runs out, delegates negotiate urgent positions "in character" with their country's interests in mind, but aiming for the global temperature goal. The facilitator enters our final numbers into the C-ROADS and we're north of 2°C, headed for a dangerously warming world.

By the end of the game, we step out of our roles as delegates at the United Nations and everyone is appealing to the Reverend to have another round. "I wanna get that number down!" the former EU delegate shouts, complaining. Heads nod in agreement across the room, faces creased in consternation. Someone formerly from the Chinese delegation says they could see this lasting all day. Participants talk about how they felt empowered or caught up by the role they were playing. The Reverend shows us in the model what it takes to get below 2°C, then guides us through a debriefing exercise that invites participants to sit and reflect on future possibilities. "When we talk about future scenarios for our climate," our facilitator dutifully says, "we spend most of the time focused on how bad the worst-case future looks or how difficult change will be. Instead, I'd like for us to spend just one minute silently considering the possibility that we could create this better future from our scenario." After this moment of reflection, we're encouraged to share what we would love about being part of this sort of future—and what we could do, with our skill sets and in our communities, to start enacting that vision. People mention the desire to start campaigning to electrify the vehicle fleet at their workplace and get involved with local environmental groups.

Although we started slowly, the World Climate simulation at Grace Cathedral had us participants riled up. People were smiley, angry, stubborn, gleefully ornery, and downright upset. A sense of urgency pervaded the room once we realized just what it would take to turn the temperature down—serious emissions reductions from not just the US and EU but "developing" countries, too. It quickly became apparent that global climate system dynamics are not straightforward. It also became clear that climate negotiations are social and political affairs in which the power relations and inequities of the real world come to the fore; they are not simply technical exercises of deliberating rational action informed by science, as the deficit model might suggest. Ideally, Climate Interactive's climate-policy simulations are meant to teach people some of this dynamic complexity of the climate-policy system, relating their own lives to broader systems and equity issues, while teaching them to connect delayed and distant climate causes and effects that are not intuitive. Climate Interactive's simulations aim to ultimately build users' capacities to do something about climate change in their own way, in their own communities—connecting knowledge to positive visions and ways to take action, creating new possibilities on climate change. They do this not through showing people research, but by allowing people to learn for themselves through interactive experiences.

While I saw first-hand the transformative and energizing impact Climate Initiative can have on people, its approach does have limitations. Climate Interactive's workshops and role-playing games have been run in 140 countries for more than 225,000 participants, but this number is paltry compared to North American or global populations. Theirs is an approach of quality more than quantity. But even this will not move everyone who participates in their learning experiences to act. However, for the people that it does, Climate Interactive's engaging learning experiences enable them to form immediate relations between their lives, the global climate, and future ways of being in the world. And, importantly, this helps people to not only understand but also feel empowered to act on climate change.

Conclusion

To conclude, let's turn back to the insight that opened this chapter: things have been different in the past, they are different in some places on Earth right now, and they can and will be different in the future. When it comes to climate change, STS and the related fields of anthropology and social movement studies teach us that global anthropogenic climate change is a combination of "natural" and "social" phenomena and scientific knowledge. Yet just as global anthropogenic climate change is, in part, constructed by humans, so are the social, political, and economic systems that have caused it—and through which we are responding to it. In other words, the problem of climate change is not simply a "natural" phenomenon caused by, yet beyond humans; it is constructed, and *it can be constructed differently*. There is possibility, wiggle room, the potential to do things differently. The current state of things is not inevitable, inescapable. Research agendas like the one introduced in this chapter can show how climate actors actively shape the space of climate science and politics to make climate change a vivid and concrete issue for people to understand and act upon. In doing so, they reveal the possibility of alternative futures.

Further Reading

Callison, C. (2014) *How Climate Change Comes to Matter: The Communal Life of Facts*. Durham: Duke University Press.

Fleischmann, A. (2023) "Fire, Ice, and Flood," *American Anthropologist*, 125(1), pp. 199–201. Available at: https://doi.org/10.1111/aman.13813.

Knox, H. (2020) *Thinking Like a Climate: Governing a City in Times of Environmental Change*. Durham: Duke University Press.

References

Beck, S., Forsyth, T., Kohler, P. M., Lahsen, M., and Mahony, M. (2016) "The Making of Global Environmental Science and Politics." In U. Felt, R. Fouché, C. A. Miller, and L. Smith-Doerr (Eds.), *The Handbook of Science and Technology Studies*. Cambridge, MA: MIT Press, pp. 1059–1086.

Callison, C. (2014) *How Climate Change Comes to Matter: The Communal Life of Facts*. Durham: Duke University Press.

Edwards, P. N. (2010) *A Vast Machine: Computer Models, Climate Data, and the Politics of Global Warming* (1st ed.). Cambridge, MA: MIT Press.

Fleischmann, A. (2022) *Possibility in an Era of Climate Change: Anthropology, Knowledge, Politics*. PhD Dissertation, McGill University, Montreal, Quebec.

Fleischmann, A. and Yip, J. (2019) "'Culture' and Climate Change: Anthropology and the Greatest Challenge of Our Time." [Conference Presentation] Changing Climates: American Anthropological Association, Vancouver.

Knox, H. (2020) *Thinking Like a Climate: Governing a City in Times of Environmental Change*. Durham: Duke University Press.

11 Skirting the Frame

Prepping and the Conservative Politics of Climate Change

Allison Ford

Introduction: Prepper Frames

Prepping, derived from the verb "to prepare," is a growing cultural movement of more than 20 million people around the world who anticipate and prepare for disaster (Saragosa, 2020). Some preppers focus on short-term disaster response through emergency preparedness. They stockpile extra supplies like food, emergency water, water filters, cookstoves, and other equipment that doesn't rely on electricity or gas lines, and construct plans to either "bug out" (escape the area where disaster hits and head to a safe, pre-planned destination) or "bug-in" (hunker down in place and plan to wait out the disaster). Some preppers stop here. But for many preppers, the goal of prepping is to survive a much bigger disaster: "the end of the world as we know it," or, as they call it, TEOTWAWKI.

In anticipation of disaster, preppers stockpile food, gather emergency supplies, learn how to communicate over ham radios, and study first aid and emergency medical interventions. Many preppers also prepare to live as self-sufficient households when *the world as we know it* ends. *The world as we know it* refers to a modern, complex, industrialized society. This means life in a 21st-century, globalized, wealthy, capitalist democracy. Life in such a society is marked by a deep division of specialized labor. This division of labor is in opposition to a self-sufficient lifestyle, in which a household meets almost all of its own needs without relying on external institutions and systems that many of us take for granted in the industrialized world, including roads and highways, running water, electricity, Wi-Fi, and sewage management. To live a self-sufficient lifestyle thus demands looking back in time to how people lived prior to the spread of these interconnected technologies.

Prepping originated in the United States in the late 20th and early 21st centuries, about the same time that the concept of anthropogenic climate change entered American public discourse. It's linked to but distinct from earlier waves of interest in self-sufficiency throughout American history, such as back-to-the-land movements. While back-to-the-land movements and prepping both orient around a critique of modernity and its human costs, prepping is unique in its fixation on disaster.

Prepping is in many respects an environmental practice. In anticipating disaster, preppers must think about how to meet their bodily needs for survival, wellbeing, and comfort, outside of complex networks of material flows that are the default way most citizens of wealthy, industrialized economies get their needs met (Schlosberg and Coles, 2015). Planning where you will get water, how you will generate heat, and what you will eat once society has collapsed, while doing away with municipal water infrastructure and service, the electric grid, and grocery stores are exercises in environmental imagination. However, as I will show in this chapter, preppers rarely think or talk about the environment, unless prompted to do so.

This chapter is based on my ethnographic observations of preppers, which I undertook between 2014 and 2018. It draws from semi-structured, in-depth interviews that I conducted with

DOI: 10.4324/9781003409748-15

20 preppers and participant observation of social sites where preppers were visible, including public events, such as expositions, fairs, in-person meetings of online clubs, visits to private businesses, and private events that I was invited to attend through my connections to the prepping community. I also conducted a digital ethnography to understand the ways preppers use the internet as a social space for exchanging information, ideas, and imagining disaster scenarios amongst "like-minded people."

I argue that prepping is a conservative cultural movement that allows Americans who are skeptical or uncertain about climate change to *skirt the frame* of climate change entirely. To skirt something means to move around its border, side-stepping it completely, whereas **framing** is a concept that social scientists use to describe the perceptual "organization of experience" (Goffman, 1974). Social movement **framing theory** explores how political actors produce and maintain meaning in order to put forward their construction of reality (Snow et al., 1986). While scientists frame climate change as real, dangerous, and anthropogenic (human-caused), and thus situate it as a social problem, conservatives offer a contradictory framing of climate change as non-problematic (Freudenburg, 2000; McCright and Dunlap, 2003). The validity of government intervention in the market to manage environmental risk is at the heart of contested climate frames.

The preppers who I researched focused directly on disasters, which they understand as real. But instead of framing such disasters in relation to climate change, they adopt a depoliticized response to environmental risk by focusing on what I call *the constituent elements of climate change*. The constituent elements of climate change refer to the effects of climate change that can be experienced, talked about, and understood without referring to climate change itself. For example, we can talk about the likelihood of experiencing a major storm without acknowledging that that likelihood increases because of a warming globe. The storm remains in the frame, while the cause of its increased frequency and severity stays outside of the frame, and thus outside of the conversation. I argue that preppers avoid engaging with the scientific discourse of climate change, which challenges their understanding of social reality. This includes both disaster scenarios and historical events that they draw on to construct an alternative framework that focuses on individual responsibility to respond to disaster and downplays government intervention as a solution to risk.

How do Preppers Feel About Climate Change?

Compared to people in other nations, Americans find it more difficult to talk about climate change, and Americans are much more likely to report climate skepticism (McCright et al., 2016). Yet it is impossible to ignore that we are living on a planet undergoing major atmospheric and ecological changes, many as a consequence of human industry. Preppers are all too aware of the prevalence of disaster. John, a prepper in Idaho, told me,

> Now, with the ability to communicate instantaneously anywhere in the world, people are seeing calamities occur within a thousand miles of their house. They're seeing fires that are killing people and destroying property, and taking everything you [sic] own. They're seeing floods, and they're seeing … everything; earthquakes, hurricanes.

It wasn't just that people were exposed to more information about disaster, John continued, but that the networks of dependency built into modern life were also more visible.

> Then you have … this is quite real … because you can see this yourself. There are political and governmental problem that occur, like [in] Long Beach, down south California.

Longshoremen went on strike, just a few years ago. "Who cares? A bunch of union commies down in there, who cares?" Except for one thing, that these guys took care of all these ships, to bring in food and stuff that comes up here, in trucks.

All of a sudden [the grocery stores], their shelves are starting to look bare, because all they have left [are] Tabasco sauce and salsa … Not only can it happen, it has twice, up here. You've seen it.

The visibility of disasters around the world combined with personal experiences of the precarity of extended networks of goods and services leave preppers feeling vulnerable—a feeling that is uncomfortable for people whose political leanings include a deep commitment to personal responsibility and self-reliance.

Despite being hyper-aware of disaster and concerned about risk, when I asked John if he was concerned about climate change he responded,

I've done a lot of study on that. That's like being concerned about sunlight because it's a natural phenomenon. The perverted concept of climate change that's been going around is a disgusting smoke screen to gain more control over the population. The climate's always changing. We have not had two years in a row here of the same weather.

John draws on familiar conservative talking points that reject climate science and attribute changes in the climate to nature. This builds on the conservative movement's attempt to construct what sociologists Aaron McCright and Riley Dunlap (2003) call the "non-problematicity" of climate change. Although John attempts to claim authority over the topic ("I've done a lot of study on that") and engages lay scientific language, his argument here is not really scientific, but social.

We often think about climate denial as a disagreement about science. But climate denial is not rooted in rational scientific debates, but rather in politics (Jacques, 2006). If we take seriously the magnitude of dangerous, anthropogenic climate change, and face the fact that human combustion of fossil fuels is its root cause—something just about all qualified climate scientists agree on (Oreskes, 2004)—then the problem is not the science. It's the political implications of what the science tells us. The predominant response that scientists have advocated for involves major governmental regulations, which conservatives and industry groups oppose on principle.

The political significance of John's rejection of climate change comes out not in his attempt to explain away the science, but in the affective salience of his reference to climate change as a "disgusting smoke screen to gain more control over the population." In this powerful turn of phrase, he betrays his distrust of climate scientists, and the institutions that support their ability to disseminate climate science and political recommendations about how to respond. Here, John reveals fear that the discourse of climate change is a political tool that is being used by the scientific elite to exert control over the population.

Hank, another conservative prepper, responded similarly to John when I asked about climate change, but he was more direct about his emotional response. The topic made him so angry that he practically shut down what had otherwise been a friendly and engaged conversation. This was especially noteworthy because the rest of the interview was markedly different in tone. By the time I brought up climate change, Hank and I had been talking for several hours already, and he had maintained a friendly and gregarious demeanor despite our common understanding of the deep political differences between us.

We had successfully remained friendly and engaged throughout discussions about contentious, politicized topics such as race, immigration, religion, guns, and the Trump Administration (our interview was in 2017). But Hank's enthusiastic willingness to explain his beliefs changed as soon as I brought up climate change. Suddenly, the conversational well ran dry:

ALLISON: Are you concerned about climate change?
HANK: [scoffs] No... not at all! Our climate changes all the time.
ALLISON: Ok...
HANK: It's bullshit. And finally there's some scientists who are finally getting the backbone to be able to say, this is all bullshit. No. Um... our climate changes on a regular basis, and has been since God first made this country—[cuts self off] this planet. So it doesn't bother me at all. So, I'm old enough to remember when they said that we got an ice age coming. No! How'd that work for you? So. No.
ALLISON: So it's not something you consider an environmental issue worth considering?
HANK: No, not at all. And it angers me! Because of what the radical climatologist, the radical supporters of climate change have done to those who have dared speak out against it. Ummm... it angers me beyond belief. So. There's my opinion about climate change. [laughs]
ALLISON: Ok! Thank you! [smiles]
HANK: So ...

As soon as I used the phrase "climate change," Hank's tone shifted perceptibly. While we had laughed together plenty over the course of the interview, my question about climate change was met with a scoff, a short burst of laughter that suggests contempt, rather than humor. Like John, Hank rejects the validity of anthropogenic climate change to argue that the climate changes all the time. Unlike John, Hank is explicit about his anger. Sitting across the table from him, I could feel it radiate off him. It was controlled but fierce, and it is reflected in the way he shuts the conversation down—"there's my opinion about climate change." The topic was closed.

Prepping as a Libertarian Cultural Movement

Most of the preppers I met during my ethnography tended toward conservative, especially libertarian, political beliefs. Libertarians value individual freedom above all else. They typically view a free market (capitalist) economy based on private property as the economic system most able to create the conditions of freedom and are therefore strongly opposed to any form of government intervention or regulation that limits the agency of either individuals or market actors.

Sociologist Matthew Schneider-Meyerson (2015) argues that individualistic responses to collective environmental problems can in part be explained by a libertarian shift in American political culture. His study on peak oil narratives explores another subset of Americans who address concerns about collective risks (the environmental effects of fossil fuel economies and anticipation of their collapse) through individualized strategies. In advocating for individuals to be self-sufficient, prepping is consistent with, and thus appealing to, those with libertarian leanings. That does not mean that all individual preppers are libertarians or conservatives. I met people across the political spectrum during my research who engaged with prepping at various levels of depth. Rather, I argue that prepping as a culture draws upon tenets of libertarian politics that are widely appealing to Americans who are wary of government, accustomed to cultural valuation of individualism, and put off by the contentious politics of climate change. In this, prepping serves as a cultural movement that advocates for individual and household scale responses to socio-environmental risk based on anti-institutionalism, especially anti-governmentalism.

Anti-Governmentalism

It became evident very early in my study that preppers did not believe that public emergency services were up to the task of the scale of disasters they anticipated. But it's always good to verify your initial observations while doing ethnography. "Do you trust the government?" I asked in interviews. "Hell no!" John responded. "You know better than to ask that." We both laughed. "I had to get it on record," I explained. "I know," he replied. "Here's the thing," and he went on to explain his views:

> Government is made up of individuals. Individuals are made up of human emotion and reaction.
>
> The federal government, this large, strange creature, is made up of individuals who have their own hates and fears and wants and perversions …
>
> You'll find the problem is that … they're trained, generally, for their position, but then if you piss them off they'll violate their training just to make you wrong. They screw with you. We see it all the time. You hear about it all the time. That's not just federal, it's city, county, state, and federal.

Consistent with libertarian beliefs, preppers deeply distrust government. While on the one hand, libertarians believe in human potential through individual freedom, this conviction disappears as soon as we consider humans in the collective, as John does. His distrust in government here is justified by a pessimistic theory of human nature, reflecting conservative skepticism about the possibility of predicting, let alone controlling social outcomes in complex societies (O'Hara, 2011). This precludes any consideration of collective solutions to risk—or in this case the construction of social problems whose foregone solution is government intervention.

Anti-governmentalism is an important feature of American conservative political thought, and it's on the rise among the population as a whole (Gallup News Service, 2019; Mettler, 2018; Pew Research Center, 2017). But preppers don't limit their distrust to government. The possibility of collective human enterprises as a whole is questionable. In anticipating societal collapse, preppers are distrustful of most big, complex social institutions (including but not limited to corporations, churches, civil society, etc.). While many of the conservative preppers I spoke with were also religious Christians, many did not attend church. Christian faith was clearly important to Max, a prepper in the Southeast region, but he told me he felt driven away from organized religion due to human fallibility and economic greed.

> I saw denominations compromising on their fundamentals, just to enlarge the offering plate. They would compromise on their principles … For my money, I don't know of any church locally that I would even participate in because there are things that they are doing, or things that they are espousing that I simply say that's not right. If that's the road they want to take, fine. I'm not going to participate. I'm not going to support them. I'm not going to go along with them. I'll go my own way.

Preppers believe that individuals are essentially responsible for their own survival and wellbeing. This extends to both body and soul.

Nobody's Coming to Help

Hyper-awareness of disaster combines with distrust in institutions. The result is the sense that disaster is inevitable, but that "nobody's coming to help." This was a common refrain amongst preppers. They take the position that the individual (with the possible extension of the family unit) is the only viable unit of action.

While this perspective engenders a strong sense of personal responsibility, it also has a dark side. I noticed that preppers regularly justify their practices through moral comparisons that involve *claims to superiority*. When individuals or institutions don't meet their expectations for independence, integrity, or rationality, it reinforces preppers' pessimism about human nature and thus the possibility of functioning, complex systems. They may look down upon people who are willing to rely on other people, or social services to meet their needs, despite the fact that interdependence is a condition of modern life.

While materially, preparations involve preparing to secure food, water, shelter, energy, and other basic necessities of survival, many prepper scenarios include a social layer that focuses on preparing to defend oneself against individuals who have not prepared. Preppers believe that in addition to the "shit hits the fan" event that triggers collapse, a major security threat will come from "the unprepared," essentially anyone who is not a prepper. As Hank explains it,

> They're trapped where they live, *I gotta get water for my kid*, this kind of thing. They're gonna resort to some nasty things, to do so. they are going to go *FERAL*.

Preppers judge those who don't have the foresight to prepare as less rational, and therefore less able to stay calm in a disaster. The unprepared will panic and target individuals who *have* planned ahead—preppers. This narrative of the unprepared allows preppers to maintain an ambivalent subject position of both moral superiority *and* vulnerability. Preppers feel both strong and vulnerable—an affect that reflects contemporary tensions around privilege and power. People who are white, conservative, financially secure, and males have generally been privileged by the same systems that marginalize people of color, women, and the poor. People with privilege may experience critiques of racial, class, and gender oppression as an attack on their valued identities, just as they experience the framework of dangerous anthropogenic climate change as an attack on their faith in the free market, political systems that have worked in their favor, and an unproblematized belief in the success of modern, Western civilization.

Conclusion: Contested Climate Frames and the Environmental Politics of Prepping

As shown through my ethnographic research, preppers' preparations for disaster are not linked to scientific assessments of the likelihood of increased frequency and intensity of disasters associated with anthropogenic climate change. In fact, climate change rarely came up unless I mentioned it. When I did, reactions ranged from disinterested to outraged. Preppers do not generally discuss climate change. Given their fixation on disaster, this posed a fascinating theoretical puzzle. While they were very grounded in the realities of disaster, they deny the scientific attribution of disasters to anthropogenic climate change. This can best be explained by observing how preppers "skirt the frame" of anthropogenic climate change by focusing on the constituent elements of climate change as independent events.

American conservatives are notably skeptical about anthropogenic climate change and prepping is a conservative cultural movement. Preppers generally distrust the government and hold strong anti-government sentiments consistent with American conservatism. Prepping as a subculture is dominated by conservative white men, the demographic group most likely to demonstrate climate skepticism (Nagel, 2016). And yet preppers, like many climate scientists and activists, believe that disaster is imminent. Both the scientific discourse of climate change and the discourse on prepping rely on risk frameworks (IPCC, 2022). Both scientists and preppers attempt to manage uncertainty about an unstable future through the use of scenarios that serve as heuristics (analytical tools) for managing and, in the case of preppers, preparing for risk.

Given its focus on individualism, rejection of government capacity to protect the collective, and denigration of people dependent on the government, prepping offers a conservative response to climate change, and environmental risk that appeals to Americans who are uncomfortable with the idea of global governmental intervention. In this, prepping holds appeal not just to self-identified conservatives, but to people who do not identify with, or are not moved by, the politics of liberal environmental*ism* which tends to see government intervention as the most appropriate response to environmental problems. To understand how people outside of scientific communities or environmentalist movements make sense of climate change, we must consider how culture, politics, and identity shape the ways people *know* about climate change. Prepping is an excellent case study that can help us consider the complex ways people respond to climate change, even when they purport not to care much about climate change at all.

Climate denial is firmly rooted in conservative politics, as conservatives have resisted the necessity of coordinated government regulation that scientists and environmental policy specialists argue is necessary to tackle a global problem like climate change. But preppers are not members of the political elite who are taking on the scientific community. Rather, they are ordinary Americans, mostly middle-class people, who are aware of environmental risk. But, due to conservative political leanings and distrust in government and scientific institutions, they are unable to accept the scientific concept of climate change, which from its introduction to the public has been paired with calls for government intervention. To be clear, I don't disagree with scientist's assessment that a global, united response to climate change is necessary, and in the current circumstances, this requires the coordination of sovereign nation states. I am concerned, however, that liberal efforts to achieve movement on this in the US have rarely tried to make sense of the failure of scientific communications about climate change to move a majority of the American public. Sociologists Philip Smith and Nicolas Howe (2015) argue that a successful, dramatic narrative about climate change that moves the public as a whole has failed to emerge. I agree, and I believe we can better understand why by looking at ways Americans are responding to environmental risk outside of scientific and environmentalist frames.

It is no coincidence that prepping has grown in popularity as a cultural movement at the same time that public debates about climate change have been contorted by disinformation campaigns. Prepping is a response to climate change and other global environmental changes that *skirts the frame* of climate change itself. In short, prepping arises as a response to climate change among a population of people who cannot ignore the prevalence and severity of socio-ecological risk, but who do not feel comfortable accepting that climate change is human-caused, and the implications that such a finding carries. If we see preppers' focus on the constituent elements of climate change—hurricanes, storms, fires, droughts, food shortages, floods, disruptions to power grids and transportation networks, and so on—preppers are in fact responding to climate change, even if they don't call it that by that name. For Americans pessimistic about government, or collective institutional enterprises more broadly, across the political spectrum, prepping is a much more culturally and politically palatable response to risk than political action.

Further Reading

Ford, A. (2021) "They will be like a swarm of locusts: Race, rurality and settler colonialism in American prepping culture." *Rural Sociology* 86(3), pp. 469–493.

Ford, A. (2019) "The self-sufficient citizen: Ecological habitus and environmental practices." *Sociological Perspectives* 62(5), pp. 627–645.

Ford, A. and Norgaard, K. M (2020) 'Whose everyday climate cultures? Environmental subjectivities and invisibility in climate change discourse.' *Climatic Change* 163, pp. 43–62.

References

Freudenburg, W. R. (2000) "Social constructions and social constrictions: Towards analyzing the social construction of 'the naturalized' as well as 'the natural'." In G. Spaargaren, A. P. G. Mol, and F. H. Buttel (Eds.), *Environment and Global Modernity.* London: Sage, pp. 103–119.

Gallup News Service. (2019) "Americans' trust and confidence in US Government on handling international, domestic problems (Trends)." *Gallup.* Available at: https://news.gallup.com/poll/246377/americans-trust-confidence-government-handling-international-domestic-problems-t.aspx (Accessed June 23, 2023).

Goffman, E. (1974) *Frame Analysis: An Essay on the Organization of Experience.* New York: Harper & Row.

IPCC. (2022) "IPCC AR6 Working Group II: Summary for Policymakers: Climate Change 2022, Impacts, Adaptation And Vulnerability." in *Implementing a US Carbon Tax: Challenges and Debates*, edited by H.-O. Pörtner, D. C. Roberts, M. Tignor, E. S. Poloczanska, K. Mintenbec, A. Alegría, M. Craig, S. Langsdor, S. Löschke, V. Möller, A. Okem, and B. Rama. Cambridge UK: Cambridge University Press.

Jacques, P. (2006) "The rearguard of modernity: Environmental skepticism as a struggle of citizenship." *Global Environmental Politics*, 6(1), pp. 76–101.

McCright, A. M. and Dunlap, R. E. (2003) "Defeating Kyoto: The conservative movement's impact on US climate." *Social Problems*, 50(3), pp. 348–373.

McCright, A. M., Marquart-Pyatt, S. T., Shwom, R. L., Brechin, S. R., and Allen, S. (2016) "Ideology, capitalism, and climate: Explaining public views about climate change in the United States." *Energy Research & Social Science*, 21, pp. 180–189.

Mettler, S. (2018) *The Government Citizen Disconnect.* New York: Russell Sage Foundation.

Nagel, J. (2016) *Gender and Climate Change: Impacts, Science, Policy.* New York: Routledge.

O'Hara, K. (2011) *Conservatism.* Wiltshire, UK: Reaktion Books.

Oreskes, N. (2004) "The scientific consensus on climate change." *Science*, 306(5702), p. 1686.

Pew Research Center. (2017) Public trust in government: 1958–2017.

Saragosa, M. (2020) "Why 'preppers' are going mainstream." *British Broadcasting Corporation.* Available at: www.bbc.com/news/business-55249590 (Accessed June 23, 2023).

Schlosberg, D. and Coles, R. (2015) "The new environmentalism of everyday life: Sustainability, material flows and movements." *Contemporary Political Theory*, 14(2), pp. 1–22.

Schneider-Mayerson, M. (2015) *Peak Oil: Apocalyptic Environmentalism and Libertarian Political Culture.* Chicago: Chicago University Press.

Smith, P. and Howe, N. C. (2015) *Climate Change as Social Drama: Global Warming in the Public Sphere.* New York: Cambridge University Press.

Snow, D. A., Rochford, E. B., Worden, S. K., and Benford, R. D. (1986) "Frame alignment processes, micromobilization, and movement participation." *American Sociological Review*, 51(4), pp. 464–481.

Part V

Climate Justice

Introduction

Tamar Law

The statue of Lady Justice is the portrayal of the moral idea of justice. It is often depicted holding scales that represent societal balance, a sword representing the power to punish injustice, and notably wearing a blindfold, representing how justice should be impartial and unbiased. When we consider climate change, however, we know that justice is not served across these three representations.

At the global scale, countries in the Global South disproportionately bear the brunt of rising temperatures and extreme weather events, even though countries in the Global North are the main contributors to greenhouse emissions. This is an example of what Sultana (2022) refers to as **climate coloniality**. Moreover, as global economic disparities worsen, the rich are better able to shield themselves from these climate change impacts while the poor are left to suffer the consequences. Some describe this as **climate apartheid**.

At the local scale, low-income communities, communities of color, Indigenous peoples, and immigrant communities are often disproportionately vulnerable and exposed to climate hazards due to social, political, and economic histories rooted in legacies of class, race, and ethnic discrimination. These frontline communities experience the most immediate and detrimental impacts of climate change.

In response to these injustices, **climate justice** has gained traction as a powerful concept and grassroots movement that acknowledges the unequal distribution of climate change impacts and the imperative for fair and equitable responses. Emerging from and in conversation with scholarship on environmental justice, social science scholarship demonstrates that climate change is not solely the result of natural processes but is informed by power structures and entangled within histories of imperialism, industrialization, and globalization. Centering on how the root causes of climate change are inherently social, economic, and political, such research reveals how climate change and climate interventions can exacerbate existing social inequalities if they are not properly understood and addressed.

Specifically, scholarship in Science and Technology Studies (STS) deconstructs social scientific understandings of climate change, revealing climate change inequalities and injustices, and showing that simply relying on technical solutions or technical fixes without recognizing and analyzing their broader social dimensions can perpetuate and produce new inequalities and injustices. STS is unique in the sense of understanding how inequalities and injustices are built into climate change-relevant infrastructure and knowledges. STS interventions involve bringing more diverse stakeholders into the design of sociotechnical solutions and climate change knowledge development and making certain they have a seat at the policy table. Work in this Part highlights the need for participation within climate decision-making processes, including both affected stakeholders and their diverse and local knowledge systems.

DOI: 10.4324/9781003409748-16

In Chapter 12, Ankit Bhardwaj draws from three discrete geographies from fieldwork across India to show the place-based nature of science, technology, and justice and highlights the scalar implications of addressing a global problem at a local level. Specifically, he points to how the just transition to the use of low-carbon energy technologies not only needs to be meaningful for emission reductions but, moreover, for the lived experiences of local workers and residents. Through his situated approach to climate justice, Bhardwaj argues that climate justice will come when the knowledge made by local people, and their actions to reduce emissions in the worlds where they live and work, are taken seriously by researchers and policymakers.

Emphasizing a just energy transition, Kendra Kintzi (Chapter 13) focuses on Jordan's national decarbonization program as it relates to renewable energy systems. Kintzi uses the concept of affordances to think through how different solar infrastructures can enable or disable different kinds of social practices, affecting who is included or excluded from this energy transition. In Jordan, many households benefit from personal rooftop solar thermal systems to diminish the effects of rising electricity costs. However, others including refugees and asylum seekers in formal camps and informal settlements are unable to access these benefits. Through this example, Kintzi reveals how seemingly highly beneficial solar sociotechnical infrastructures can nonetheless continue to shape the uneven distribution of the burdens of climate change.

Attending to how researchers might address contemporary climate injustices in practice, Roopali Phadke (Chapter 14) focuses on the politics of public participation within the infrastructural re-design and removal of locks and dams along the Mississippi River. Drawing from Phadke's undergraduate lab research with the U.S. Army Corps' implementation of a place-based model in infrastructure design, the chapter offers a methodological lesson on how to engage local communities. Implementing "upstream engagement," Phadke illustrates how government agencies can reconstruct notions of accountability and responsibility to engage the public in technology design.

Finally, in Chapter 15, Shangrila Joshi reviews several key forms of climate justice: distributive, participatory, epistemic, and transformative justice. Through the example of Nepal's *Guthi*, an Indigenous commons governance institution, she takes seriously the role of the commons for climate justice showing how reclaiming the commons may be transformative in enabling climate justice across various contexts and scales. Essential to reclaiming the commons is a respect for and empowerment of Indigenous knowledges and how that services the different elements of climate justice.

References

Sultana, F. (2022) "The Unbearable Heaviness of Climate Coloniality," *Political Geography*, 99, p. 3.

12 Postcards from Small Town India

Situated Climate Justice, Science, and Technology

Ankit Bhardwaj

Introduction

In this chapter, I present to you brief accounts from three vastly different towns across India. These stories are about the changes these places are undergoing due to the climate crisis. Without action, the towns will come apart at their seams. People who reside and work in them are facing dire consequences but are also forging creative responses. Their stories point to a pattern of how we all relate to climate science and technology. With these accounts, I hope to convince you

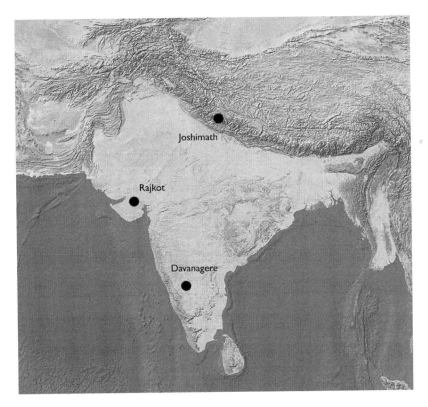

Figure 12.1 Approximate locations of the small towns in the Indian subcontinent. Joshimath sits in the majestic but climate-vulnerable Himalayas, Rajkot in the dry, shrubby Saurashtra region in the Gujarat peninsula, and Davanagere, southernmost, in semi-arid central Karnataka. Each town is more than two hundred kilometers by road away from the nearest major city, Dehradun, Ahmedabad, and Bengaluru respectively.

DOI: 10.4324/9781003409748-17

that building a world of lower greenhouse gas emissions and protections from climate impacts, which is also just and equitable, involves a situated understanding of the unique harms and priorities facing the workers and residents of each place. The map, below, is to help locate your imagination.

Joshimath, Uttarakhand, January 2023

Cracks appeared in homes across the small ancient town of Joshimath which had stood for more than a thousand years. Many residents, feeling unsafe, soon evacuated their dwellings. There was much speculation about what caused the cracks. Environmental activists took to social media to lay blame on climate change. They pointed to a flood that had affected the region just two years earlier. A glacier had burst, overwhelming dammed rivers and saturating the soil with liquid, weakening the land the town sat on. Scientists agreed in part, stating that though the flood was triggered by a landslide, the glacier was more precarious due to the increasing average temperature caused by climate change.

Critics of state infrastructure development argued that it was not climate change, but the rapid construction of roads, buildings, and hydropower dams in the Himalayas that led to the crisis. Local residents, alongside allied geologists and environmentalists, blamed tunneling for a nearby dam project which they posited ruptured an aquifer and destabilized the land under the town. Activists put up signs in the town's main market, blaming the government agency in charge of the dam's construction. They pointed to a report released by the national government's space agency which used satellite data to show rapid subsidence of land in the last two weeks.

However, the government quickly disputed that the land was sinking due to their hydropower projects. The Minister of Power defiantly argued that a committee of experts showed that dams led to fewer landslides and more greenery. Engineers testified that the construction was perfectly safe. Just a day after the space agency's report that showed rapid subsidence was released, it was no longer accessible; the government had taken it down.

Davanagere, Karnataka, March 2017

I had been walking about five minutes on a dusty road, off a local highway on the edge of Davanagere, when the air became noticeably smokier. I came up to a tight row of one-story huts, on the roof of each were slim, charred chimneys, emitting a sooty smoke. That must be black carbon, I thought to myself, one of the worst offenders of the greenhouse gas effect causing climate change. In front of the huts, men were carrying sacks filled to the brim; on the ground, rice was drying. I was in Mandakki Bhatti, an area known for making puffed rice. I asked a worker if I could go inside, curious about what caused the smoke. He gestured me in.

The room was stiflingly hot, not only from the baking sun but also from a blazing earthen oven in the center of the room that was heating a bowl of what appeared to be black sand. In a matter of minutes, I was sweating profusely, and my throat began itching. I could not imagine how the two men crouched around the oven felt. One of them sprinkled what looked like sawdust into the oven, bolstering its heat. The other took a bowl full of husked, dried rice and dropped it into the heated sand. He pulled a string that dropped a large whisk that churned the rice and sand. In a matter of seconds the grains of rice puffed up, which one worker sieved out and tossed into a huge pile in the corner. He immediately dropped another batch of raw rice in, working in a practiced and quick rhythm. The manager peered in, curious about who I was. I pointed at the oven and asked him about the temperature, shocked at how quickly the rice puffed. "500, 500 degrees," he muttered.

Mandakki Bhatti was the subject of grand designs by the municipal government to reduce emissions in Davanagere. Bureaucrats wanted to address air pollution from the traditional puffing process by outfitting roofs with solar panels and subsidizing electric ovens to replace the existing earthen ones. They commissioned engineers at a government university to design an electric oven. Yet, workers rejected prototypes as the ovens could not reach the high temperature necessary to puff rice quickly. Nor did the electric oven impart the characteristic smoky taste their puffed rice was known for. Bureaucrats did not hide their frustrations and confusion from me that workers rejected a technology that reduced emissions.

Rajkot, Gujarat, October 2017

Engineers were busily shuffling back and forth in a large, fluorescent tube-lit room in Rajkot, looking over plans for a national government scheme to provide "Housing for All." The town's chief engineer proudly announced that they were on track to build 6000 subsidized homes that year. She introduced me to her right-hand man, who then sat me down next to his computer to show me the blueprints in more detail. I was familiar with the basic template of government-built housing in India: a rudimentary block of eight flats, two on each floor separated by an open stairwell, often built with shoddy concrete. Its paint would peel in a few years, and few would choose to live in them. Yet, I noticed some small differences in the drawings he was showing me. The windows were slightly more set back into the walls and the openings into the stairwell slightly larger than usual, both of which I remarked to the engineer. He smiled and asked: "Building dekhna hai?" ("Do you want to see the building yourself?")

The next thing I knew I was on the back of his motorbike speeding through Rajkot's traffic on a hot and dusty day, wearing a spare, musty helmet he kindly lent me. After passing some fields, where cows were grazing, we arrived at a set of flats, about 40 in total, freshly painted in shades of white and earthy yellow. He proudly guided me through the flats, showing me the windows designed to keep the shade out, and the openings to ventilate the buildings. He also pointed out the system of pipes and tanks to harvest and store rainwater, a precious necessity in dry Rajkot where many households only received tap water for 20 minutes a day. He took me to the roof, painted a stark white to reflect heat. "Cool roof," he said, proudly puffing out his chest, and proceeded to tell me that they were installing solar panels soon.

Over the rest of the afternoon, he took me to another set of flats five minutes away, and then another, where people were already living, milling about the public area. There, a sign read "Grow Trees, Go Solar." I returned later without the engineer in tow, to ask residents how they felt about their experience living in the flats. Many expressed happiness with their new homes, barring the distance from the city center. The government was going to build six thousand more of these this year, and perhaps more the year after, and yet more after that.

Climate Change as Local, Controversial, and Uncertain

The three accounts challenge an influential way to think about science, technology, and justice. Specifically, we consider science and technology to be universal, or constant across the world. There is good reason to believe so. The science that establishes the reality of climate change is based on the construction of models of geophysical phenomena of the whole Earth, building on climate data collected across many countries. A solar panel works the same in the hut in Davanagere where workers puff rice, state-built flats in Rajkot, or on the roof of your home. The scientific measurement of sinking land is based on common principles whether in the Himalayas or the Rockies. Believing climate change is about trusting this globally uniform science; those that reject these globally stable facts are labeled climate deniers. But some of those who study science and technology have come to understand them as not globally uniform and context-less but **local** or **situated**. Science and technology, they argue, are the product of many small decisions people make together in specific settings such as laboratories, field sites, homes, and workplaces. What comes to matter is not abstract principles or the technologies themselves, but how people adapt them to local settings and problems. This has profound consequences for how we understand climate change, a phenomenon we tend to think of as impacting the whole planet uniformly.

Justice too can neatly map onto patterns of global inequality. **Climate justice** can be defined across several dimensions: who gains from or is harmed by climate change (distribution), who should bear the cost of the action (responsibility), who influences decision-making (procedural), whose interests are seen as legitimate (recognition), and who deserves compensation for past harms (reparation). It is undeniable that the rich have profited more from the consumption of fossil fuels, and therefore bear more responsibility for the crises, while the poorest will not have the resources to survive the harms of climate change, furthering their vulnerable suffering. At the global scale, colonialism and globalizing capitalism have structured a stark inequality. Many of the poorest are in the equatorial global South, heavily populated regions of postcolonial Africa, Asia, and Caribbean and Latin America. This is a cruel fact, as much of climate change's most vicious impacts including heat, rains, droughts, and storms will be most pronounced in this region. The poor also have little representation in the halls of power and control over decision-making. Their interests are seen as less legitimate. As a result, if things do not change, they will systematically have less influence on actions to respond to climate change.

Climate justice is debated on the global stage in a diplomatic rift between the richest countries in the global North and the poorest countries in the global South and is often characterized in terms of a conflict between the haves and have-nots. Yet, such a global understanding of climate justice, while a useful guide, provides little detail on how people's lives will change in response to climate change. Those impacted by climate change will likely have their own definitions and pursuits of justice formed depending on their situation. The challenges they face due to climate change, the priorities they have in response and the capacity to achieve their goals, will likely vary locally. Climate justice is thus not only a global pursuit but also a local one. I term this **situated climate justice**, the locally uneven ways in which people experience the consequences of climate change, understand the crisis, get involved, prioritize demands, and devise projects for a just future.

The three sketches I outline above indicate that climate science, technology, and justice will have different meanings in different situations. The first two examples show how political discussions on climate change tend to go. They are controversies, situations in which people disagree over the explanations of events or the solution to problems. Controversies are common in science, as they involve skepticism of explanations and solutions which are still uncertain (when the likelihood of correctness is unclear) and indeterminate (when relevant causes are unclear). Controversies protract debates, using up precious time necessary to avoid the most dangerous consequences of climate change. But they can also generate possibilities for the many affected to raise their concerns. In my first sketch regarding the land subsidence in Joshimath, we see controversy around the causes of a seemingly "natural" geophysical event that harmed people. Is climate change to blame? In the case of the plans to decarbonize traditional puffed rice production in Davanagere, we see controversy around reducing greenhouse gas emissions, known as either mitigation or decarbonization, a necessity if we are to avoid dangerous climate change. Are technologies to reduce emissions suitable? The third sketch about low-carbon, public housing development in Rajkot points to a means for reducing emissions that while small, was not controversial. Does it provide a lesson on how climate action can be pursued without conflict?

Contesting Scientific Frames Beyond the Popular

The debate over what caused the cracking homes in Joshimath reveals that struggles over what we think is happening, or knowledge about the crisis, will set the grounds for justice in climate controversies. When events have multiple possible causes, groups will seek to establish knowledge so that it bolsters explanations suitable to their positions. Sociologists call this organization of knowledge a frame: what is considered important, how it is portrayed, what is explained, and what is ignored. Climate controversies often involve multiple, conflicting scientific frames. Power, in part, is the ability to impose one's frame on others.

At the global scale, vast amounts of planetary-scale models have underpinned a settled scientific frame that climate change is real and dangerous. The frame is settled through consensus – an agreement amongst a vast majority of scientists. But these planetary-scale data and models have yet to capture intricacies at the scale most familiar to us, such as the places we live and work, leaving local climate science unsettled.

Even when pertinent, climate change might not be the first cause on people's minds when they face a crisis. For example, the ire of Joshimath's residents focused on the government agency for building a nearby hydropower dam and not necessarily those responsible for climate change, the position emphasized by environmental activists. It would be a mistake, though, to consider the residents' cause as different from one for climate justice. They rightly note that government construction puts them in harm's way, and this harm will only be further intensified due to climate change. Climate change acts as a threat multiplier, exacerbating harms people are already familiar with. But as such harms can have other relevant, and more locally familiar causes, calls for climate justice do not have to be framed in the name of climate change, but the threatful avatars it will unleash.

In a crisis, no single cause may readily stand out to blame. Instead, different stakeholders wield science to defend their preferred cause. What is considered truth is established in a public debate between alternative explanations. In classical accounts of controversies of environmental knowledge, scholars have often counterposed the expert scientific knowledge used by states and firms with citizen science. Popular science is knowledge made by a local community, based on their cultural tradition, made credible by alignment with lived experience, and communicated through stories. In Joshimath, residents drew on their own experiences and local knowledge of their region to blame new hydropower construction for the harms they faced. But they also drew from expert scientific reports by geologists and satellite data from government agencies that bolstered their claims. The government also cited geologists and engineers who argued that the dam was safe. Climate activists leaned on global scientific data, and computational models, to point toward the role of climate change. The controversy was not over those who believed "The Science" versus those who did not. There was no single scientific position on the issue. It was instead a public contest between different scientific frames to explain the crisis.

If climate justice is about forwarding community perspectives and welfare, then it seems that citizen science aligns with movements for justice. But in Joshimath, calls for justice did not just involve experience and tradition-based knowledge but also professional science that aligned with their claims. It involved government agencies disagreeing openly over causes, leading one to censor the other in an attempt at controlling the narrative. Alongside traditional wisdom, local communities can find professional scientific data such as from government satellites useful in bolstering their call for justice. But those opposed to climate justice too can marshal their own narratives and data. The climate crisis then poses a problem of how we come to know, and make decisions about, the causes of harms we face.

The Unavoidable Weight of What Exists for Just Transitions

The second sketch shows the influence of a community's traditional knowledge on climate action. In Davanagere, workers committed to existing traditional practices for making puffed rice, even as it polluted the air they worked in and worsened their health. While open to the use of solar panels, their experience with the prototype electric ovens made them skeptical that it would provide the temperature necessary for production. They came to believe that the electric ovens would make their lives harder and therefore continued to raise concerns.

STS scholars have studied how we change our use of technologies over time or technology transitions. To reduce greenhouse gas emissions, societies will have to undertake the mind-boggling task of shifting from using fossil-fuel energy to sources that do not emit carbon. We have undertaken such vast energy transitions before. Early societies used biomass (wood or other organic matter such as peat) for energy, then increasingly added more fuels that were more efficient such as whale oil, coal, and eventually fossil oil and gas. As of today, according to the Statistical Review of World Energy, we meet around 80% of our energy needs with fossil fuels. The transition to avoid dangerous climate change will involve replacing fossil fuels with low-carbon sources of energy. This will be a profound social change. At stake is how people make, work, move, and live in comfort, and the costs they bear to do so. Labor and social movements calling for a just transition argue that this shift to low-carbon energy should address the needs of workers and citizens such as gainful employment, social welfare, and clean environments. It should shift the costs of the transition not to the poorest but to those most responsible for the high consumption of fossil fuels.

The case of Davanagere shows that a transition to low-carbon energy will have to change long-existing traditions and labor practices of using carbon-intensive fuels. Just transitions will involve an unavoidable reckoning with how reducing emissions will involve transforming livelihoods, or how people secure the necessities of their lives. Even when the government subsidized the economic cost of the ovens themselves, the workers argued that the electric ovens would reduce the productivity of their labor, likely extending hours in a stifling workplace to produce the same amount, while taking away a preferred taste.

The government's climate-friendly aim to decarbonize the traditional practice faltered, not because the workers did not believe in climate change but because they saw the proposed technology as insufficiently aligned with their existing lives. The uptake of low-carbon technology was not a technical but rather a social and cultural problem. What came to matter was whether the specific, cultural practices in Mandakki Bhatti could be decarbonized with existing technology without disrupting the priorities of laborers. This indicates hard but unavoidable local choices for achieving a just transition: how will we change the many culturally specific practices that have come to rely on fossil fuels? And if it cannot be done without costs to the most vulnerable, is it worth it to rapidly reduce emissions?

Styles of Climate Justice

In the second and third cases, government bureaucrats aimed to reduce emissions, but they adopted different approaches. In Davanagere, bureaucrats bargained with workers over the suitability of low-carbon technology but faced opposition when workers framed the transition as disruptive to their work. In Rajkot, bureaucrats did not talk to the residents they were building for, often an indicator of climate injustice. But in their project, residents continued life uninterrupted and did not see any negative impacts of a project to reduce emissions on their lives. The city engineers had built low-carbon housing for the city's poorer residents with only small changes to familiar housing.

Due to the gravity and global scope of the crisis, responding to climate change will require vast, disruptive social changes. But the transition to using less energy and more from lower carbon sources can also involve small changes. They just need to be widespread. In Rajkot, building apartments that used less energy meant only tweaking the status quo: slightly changing the size and position of windows, choosing the right paint, and procuring increasingly affordable solar panels. Rajkot's bureaucrats creatively added these climate-friendly features to their existing practices. It required no new money or major policy on the part of the government. It did not disrupt their lives or pose unbearable costs to lower-income residents. For better or worse, responding to climate change can look like continuing as-is.

In contrast to Davanagere, emission reduction in Rajkot's low-carbon homes aligned with social welfare. Residents benefited because their homes would be slightly cooler and their electric bills slightly lower than usual. Action to reduce emissions need not be disruptive. It can have social co-benefits. These benefits include more than better homes. Other low-carbon projects such as public transit, electrified and efficient appliances, new green public spaces, and easily accessible services can mean healthier, and more equitable cities. The focus on co-benefits turns our attention to how climate action addresses the other pressing social and environmental needs of people.

Studying such a situated justice involves analyzing how climate actions are related to factors of situational importance such as housing, potable water, affordable electricity, livelihoods, and culture. The importance of each aspect of justice shifts in relevance, depending on the concerns of the actors involved. For example, the procedural justice of involving workers in discussions around low-carbon technologies in Davanagere rightly gave opportunities for workers to raise their concerns but stymied the distribution of low-carbon technologies.

There was a lack of procedural justice in Rajkot but the distribution of solar panels and cool housing to the city's low-income residents involved recognition of their climate vulnerability, and distribution to alleviate harms. Justice is not one-size-fits-all. It involves consideration of local cultures and situations. People create climate justice in their own style by coming together to define it in ways consistent with their needs.

Conclusion

Climate change will bring extraordinary changes to ordinary places. But its changes will not be uniform, and neither will climate science, technology, and justice. Climate change will multiply existing threats of floods, heat, and drought in cities. It will also mobilize groups with differing worldviews and complementary scientific findings. Culture will shape how people frame the crises they face and the solutions they support, as will professional scientific findings and innovative technologies. Disruptions to what people hold dear, such as how we traditionally make food, will invite opposition. But the changes that will better our lives will quietly find a place in our homes.

Rather than focus on the big disagreements over abstracted facts and global principles, a situated approach helps us see that both knowledge about climate change, and actions to reduce emissions will be made by people in the local worlds where they live and work. Causes for climate justice that aim to reduce emissions and increase welfare will have to consider how people make a living, and what they treasure about the world around them. The situations in Joshimath, Davanagere, and Rajkot are unique, but they reveal a truth. The success of climate science, technology, and justice will not be won in a global debate, or descend by international fiat, but be built piece by piece, in every place, including around your corner.

Further Reading

Angelo, H. (2022) "Boomtown: A solar land rush in the West," *Harper's Magazine*, December 12.

Araos, M. (2023) "Democracy underwater: Public participation, technical expertise, and climate infrastructure planning in New York City," *Theory and Society*, 52(1), pp. 1–34.

Bhardwaj, A. (2022) "Styles of decarbonization," *Environmental Politics*, Taylor & Francis, pp. 1–23.

Boyer, D. and Howe, C. (2019) *Wind and Power in the Anthropocene*, Durham, NC: Duke University Press.

Castro, B., & Sen, R. (2022). Everyday Adaptation: Theorizing climate change adaptation in daily life. *Global Environmental Change*, 75.

Dubash, N.K. (ed.) (2020) *India in a Warming World: Integrating Climate Change and Development*, New Delhi: Oxford University Press.

Elliott, R. (2021) *Underwater: Loss, Flood Insurance, and the Moral Economy of Climate Change in the United States*, New York: Columbia University Press.

Rice, J.L., Cohen, D.A., Long, J., et al. (2020) "Contradictions of the climate-friendly city: New perspectives on eco-gentrification and housing justice," *International Journal of Urban and Regional Research*, 44(1), pp. 145–165.

Schlosberg, D. and Collins, L.B. (2014) "From environmental to climate justice: Climate change and the discourse of environmental justice," *WIREs Climate Change*, 5(3), pp. 359–374.

Táíwò, O.O. (2022) *Reconsidering Reparations*, New York: Oxford University Press.

13 Solar Affordances and the Struggle for Climate Justice in Southwest Asia

Kendra Kintzi

Introduction: A Rooftop View of Solar Thermal Systems in Jordan

Summer days are hot in the northeastern corner of Jordan, near the border with Syria. The sun beats down across the golden hillsides, penetrating the thin walls of many of the houses that populate the towns and villages along the border. On a bright spring day in 2022, I drove out to a Jordanian village near the city of Irbid to meet with Maryam, a woman who had installed a solar thermal water heater on her rooftop to help conserve energy and reduce her monthly electricity bill. Maryam's family migrated to this village from Damascus over a century ago, and generations of her family have built homes and gardens that made use of the rich ecological diversity of this region. As I walked up to her house, the sun glinted off the golden stalks of barley and flowering fruit trees that interlaced the old stone buildings. Sitting on the veranda of Maryam's house, we drank mint lemonade and ate fresh loquat, a small orange fruit, that Maryam had just harvested from her garden.

We talked about the challenges and opportunities for household renewable energy systems in the village. She explained how electricity prices had become increasingly unaffordable amidst nation-wide energy sector reforms, with monthly electricity bills becoming an unbearable burden for many households in the village. In addition, she had been noticing shifts in the local climate in recent years, as climate change brought bitterly cold winds and ice in the winter and extended periods of drought in the summer. Maryam took me up to the roof to show me the small solar thermal system that she installed. It uses the thermal rays of the sun to heat tubes filled with water, which is then piped directly into the house. As we explored the form and function of the system, Maryam shared how this simple array had reduced her electricity bills and her reliance on the local electricity distribution company. This renewable energy system gave her more economic independence and direct control over the comfort of her home.

Around the world today, many individuals, communities, and governments recognize the urgent need for rapid decarbonization. **Decarbonization**, or the reduction of global greenhouse gas (GHG) emissions, can take many different forms. Renewable energy transition is a key piece of decarbonization and can include simple renewable energy technologies like Maryam's rooftop solar thermal system, as well as more complex renewable technologies like large-scale wind farms with battery storage systems. Decarbonization also includes making changes to the built environment, which can take the shape of small interventions like installing better household insulation and switching to more efficient light bulbs, as well as large-scale interventions like transforming urban and rural transportation and agricultural systems to reduce the use of hydrocarbon-based inputs like gasoline and petrochemical fertilizers.

The government of Jordan, among others around the globe, has set ambitious decarbonization targets and developed national programs to reach these goals. Jordan is a middle-income country that faces multiple, pressing development challenges, including high unemployment

DOI: 10.4324/9781003409748-18

rates, high inequality, and high public debt. In 2012, Jordan created a national scale Renewable Energy and Energy Efficiency program, which has already led to the development of over two gigawatts of new renewable energy to power 30% of the country's electricity needs. Jordan was one of the first countries in Southwest Asia to adopt such an ambitious decarbonization program. Part of the reason Jordan made this aggressive shift toward renewable energy is that, in contrast to many of the countries in the region, the country has very limited domestic hydrocarbon resources. As a result, it imports most of its oil and gas, which creates a sizable economic burden for individual households and for the country. Most households across the country have access to electricity and cooking gas, but the challenge of energy poverty is acute, meaning that many households are not able to afford their monthly electricity and gas bills.

Like Jordan, a growing number of countries, states, and cities around the world have taken up the challenge over the past decade to transition to renewable energy, reduce dependence on fossil fuels, and decarbonize their infrastructure. National decarbonization programs provide a potential pathway for countries like Jordan to meet international climate goals, reduce dependence on imported hydrocarbon fuels, improve household and national level energy security, and boost domestic production of renewable energy technologies. Many of these programs are designed as sociotechnical solutions that involve a combination of incentives and regulatory changes at the national, state, or city level. For example, they can include individualized incentives like tax credits and rebates to help encourage users to adopt renewable energy technologies. They can also include broader policies to disincentivize the use of hydrocarbons across commercial and industrial systems or rebuild and restructure public infrastructures.

While decarbonization is an urgent priority for mitigating the impacts of climate change, implementing large-scale decarbonization programs in a just, equitable, and affordable manner has proven to be a daunting challenge in practice. This challenge is particularly acute for low- and middle-income countries where many communities do not have consistent access to affordable power. Over a billion people around the world do not have access to electricity; and in many low-income countries, over 70% of the population does not have consistent access to reliable, affordable power. Implementing decarbonization policies can also create conflicts between different groups of interests, as state governments, electric utilities, regulators, and communities around the globe may have competing needs and priorities. Understanding the uneven distribution of the benefits and burdens of energy systems is a key focus of **energy justice** scholarship. An energy justice framework enables us to begin thinking through the different needs of diverse groups of users to work toward more just and equitable energy systems. Energy justice is a critical piece of broader struggles for climate justice. Around the world, climate justice scholarship and activism have brought critical attention to the uneven global impacts of climate change and the powerful connections between environmental and social change.

This chapter provides a brief introduction to the key concepts of sociotechnical systems, infrastructure, and affordances as they relate to ongoing struggles for energy and climate justice in Southwest Asia. These concepts have been developed in Science and Technology Studies (STS) and offer key insights to understand why decarbonization efforts often generate uneven impacts that benefit some while disadvantaging others. This chapter draws from STS and climate justice scholarship, as well as my own ethnographic research on Jordan's national decarbonization program. This ethnographic research consisted of interviews and site visits with renewable energy developers, manufacturers, and residents in Jordan to examine how renewable systems are woven into the country's changing social and environmental landscapes. By exploring ongoing struggles for affordable energy and community-based renewable energy technologies, this chapter provides insights into affordances of solar technologies and the possibilities for just decarbonization around the world today.

The Social Life of Renewable Energy Technologies: Affordances, Infrastructures, and Sociotechnical Systems

Why is decarbonization so difficult? Despite decades of scientific study, careful intergovernmental planning and coordination, and the development of ambitious national, state, and city-level targets, GHG emissions continue to rise. One of the key reasons for this is that energy systems are not just technical, but also social. What does it mean for a system to be both technical and social? In STS, a vast body of scholarship explores the meaning of **sociotechnical systems** to help us understand the ways that material objects come together with social practices to shape the world around us. A sociotechnical systems perspective requires investigation of both the social context in which specific technologies were designed and the social relations that shape how they circulate and are reproduced (Bijker, 1994).

Transitioning global energy systems to renewable and low-carbon sources requires attention to the sociotechnical dynamics of design and use, which may include rethinking the kinds of calculations that go into system-wide energy planning and transforming everyday practices in residential and other settings. Decarbonization demands more than a simple technical fix; it requires the transformation of intertwined sociotechnical processes, from changing the way we heat our homes and the ways we cook our food to the ways that we travel and interact with each other. In short, it means changing the way we live.

STS approaches to infrastructure emphasize how they accrete, or build up, over time and their diverse uses and adaptations by different actors (Edwards et al., 2009). Susan Leigh Star writes that infrastructures are relational. They are not the same in all times and places. Rather, they are experienced in different ways by different constituencies (Star, 1999). For example, in the case of renewable energy, the construction of a large-scale solar farm on communal grazing lands in Southwest Asia may generate electric power that enables certain groups of people to power cell phones, heat their homes, and plug in a refrigerator, while also limiting access to a key resource (grazing lands) for local goat and sheep herders and their flocks. In this way, infrastructures shape how resources flow within and between communities, often distributing benefits and burdens in uneven ways.

A variety of actors make up infrastructural systems, which enroll diverse practices, norms, and values. As infrastructures grow and evolve, these practices and values are translated and adapted to new contexts with different conditions and needs (Larkin, 2013). This process of translating infrastructures from one locale to another is not always easy; in fact, it is often messy, complicated, and controversial. To untangle this messy and complicated picture, it is helpful to draw on the concept of **affordances** to examine the possibilities of infrastructural arrangements. Affordances are the possibilities for action that arise from the way that objects are configured. In STS scholarship, affordances provide a way of thinking through how different sociotechnologies can enable or disable different kinds of social practices. For example, solar photovoltaic panels that were designed for the temperate atmospheric conditions of Northern Europe may require new practices of maintenance and repair when they are installed in the hot, dry, and dusty climates of Southwest Asia. As particles of sand and dust cover the photovoltaic panels and block the sun's rays, the efficiency of the panels declines, prompting new labor practices to clean and repair them on a regular basis.

These concepts help us understand how renewable energy systems assemble disparate technological, political, and cultural histories. As sociotechnical systems, renewable energy infrastructures bring together a vast network of diverse technical and social practices as intimate as reading by lamplight and as distant as lithium mining. Specific renewable energy technologies have affordances, and as these technologies are brought together through broader decarbonization efforts, they both shape, and are shaped by, diverse social practices around the globe.

Unpacking the sociotechnical dimensions of decarbonization requires us to be attentive to uneven power relations. Infrastructures are more than simply the sum of their parts; critically, they also enable important flows – not only of energy, but also of power, resources, and information. Renewable energy infrastructures are not simply systems of technologies and practices; but structures that enable flows of capital and sustain particular social relations. These flows can both reinforce and transform existing governance arrangements and power imbalances between different social groups (Rutherford and Coutard, 2014).

These concepts are key to understanding the possibilities for energy and climate justice around the world today. As climate justice scholars have long shown, processes of technical and environmental change are never free of politics but take shape in a highly uneven global landscape marked by profound social inequality. Energy and climate justice scholarship helps us see the ways that seemingly neutral regulatory and governance processes, such as the siting of energy infrastructure and the development of urban climate resilience plans, are deeply entangled with situated struggles for access and inclusion (Ranganathan and Bratman, 2021). In the following section, I examine how the affordances of solar thermal technologies take shape in the context of Jordan's decarbonization program.

Solar Thermal Affordances and the Uneven Infrastructures of Decarbonization

Rooftop solar thermal systems, like the one that Maryam uses in her home, are simple technologies assembled from glass, steel, and fiber insulation. These systems are the most common form of renewable energy technology in use in Jordan today, and they can be found in homes and apartments across the entire country. Examining the affordances of these systems reveals how the existing social and environmental landscape has shaped the development of localized forms of renewable energy infrastructure, while shedding light on the possibilities for community-centered and just decarbonization.

Although solar thermal water heaters typically do not produce grid-tied electricity, they channel the thermal energy of the sun to directly heat water that is then piped into residences. There are a few different models for these systems, including ones that use photovoltaic cells. However, the two simplest and most prevalent models in Jordan use evacuated tubes or flat plate collectors. In the evacuated tube model, water is kept in a galvanized or stainless-steel tank that is positioned at the top of a panel of glass tubes, which are typically lined with copper or bronze. Cold water from the tank flows down to the base of the pipes, and as the sun's rays hit the tubes, the water heats up and rises upwards back into the holding tank. This relatively simple process creates a dynamic, circular flow of water and heat. During peak sun hours, hundreds of liters of water can be heated. Similarly, the flat plate collector model uses a panel of tubes set within a flat surface, typically painted black, which absorbs the sun's energy and creates the same dynamic flow of heated water. The tanks are often insulated so that the water stays warm long after the sun sets (up to 72 hours for some models).

The reason that these systems are so ubiquitous across Jordan is due to both the technical affordances and the social and environmental landscape where they are developed and adapted for different uses. A crucial part of this landscape is the lack of consistent water supply. Jordan is one of the most water-scarce countries in the world. Across Jordan's rural and urban landscapes, most buildings have piped water infrastructure. But the water service is erratic and limited, resulting in frequent water shutoffs and supply disruptions, particularly in the summer and fall dry seasons. Nearly all homes and apartment buildings have installed water tanks made

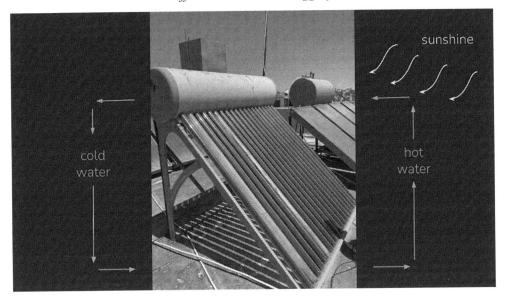

Figure 13.1 Diagram of a rooftop solar thermal system.
Source: Photograph by the author.

of galvanized or stainless steel or polyethylene to ensure water supplies amidst these frequent disruptions. The tanks typically hold 200 to 500 liters of water and are usually housed on the roof, due to space constraints inside most buildings and in order to use gravity to assist with water pressure. Most buildings are constructed with flat rooftops to accommodate rows of dozens of tanks linked up to individual apartment units. This landscape lends itself to the adoption of rooftop solar thermal systems, since the tanks can be reconfigured to accommodate solar thermal technologies.

Another key affordance of solar thermal systems is its simplicity and affordability of readily available component parts (costing $200 to $600). Unlike solar photovoltaic systems, which require minerals and materials sourced from specific locations and intensive, industrial production processes, solar thermal systems require a relatively low level of resource inputs and production capacities. In addition, solar thermal systems can be recycled and repaired relatively easily compared with solar photovoltaic panels. In Jordan, over a dozen local manufacturers can produce these systems by hand, using low-tech fabrication and materials sourced either locally or from nearby countries. The simplicity and adaptability of these systems also means that they are easily scaled up or down in size and volume, based on users' needs. They are relatively affordable and accessible across Jordan, making them both the most common form of renewable energy and the most impactful on household wellbeing.

There are also limitations to this renewable energy infrastructure. While they are highly efficient at heating water, they do not produce electricity. A solar thermal system cannot power a computer, car, or lightbulb. While these systems are accessible to many, they are not accessible to all. Even with the added support of government incentives that subsidize their cost, the systems are still unaffordable for the millions of residents living below the poverty line. The requirement of ample rooftop space for these systems is a further barrier to households living in temporary dwellings or facing housing insecurity. This challenge is particularly acute for the three million refugees and asylum seekers currently living in Jordan.

During my interviews with residents across the towns, villages, and cities of Jordan, I encountered many households that benefitted daily from personal rooftop solar thermal systems. But I also encountered many households, particularly those in formal camps and informal settlements, that were unable to access them. Nearly all households struggled with the rising electricity costs that accompanied the national energy transition program. But while some households were able to invest in renewable technologies to offset their electricity bills, other households had little choice but to disconnect, turning instead to time-tested techniques like rolled-up blankets and rugs to help insulate the home and keep out both blistering summer heat and bitter winter cold (Faxon and Kintzi, 2022). For these households, the struggle for energy and climate justice is not just about renewable energy technology, but also the daily struggles for better insulation, more accessible infrastructure, more secure and supportive housing, and thriving community spaces.

Conclusion: Toward Just Decarbonization

A key insight and goal of climate justice work is to bring attention to the lived realities and practices already shaping climate realities around the world today, from the knowledge that already exists, to the ways that community spaces and resources are already used, and the ongoing struggles for access and inclusion that matter to peoples' everyday lives. Simple technologies like rooftop solar thermal systems form a prime example of the kinds of climate change mitigation technologies that already exist in the world today, providing powerful alternatives for communities facing rising energy costs and an increasingly volatile climate. But there is no one-size-fits-all solution to decarbonization. While rooftop solar thermal systems offer affordances particularly conducive to Jordan's built environment, they would require translation and adaptation in other environments. There also are likely other sociotechnical alternatives already in place in other settings better adapted to localized conditions.

In analyzing the linkages between diverse decarbonization efforts around the world, STS approaches to infrastructure, affordances, and sociotechnical systems can help us think more flexibly about how we design, build, and extend energy systems and how we engage place-based needs and priorities. The affordances of rooftop solar thermal systems highlight the transformative potential of technologies that incorporate local, sustainable supply chains to facilitate community-based production and local ownership of infrastructure development and implementation. At the same time, analyzing the limitations of these sociotechnical systems reveals the ways that place-based inequalities and struggles for access and inclusion continue to shape the uneven distribution of benefits and burdens of climate change. Deploying the analytical tools of sociotechnical analysis brings focus to these dynamics and can help clarify the terms of ongoing struggles for energy and climate justice in Jordan and elsewhere.

The proliferation of rooftop solar thermal systems in Jordan also provides key insights into the social dimensions of decarbonization. For low- and middle-income countries like Jordan, broadening access to simple technologies like rooftop solar thermal systems can address energy poverty and improve daily living conditions for individual households while also helping national governments meet their decarbonization and economic development goals. However, these efforts are never just technical. They are also political. STS approaches call attention to how the affordances of technologies like this interact with the social dynamics of each place. The advancement of low-carbon technologies requires governance decisions about how resources should be allocated, which can advantage some groups of actors while disadvantaging others. An energy and climate justice approach calls us to question how existing social dynamics shape who benefits and who is burdened by new decarbonization efforts. How do the affordances of a particular technology intersect with the existing landscape? How do ongoing struggles for energy access, or acute challenges of energy poverty, shape access to different decarbonization technologies? How do existing infrastructures channel flows of resources toward, or away from, different groups of actors? Addressing these questions is key to energy and climate justice.

Further Reading

Dawson, A. (2020) "We need a national People's Power campaign," *Verso Blog*. www.versobooks.com/blogs/news/4677-we-need-a-national-people-s-power-campaign.

Sou, G., Risha, A., and Ziervogel, G. (2022) *Everyday Stories of Climate Change*. Manchester: RMIT University and The University of Manchester.

Sze, J. (2007) *Noxious New York*. Cambridge: MIT Press.

References

Bijker, W. (ed.) (1994) *Shaping Technology/Building Society: Studies in Sociotechnical Change*. Cambridge: MIT Press.

Edwards, P., Bowker, G., Jackson, S., and Williams, R. (2009) "Introduction: An Agenda for Infrastructure Studies," *Journal of the Association for Information Systems* 10(5), pp. 365–374. DOI: 10.17705/1jais.00200.

Faxon, H. and Kintzi, K. (2022) "Critical Geographies of Smart Development," *Transactions of the Institute of British Geographers* 47(4), pp. 898–911. DOI: 10.1111/tran.12560.

Larkin, B. (2013) "The Politics and Poetics of Infrastructure," *Annual Review of Anthropology* 42(1), pp. 327–343. DOI: 10.1146/annurev-anthro-092412–155522.

Ranganathan, M. and Bratman, E. (2021) "From Urban Resilience to Abolitionist Climate Justice in Washington, DC," *Antipode* 53(1), pp. 115–137. DOI: 10.1111/anti.12555.

Rutherford, J. and Coutard, O. (2014) "Urban Energy Transitions: Places, Processes and Politics of Sociotechnical Change," *Urban Studies* 51(7), pp. 1353–1377.

Star, S. (1999) "The Ethnography of Infrastructure," *American Behavioral Scientist* 43(3), pp. 377–391. DOI: 10.1177/00027649921955326.

14 Upstream Engagement in the Era of Climate Change

Roopali Phadke

Introduction

On a frigid night in February 2021, thousands huddled on the railings of Upper St. Anthony Falls in Minneapolis, the first major lock and dam on the Mississippi River, to peer down at Indigenous artist Moira Villiard's light and sound show, *Madweyaashkaa: Waves Can Be Heard.* Our attention was focused deep inside the 400 by 50-foot lock chamber at floating images of Grandmother moon, fire, earth, water, and the jingle dress dance entwined in swirling colors. Challenging the histories of violence and racism that have defined our relationship with the river, this piece asked us to imagine how physical and social infrastructures can promote healing amidst pandemic and racial uprisings. The performance was surprisingly sanctioned by the Army Corps of Engineers, the agency that has forcefully held the Mississippi River in place for more than a century.

Like many water agencies across the US and the world, the Corps' mandate is rapidly shifting away from its 19th-century mission to harness rivers for commercial exploitation toward a 21st-century need to repair rivers facing climate change impacts such as biodiversity loss. For example, Corps projects have been impacted by invasive species, unpredictable flooding, and federal mandates to restore destroyed fish habitats. In 2018, the agency was directed by the US Congress to complete a multiyear, multimillion-dollar study to determine what "value" the American public derives from the operation of infrastructure on the Mississippi, particularly in the upper basin where commercial navigation is no longer economical. It is likely that the Corps

Figure 14.1 Mockup of *Madweyaashkaa.*

DOI: 10.4324/9781003409748-19

will recommend divestment, as they have done elsewhere, and perhaps even pursue dam removal and massive river restoration.

There is no precedent for undamming a river system the scale of the Mississippi, encompassing the fourth-largest watershed in the world. The mere act of imagining this future will have major consequences for how the Corps operates around the nation and ripple effects for millions who live downriver. With massive injections of public money coming for infrastructure revitalization in an era of climate change and green energy development, this is an extraordinary time to consider how 21st-century river management can become a force for climate justice. Herein, achieving climate justice requires us to acknowledge and repair the damage done to ecosystems and communities.

I begin the chapter with a historical look at why the Mississippi was dammed and dredged, and the long-term social and environmental implications of these efforts. I then describe how STS-informed approaches to participatory research and critical infrastructure studies can be applied to thinking about the future of the Mississippi River. The third section describes the community-based research conducted by my research lab over the last several years and how the lessons we've learned can apply to future Corps' actions. I end with reflections on the implications of this case study for thinking about the role of water infrastructure in an era of climate changes and climate justice.

The Mississippi's Transformation

The Mississippi, the second longest river in North America, travels from its origins in Lake Itasca in northwestern Minnesota 2,350 miles south to the Gulf of Mexico. The river is hydrologically and administratively divided into three sections: the Upper Mississippi, from its headwaters to the confluence with the Missouri River near St. Louis; the Middle Mississippi, which is downriver from the Missouri to the Ohio River; and the Lower Mississippi, which flows from the Ohio to the Gulf of Mexico.

The Corps built and operated the Mississippi's infrastructure with a battle general's intent for over 150 years (Frankel, 2018). Founded in 1755, the Corps is a branch of the US Department of Defense. It is one of the world's largest public engineering, design, and construction agencies. Congress first authorized the Corps to construct six dams in the headwaters area between 1880 and 1907 to support a vocal and powerful flour milling and timber industry. Figure 14.2 shows the state of the river in the early 1900s when the Meeker Dam was constructed. This was the first and northernmost lock and dam on the Mississippi River. In 1910, the Corps built America's first national dam with a hydroelectric plant at Lock and Dam 1 in St. Paul.

Today, the Corps operates a "stairway of water" that consists of nearly 30 locks and dams on the Upper Mississippi River between Minneapolis and St. Louis to permit barge traffic and protect farms and cities from flooding. The Corps also maintains a nine-foot-deep channel to enable navigation through this stretch. Between the first and last lock, the Mississippi drops 420 feet over the course of 670 miles. Before channel construction, the depth of the Upper Mississippi averaged approximately three feet and was as low as one foot deep near St. Paul in dry seasons.

The Upper Mississippi's industrial past and imagined futures cannot be separated from histories of native dispossession. The locks and dams on the Upper Mississippi effectively drowned the waterfalls, sand bars, islands, and gorges that once covered this territory. The Dakota people, who occupied the region for nearly 10,000 years before the onset of settler colonialism in 1680, were disregarded and abused when the river was claimed as a commercial resource. The confluence of the Mississippi and Minnesota Rivers, sacred grounds known to the Dakota as *Bdote*, was made into a military base where the Dakota were imprisoned during the War of 1862. The

Figure 14.2 Meeker Dam Construction.

Source: Army Corps of Engineers St Paul District.

Dakota still claim this territory and are deeply connected to the politics of infrastructure development in the region.

Moreover, the river's industrial past is also inseparable from the violence of slavery, anti-Blackness, and the ongoing dispossession of Black communities. Historians have described how the Mississippi served as both refuge and oppressor for those enslaved peoples who toiled the crops and labored on the boats of the antebellum economy (Zeisler-Vralsted, 2019). Zoning and redlining policies in the 20th century further segregated Black populations, simultaneously barring them from the river while exposing them to the impacts of toxic industries that took up residence on the banks (Miller, 2020).

In 2015, the Corps was forced to close the uppermost lock to block the spread of invasive carp further upstream. This was the first time a navigable waterway in the US was closed to stem the tide of an invasive species. Soon after, the US Congress passed the Water Resources Development Act of 2018 (PL 115–270), which authorized the Corps to conduct a disposition study that covers the three uppermost locks and dams in Minneapolis and St. Paul. This includes Upper St. Anthony Lock and Dam, Lower St. Anthony Lock and Dam, and Lock and Dam 1. The goal was to determine whether it was still in the public's interest for the Corps to continue owning and maintaining them (at a cost of $1.6 million per year) now that they no longer meet a navigation purpose.

The Corps completed their report on the first structure, Upper St. Anthony Falls, and have moved on to examine the two other structures in a second study scheduled to be completed by

2024–25. The disposition study process examines three options: 1) no action, 2) partial disposal (deauthorize and retain some flood mitigation features), and 3) full disposal (deauthorize and completely dispose). After disposal, the facility is up for sale or transference to another public or private entity. It is also possible that the structures can be removed. Congress also directed the Corps to consider other measures to "preserve and enhance recreational opportunities and the health of the ecosystem" and "maintain the benefits to the natural ecosystem and human environment" (US Congress, 2018). The Corps is already involved in restoration projects in the upper basin, including dredging pools along the floodplain and using the sediment to construct islands and restore wetland fish and waterfowl habitat.

Interest in restoring the Mississippi River is representative of dam removal and river restoration movements cropping up across the US and Europe. Over 1,340 dams have been removed across the US, with 930 removed since 1999 (American Rivers, 2017). On the Upper Mississippi, river restoration will be a highly complex act of engineering through which new islands and channels would be created to restore whitewater conditions. Removing the dams would mean that in the dry season (July–August) it would be possible to wade across a two-foot river in the city, which today the Corps holds at a minimum nine-foot depth. Those who live, work, and play in and along the river have many ideas about what the river is and what it should be. The Corps' consideration of dam removal and river restoration has drawn impassioned and conflicting responses. Some advocate to keep the dams while many want to free the river and give land back to Native peoples.

Scholarly Inspirations from STS

I have engaged with STS scholarship on participatory technology design and critical infrastructure to research this case. STS scholars have described, analyzed, and experimented with citizen-led technology assessment and community-based research. In the realms of environment and climate, STS participatory experiments have included consensus conferences, citizen juries, and deliberative polls on topics that range from assessing biotechnology to geoengineering the planet. My environmental scholarship models a form of STS action-research, which I've referred to as "place-based technology assessment," that connects local policy actors with social movement demands through multisite, multiscale research collaborations (Phadke, 2014).

I evoke **place-making** to underscore the importance of designing technology, such as lock and dam infrastructures, in relationship with those who live, work, and play in the very environments we seek to change. Social scientists and humanists have come to define **infrastructure** as a complex and changing set of objects, laws, and knowledge practices that manage, maintain and repair places (Anand, Gupta, and Appel, 2018; Jackson, 2014; Edwards, 2003; Mitchell, 2002). Neglected and abandoned infrastructures, like lead water mains and oil pipelines, are sites for intense political conflict in the era of climate change because they may fail to safely deliver resources. Infrastructural failures tend to affect people unequally. Thus, their management and repair allow us to consider how issues of gender, race, colonialism, and classism relate to place-making.

Focusing on "upstream" engagement allows me to connect critical infrastructure studies with place-making design. **Upstream** refers to public engagement with potential problems and solutions at an early stage of the research and development process, rather than downstream when problems and social controversy are experienced (Rogers-Hayden and Pidgeon, 2007). Upstream engagement has been used across a wide set of technology contexts from nanotechnology to genetically modified organisms. While "upstream" here refers to process, the concept is particularly useful for thinking about governing rivers that move across complex cultural and physical terrains. In the context of the Corps' disposition studies on the Mississippi, the above wisdom helps us consider the limitations of conventional public engagement practice and how we might intervene toward greater inclusive, reflexive, and creative engagement.

Working in collaboration with several local and national river organizations, my undergraduate student research team has documented, interpreted, and experimented with public engagement processes. By examining which modes of engagement enable or shut down imaginings of a future river, we've aimed to create upstream opportunities for reflection among publics who have been excluded and harmed by past approaches. Our work has included archival document analysis, public surveying of hundreds of people, expert interviews, and focus group river tours. The next section describes our approach and findings.

Modeling Upstream Collaborative Research

This project began in 2018–19 when the Corps' launched its first public hearings about the future of the Upper St. Anthony Lock and Dam. Between 2000–23, our research group reviewed all the public comments received by the Corps and monitored news accounts and social media. We also interviewed local and national experts and conducted surveys and focus groups. The work culminated with a research report delivered to the Corps with a set of recommendations about how to improve their engagement efforts. I describe below our findings and reflections from the Corps about them.

Public comments received by the Corps offered insights into how residents perceived the importance of the locks and dams to their sense of place. Of the 114 submissions, 84 comments opposed full disposition and 17 comments favored partial disposition of the Upper St. Anthony Lock and Dam. Most respondents asked the Corps to stay, citing trust in their ability alone to maintain the river infrastructure that matters most to quality of life. For example, one resident wrote: "We are extremely skeptical that there is any other organization with the financial or organizational capacity to manage this critically important facility." Minneapolis Mayor Jacob Frey, along with city council members, wrote that the Corps "are the only entity that can continue to manage these structures with the expertise and oversight that considers its effect on the system of locks, dams, cutoff wall, bridges, flood mitigation, municipal water, industrial uses, transportation, and recreation that impacts millions of Americans." While the public comments overwhelmingly told the Corps to stay, the agency's disposition report on Upper St. Anthony Falls concluded that "there is no federal interest in continuing to own and operate the project, and recommends full disposal, combined with offering a monetary incentive to expedite the disposal".

Community groups stepped in to find a way forward given the Corps' desire to abandon the facility. This process, now part of "The Falls Initiative," created a process to transform the deactivated, concrete lock into "an iconic destination" honoring the site's Indigenous history. As an act of place making, the Falls plan emphasizes how infrastructure can be renegotiated to address settler-colonial injustices related to land and water dispossession. The Native Leadership Council of Friends of the Falls, the nonprofit steering the efforts, writes on their website that "We have an opportunity to create a place of healing at Owámniyomni (meaning 'turbulent waters' in the Dakota language), or St. Anthony Falls, that acknowledges the past and advances a more equitable and inclusive future." Early conceptual drawings include native landscape, walking paths, and places to gather and connect to the river. The Corps are currently in negotiation with the City of Minneapolis to convey the land over to the city, and then back to the Dakota people.

Our analysis of public comments in the first part of the study indicated that while the public trusts the Corps' technical expertise, there is great unease with the framing of the public process.

The Corps first study was narrowly focused on the locks and dams, and not the future of a river that is the lifeblood of the region. The Corps routinely responds to questions asked by members of the public by saying "that's not in the scope of the project". Members of the public often respond by asking how the scope of the study can be expanded so infrastructure goes beyond concrete and steel.

The low levels of public engagement with the Corps' first study begged asking how upstream forms of public participation might aid the policy process. We wondered how the Corps could better understand public concerns before they began the second part of their disposition process aimed at the next two locks and dams: Lower St. Anthony and Lock and Dam 1. This was particularly important because there are already strong calls from across the nation to unlock this seven-mile stretch of the Mississippi to restore Big Rapids habitat.

To increase more inclusive and intentional upstream public engagement, we helped create a partnership called the Future Mississippi Collaborative, made up of local and national river guardian organizations with expertise in engaging diverse communities.[1] We began by collecting and analyzing 270 hand-written surveys at different sites along the river near the locks and dams in question. The surveys found that only 3% of respondents had participated in the Corps public input process. Only .7% (two people out of 270) reported they had attended a public meeting and only 2.6% said that they participated through public comments. The majority of respondents did not know the primary purpose for the locks and dams. This general lack of knowledge and engagement indicated a need for expanded outreach and education.

Acknowledging this lack of public awareness was a problem, the Corps partnered with us to increase public engagement. In summer 2022, we designed and hosted 13 interactive tours, which also served as focus groups. Our tour guides described the history of the locks and dams and the questions the Corps was considering. Corps staff joined us on these tours and provided access to the lock and dam facilities so participants could get up close to the river infrastructure. Figure 14.5 includes images from our walking, biking, kayaking, and boating tours that included three community council tours, two BIPOC (Black, Indigenous, and people of color) tours, and one youth-centered tour. Over 400 people applied for these groups, and 250 people of all ages and abilities ultimately gathered at the water. We collected 233 more surveys from tour participants to gather their opinions on the future of the river.

Among our findings, we learned what types of information survey participants thought would be useful to know prior to participating in a Corps disposition study. Many participants requested information about ecological impacts (26.4%) and social and community impacts (21.3%). Tour participants also wanted to learn more about sediment toxicity, current and future costs of lock and dam maintenance, and the cost of dam removal.

Our focus group tours also asked participants: "What do you wish for the Mississippi River?" Among other sentiments, many respondents emphasized giving land back to Native peoples. Comments included: "I want Indigenous folks to have the most say—they care for better standards of the land, and water, have ancestral ties and are owed some sort of reparations for their forcible exile/expulsion from the place," "To decolonize this river, the infrastructure, the narratives, allow humans to connect with a more natural river corridor, to heal and love the river so that we may heal and love ourselves," and "To be returned to the communities they impact most and historically have been their caretakers."

We provided the Corps with a summary set of recommendations based on our findings. We advised them to offer educational resources and public tours, present community members with visual representations of how a future Mississippi River might look, and collect basic demographic information about who submits comments and attends public events. We also urged the Corps to build partnerships with youth and youth-led organizations. We found that youth are

Figures 14.3, 14.4, 14.5 Local residents on focus group tours.

Source: Macalester research team.

deeply interested in thinking about the future river and have innovative ideas about improving engagement with diverse publics.

The Corps launched its official scoping process for the second half of the disposition study in November 2023. Over 300 people submitted comments, which is more than double the number they received the first time around. We believe this is evidence of the impact of our work in stimulating public interest and engagement. The scoping report released by the Corps also called out our efforts in helping to model how they might overcome barriers for participation. This is key for climate justice. Agencies like the Army Corps are often perceived as using public consultation to simply maintain the status quo—to essentially check the box that they "engaged the public" without giving careful attention to what the public wants. By shining a light on how upstream engagement works, we hope to shift how and when the public gets involved in Corps processes and create place-making outcomes that are well supported by local communities.

The Corps' disposition study is still unfolding. The findings of the study may trigger a federal environmental impact statement review, which could take a decade to complete given the complex socio-economy, hydrology, toxicology, and ecological dynamics of river restoration. The decision-making processes around this project will likely unfold over a generation—a generation that is likely to see profound changes in the regional and global climate. For these reasons, it was important to model inclusive and intentional upstream public engagement to enfranchise diverse community members in caring for and about their river long into the future.

Conclusion

This case study has applications well beyond the Mississippi and water projects. Climate justice issues are implicated in nearly every infrastructure decision being made to engender a green economy. From wind farms to carbon capture projects, communities need to engage upstream of project designs. They need to be asked to weigh in on where things should be located, how big we should build, and how economic costs and benefits are distributed. This kind of community-oriented approach is best done through deep collaborations between state agencies and locally situated organizations who know how to engage neighbors with respect and at a speed that instills trust. Our work with the Corps suggests that such partnerships are possible.

Climate change affects infrastructure in predictable and unknowable ways. We expect disruptions from extreme weather events will bend or break some transportation, electric transmission, and water systems. This will have profound impacts on communities, and disproportionately burden those most vulnerable within those societies. Other incremental changes, such as species migrations, will be harder to predict. Adapting to climate change will require building resilience through the strengthening of literal and metaphorical walls and bridges. Place-based approaches to infrastructure design provide opportunities for diverse groups to consider the futures they want to protect and enact.

Acknowledgments

This project has been funded by the National Science Foundation (SES#1947152).

Note

1. This collaborative includes the Friends of the Mississippi River, American Rivers, and the National Parks Conservation Association and the University of Minnesota.

References

American Rivers. (2017) "Restore the Gorge." www.americanrivers.org/2015/03/restore-the-gorge/

Anand, A., A. Gupta, and H. Appel (eds). (2018) *The Promise of Infrastructure*. Durham: Duke University Press.

Edwards, P. (2003) "Infrastructure and Modernity," in *Modernity and Technology*, eds. T. J. Misa, P. Brey, and A. Feenberg. Cambridge, MA: MIT Press.

Frankel, T.C. (2018) "Taming the Mighty Mississippi," *Washington Post*. March 14. "www.washingtonpost.com/graphics/2018/national/mississippi-river-infrastructure/?utm_term=.1396c970bc8a

Jackson, S. (2014) "Rethinking Repair," in *Media Technologies: Essays on Communication, Materiality, and Society*, eds. T. Gillespie, P. Boczkowski, and K Foot. Cambridge: MIT Press.

Miller, G. (2020) *When Minneapolis Segregated. New York: Bloomberg City Lab*. www.bloomberg.com/news/articles/2020-01-08/mapping-the-segregation-of-minneapolis

Mitchell, T. (2002) *Rule of Experts: Egypt, Techno-politics, Modernity*. Berkeley: UC Press.

Phadke. R (2014) "Green Energy, Public Engagement and the Politics of Scale," in *The Handbook of Science, Technology, and Society*, eds. D. Kleinman and K. Moore. New York: Routledge Press.

Rogers-Hayden T. and N. Pidgeon (2007) "Moving Engagement 'Upstream'? Nanotechnologies and the Royal Society and Royal Academy of Engineering's Inquiry," *Public Understanding of Science* 16(3): 345–364.

U.S. Congress Public Law 115-270. (2018) America's Water Infrastructure Act of 2018. www.congress.gov/bill/115th-congress/senate-bill/3021/text

Zeisler-Vralsted, D. (2019) "African Americans and the Mississippi River: Race, History and the Environment," *Thesis Eleven* 150(1): 81–101.

15 Climate Justice

Taking Back the Commons

Shangrila Joshi

Introduction

During fieldwork in Nepal, a government forest official revealed to me that Nepal's esteemed community forestry model for forest governance was predicated on the *Guthi*, an Indigenous commons governance institution of Kathmandu's Newa. This was intriguing to me because I hail from that community. My family has always been part of various *Guthi* associations comprised of extended kin networks that continue to make important decisions about community resources related to food, life cycle rituals, and other cultural practices. I did not know the *Guthi* had inspired the community forestry model I had studied as an undergraduate student and later post-doctoral researcher.

Community forestry is an approach to forest management that recognizes the self-governance capabilities of local forest user groups, as opposed to leaving forest management entirely to government agencies. The ushering in of community forestry as government policy in the 1970s was the outcome of decades of struggle over territory between local forest users and the Nepali state. As I describe in detail in my book, *Climate Change Justice and Global Resource Commons* (Joshi, 2021), these struggles are ongoing, and the ushering in of carbon offsetting programs could erode the self-governance powers that community forest user groups won after much struggle. Meanwhile, the *Guthi* which is an integral part of Newa cultural identity and autonomy has also been threatened by the Nepali state and commercial interests. Both forms of commons governance are at risk of commodification enabled by the state. This tug-of-war between commoning and commodification represents a defining struggle of our times. In this chapter I argue that efforts to take back various commons around the world are examples of climate justice in action.

Climate justice is a multi-faceted concept and there are numerous ways to understand and implement it. Scholars have identified distributive, participatory, and transformative justice dimensions (Schlosberg, 2007; Walker, 2012). **Distributive justice** takes issue with the highly unequal distribution of greenhouse gas emissions and climate impacts in the world, within and across countries. **Participatory justice** recognizes that those who are most affected by and vulnerable to climate impacts tend not to have the power to influence decision-making, and strives to counter this injustice. Beyond these distributive and participatory justice concerns, scholars call for **transformative justice** to dismantle structures such as capitalism and colonialism that they argue form the root cause of the climate crisis. I suggest that the act of taking back the commons in various forms – as in the context of forest governance in Nepal – contributes to transformative justice by enacting structural changes in how resources are governed, facilitating a more equitable distribution of resources and a greater degree of participation within communities. Thus, the scaling up of these efforts to take back the commons is important for climate justice.

DOI: 10.4324/9781003409748-20

In the following sections, I discuss approaches to theorizing climate justice. I then offer my arguments to extend and deepen these approaches in new directions, that is, to facilitate autonomy and self-governance of Indigenous and other marginalized communities, and to support their traditional ways of knowing and being. I argue that ways of knowing, being, and governing are inherently connected. For students and researchers, commons governance and Indigenous knowledge are compelling new areas for climate justice research and scholarship.

Climate Justice – Prevailing Approaches and their Limits

Distributive Justice

The impacts of climate change – sea level rise, hurricanes, heat waves, droughts – are experienced unevenly by members of our global community. For those with limited resources, climate vulnerability can be more acute. Ironically, their contributions to the problem are the smallest. This observation led Roberts and Parks (2007) to articulate the notion of the **triple injustice of climate change**: those who are least responsible for problem creation often have higher vulnerability and weaker financial and technological capacity for adaptation. Beyond this triple injustice, climate solutions introduced with good intentions might create further injustice. For example, climate solutions and resources might be unequally distributed within a community. Solutions imposed by outside agencies without understanding the local context may backfire, or worse, damage the community if their autonomy is compromised. Climate solutions need to be just, reducing rather than exacerbating prevailing inequalities.

When discussions about climate justice began in the early 1990s in the context of UN climate negotiations, the focus was on reducing the unequal distribution of greenhouse gas emissions between the Global North and South. The distribution of emissions was considered unequal in two key ways: the cumulative emissions of the Global North were greater than that of the Global South, culminating in the North's greater historical responsibility; and the average emissions of Global North countries were much higher than those in the Global South, leading to demands for per capita equity. Industrialized countries such as the US were found to have high levels of responsibility relative to others both in terms of cumulative emissions and current per capita emissions. For example, according to the World Resources Institute, comprehensively measured per capita greenhouse gas emissions for the US were ten times that of Nepal in 2019. The same year, US fossil fuel-based carbon emissions were 25.5 times that of Nepal, according to the EDGAR database (Climate Watch, 2022; Crippa et al., 2021). Of course, national averages often conceal disparities within countries, but the cross-country comparisons do offer a measure of global inequality in how responsibility for and suffering from climate change are distributed among the global population.

In response, scholars argue that the Global North owes the Global South a **climate debt** to redistribute the earth's resources. This debt is accrued due to relations of unequal ecological exchange established during the colonial era that imparted disproportionate advantages to the Global North, such as a greater share of resources including atmospheric space to store disproportionately more greenhouse gas emissions (Goeminne and Paredis, 2010). Climate debt arguments demand that the debt be repaid through monetary and technological transfers to mitigate climate change, and by offering compensation for losses and damages incurred due to climate disasters.

Participatory Justice

While distributive justice is mostly concerned with fair distribution of outcomes, **participatory justice** is concerned with fair process. The emphasis is on the rights of affected parties to participate in decision-making processes. It is not just the distribution of emissions responsibility that matters. It also matters that those who are most affected by climate change have meaningful opportunities to participate in deliberations and decision-making on climate policy and action. My research in Nepal reveals that climate solutions such as REDD+ (Reducing Emissions from Deforestation and Forest Degradation) were introduced without heeding principles of free, prior, and informed consent. These principles were designed to protect the rights of Indigenous Peoples to exercise their agency in granting permission to externally imposed programs after receiving adequate and timely information about them (Joshi, 2021). The right to give consent is a participatory justice issue, as is adequate representation of marginalized groups (such as Indigenous, women, youth, and the disabled) in climate policymaking.

Environmental justice policy increasingly prioritizes participatory justice but critical scholars draw attention to the tendency of organizations to engage in superficial measures, such as when the participation of underrepresented groups is secured through coercion (Dahal, Sanjay, and Michael, 2013), or when the participation is tokenizing or formalistic (Satyal et al., 2019). Genuine participation must be meaningful to participants. Their involvement must come when programs are designed and planned, rather than as an afterthought when most critical decisions have been taken. Participatory environmental justice requires that underrepresented stakeholders have meaningful opportunities to shape the agenda of decision-making processes that have a bearing on their livelihoods and life chances. Meaningful participation entails taking seriously the agency as well as the worldviews – ways of knowing and ways of being – of those affected by climate change or climate solutions.

Another key element in environmental and climate justice is **recognition**, that is, the acknowledgment of the unique attributes of communities that are particularly affected, vulnerable, underrepresented, and/or marginalized. Globally, island nations and coastal and other communities that face food/water/livelihood security threats are identified as particularly climate vulnerable, facing an existential crisis. Climate change displaces people within and across international borders, yet there is no formal recognition of, or resources earmarked for, climate refugees and migrants. In the Pacific Northwest of the US, Indigenous tribes such as the Quinault and Quileute are vulnerable to sea level rise, and their food sovereignty is also threatened by climate change. Recognizing these particular vulnerabilities is necessary, but not sufficient. Recognition and validation of Indigenous ecological and cultural knowledge is also necessary. This will be discussed further below.

Transformative Justice

While achieving equity in distribution and participation is an important starting point, it does not sufficiently address oppressive systems and the structural drivers of climate change injustices. **Transformative justice** focuses on the root causes of climate injustice and seeks to transform societal structures and systems to alleviate it. Problematic structures include capitalism (Low and Gleeson, 1998), environmental racism (Mohai, Pellow, and Roberts, 2009), racial capitalism (Pulido, 2016), patriarchy (Buckingham and Kulcur, 2009; Chiro, 2008), and North-South relations (Joshi, 2021). These structures are deeply ingrained in our institutions and cultures. They are recognized as the root causes of the climate crisis, as well as the social inequalities that create differential vulnerability to climate impacts.

Recognition of root causes does not, however, effortlessly lead to structural change. Dismantling structures of capitalism, colonialism, racism, and patriarchy, while important, can seem daunting. Instead, we can target structural change in specific sectors, such as forests, agriculture, transportation, or energy, by introducing stronger regulations, institutional reforms, and by scaling up current and effective voluntary actions, both individual and collective. Some scholars argue for transformative structure change – such as replacing capitalism with socialism – often dismissing these approaches as incremental and positing an oppositional framing between reformist and transformative approaches (Foster, 1999; Hopwood, Mellor, and O'Brien, 2005). However, such a binary distinction is not necessarily helpful. Meaningful reform – including genuine participatory and distributive justice measures – may in fact constitute the building blocks for transforming society for the better. A more expansive view of the structures that need changing may help us move beyond this false binary.

An example of meaningful structural change in the forest sector comes from community forestry, hailed as a radical reform (Thwaites, Fisher, and Poudel, 2018). A detailed discussion of how community forestry can serve as an example of successful structural change follows.

New Directions for Climate Justice Research

Taking Back the Atmospheric Commons, One Local Commons at a Time

The climate crisis is often described as a tragedy of the atmospheric commons, a compelling but problematic metaphor. In a 1968 publication in *Science*, Garrett Hardin wrote about the **tragedy of the commons** as inevitable owing to the selfish nature of humans seeking to maximize individual gain at the cost of the common good. Hardin had asserted that the only way to avert tragedy is to either reduce human numbers or force enclosure of the "commons" in a top-down manner. When Hardin used the word "commons" he was actually referring to open-access resources with no rules for sustainable use. A **commons** by definition is commonly owned or governed by a well-defined community. He later published a less well-known article acknowledging his error. But in his widely read 1968 article Hardin's solution for avoiding "commons" tragedies was to privatize the resource in question.

We see this logic at play in the climate arena with the mitigation solution of buying and selling carbon credits. This approach involves commodification of the atmospheric commons. But carbon trading is not a just solution even if the financial benefits that countries such as Nepal receive in exchange for sequestering carbon in forests are welcomed by the government and forest-based communities, at the rate of USD 5 per ton of carbon equivalent. The low carbon price coupled with lack of free, prior, and informed consent indicate distributive and participatory justice concerns in carbon trade policies. Notably, for Hardin, injustice was inevitable and preferable to the ultimate tragedy. But is there truly no alternative to these tragedies?

Nobel Laureate Elinor Ostrom demonstrated the falsity of Hardin's claims by documenting communities around the world where people did not always act selfishly to destroy their local commons. Ostrom provided empirical evidence that community members do often come together to create sustainable solutions without either privatizing the resource or requiring top-down solutions. If given the opportunity and institutional validation, communities can be trusted to govern the resources on which they depend utilizing local and traditional knowledge.

A Successful Example of Commons Governance: Community Forestry

A successful example of commons governance is community forestry which is a key way that forest health is maintained in Nepal. Forest-based communities are delegated to govern local forests through the creation of formal institutions, that is, governing organizations with rules officially recognized by the state. This arrangement was developed by local community leaders and their allies to reassert control over forest commons from state authorities and wealthy landowners. This example can be understood as climate justice in action since forests play a key role in sequestering carbon and mitigating climate change. Community forestry promotes justice through local self-determination over the commons bringing structural change at a local scale.

The climate crisis is more than a problem of market failure that can be fixed with economics. It is a moral and civilizational crisis, and a big battle of our time is one between forces seeking to commodify nature versus preserving the integrity and practice of the commons. Not everything should or can be bought or sold. While just compensation for labor is an important pursuit, where life and self-determination are at stake, communities can and should rise above market logic and create ground rules that seek to preserve the integrity of life. The UN deliberation process for addressing climate change since 1992 is a space where the international community seeks to create these ground rules with the participation of all countries. When this process works, and countries cooperate effectively, it means the atmosphere is treated as a commons rather than as an open-access resource where any country can do as it pleases. Many nations have often refused to cooperate with international efforts toward global commons governance. Some did not ratify the Kyoto Protocol with its binding commitments to reduce emissions and the Paris Agreement is weaker due to voluntary targets. In the US, for example, an Executive Order declared that "Americans should have the right to engage in commercial exploration, recovery, and use of resources in outer space... and the United States does not view it as a global commons" (Executive Order 13914 of April 6, 2020, 20381). This is a blatant example of a highly individualistic approach to resource use that is neither just nor sustainable.

A silver lining for prospects of successful commons governance lies in the diverse spheres of life and resources (forests, fossil fuels, water, land, wind, sun) crucial for climate mitigation and adaptation, coupled with the difficulty of global atmospheric commons governance. These variables open spaces for meaningful local scale commons governance for climate action. Dependence on local governance also supports distributive and participatory justice because it recognizes and validates the power of communities to regulate the commons.

The practice of the commons is essential for addressing the climate crisis. When communities own or govern the means of production, whether forests, land, or water systems, they are better able to withstand climate instability because they can be more resilient financially while being more food and water secure as well. That is why Indigenous rights activists in India and Nepal refer to their struggles as *Jal, Jangal, Zameen* (water, forest, land), because they seek self-determination over the water, forest, and land commons. When local communities everywhere are able to exercise such autonomy, they have beneficial ramifications for the global atmospheric commons.

An Endangered Example of Commons Governance: Guthi

The example of community forestry described earlier is an example of a just and effective form of commons governance. Another example in Nepal comes from the Newa *Guthi*, which is a system of communal resource ownership and governance that has survived historical encroachments of settler colonialism. In 2019 thousands of Newa took to the streets of Kathmandu to successfully protest the Guthi Bill proposed in parliament, threatening to commercialize communally owned *Guthi* lands that form the foundation of the Newa civilization and culture of the Kathmandu Valley. The Indigenous institution of *Guthi* has been systematically obliterated by the Nepali government, even as it signs international declarations to uphold Indigenous rights and knowledge systems. Historically, *Guthis* facilitated the investment of agricultural surplus in the arts, architecture, and urban infrastructure to build Kathmandu's cultural heritage as we know it today. The tradition of the *Macchindranath Jatra*, an annual tradition of praying for rain and freedom from disasters, continues to this day due to the work of various *Guthis* performing their traditional rites and duties. *Guthis* are also integral to the maintenance and preservation of traditional waterspouts, called *hiti*, that are critical for water security, particularly for the urban poor.

Although not the focus of my own research, numerous examples of taking back the commons exist in the US, whether in efforts to advance Indigenous food sovereignty or community-driven acequias (a form of irrigation ditch) for water governance. In disaster management, as well, there are opportunities for local commons control despite the tendency to divert government resources to private contractors. After the 2015 earthquake in Nepal, for example, public spaces such as Tundikhel served as a temporary refuge for disaster-affected communities. Due to the historical recognition of the importance of these spaces for the public good, Kathmandu residents launched an Occupy Tundikhel movement to protect those commons from encroachment by the Nepali military.

Taking back the commons for disaster recovery entails recognizing these bottom-up support structures and validating them through formal recognition and allotment of financial resources for resilience planning and building. In the Nepalese context, taking back the commons for disaster resilience means working against the erosion or weakening of institutions such as the *Guthi* and community forestry that serve not only as resource governance structures but also as a social safety net during disasters.

Epistemic Justice

Within climate justice scholarship, there is understandably a strong emphasis on vulnerable populations victimized by social and economic structures and then again by the climate crisis. This emphasis becomes problematic when these populations are characterized *only* as victims that need to be rescued by would-be saviors. Capacity-building, too, is often conceptualized as one-directional, when it should involve reciprocity between the vulnerable and agencies and governments providing help. Those who offer help should also practice humility and be willing to follow the lead of those they wish to help, possibly learning from them in the process.

Climate justice should therefore not be limited to supporting subordinated groups through financial reparations, or including them in predetermined deliberative and decision-making processes, although these are a good start. Vulnerable and subordinated communities should also be recognized for their agency, contributions, and leadership. Indigenous people and their ancient knowledge systems ought to be recognized as an important resource in designing sustainable and just solutions to the climate crisis. This is what I mean by **epistemic justice** – the empowerment, recognition, and validation of Indigenous knowledge systems or ways of knowing.

The value of local knowledge and ways of knowing can be seen in the following example from my research. When studying community forestry in Nepal, as I mentioned above, I was surprised to learn from a bureaucrat in the Ministry of Forests and Environment that Nepal's community forestry institutions were inspired by the *Guthi* system of the Newa. In my own Western-educated upbringing, I had been conditioned to devalue the traditional practices of my own cultural group by relegating them to the space of ritual and culture. But in recent years a careful study of *Guthi* and related traditions such as the *Macchindranath Jatra* and urban infrastructure such as *hiti* indicate they have clear implications for climate resilience. The *Jatra* is an annual months-long festival that Kathmandu's Newa have been celebrating for more than a thousand years. This festival signifies worship of the deity that the Newa call *Bung Dyah*. The festival is associated most directly with the monsoon rains, but the worship of the deity is also tied to larger societal objectives including ensuring a bountiful harvest and the prevention of disasters. According to Newa mythology, the festival was initiated in response to a 12-year drought in the 7th century AD. Tantric priest *Macchindranath* was brought to the Kathmandu Valley from Assam, India to cure the valley of its prolonged drought. The drought was reportedly cured and the tradition of worshipping the deified priest has continued.

The successful execution of this annual festival is made possible by a tremendous degree of collaboration and sophisticated planning among a wide range of social groups, from astrologers to wood carvers, each performing their duties with precision to pull off the gargantuan task of constructing a sacred wooden chariot housing the deity that is taken on a procession through the historic town for the benefit of hundreds of thousands of devotees who pay their respects and perform their own community functions in tandem with the *Jatra*'s timeline. Newa scholar Hari Ram Joshi suggests that the *Jatra* is meant to embody a successful and sustainable society, the foundation of which is unity and collaboration amidst diversity, a testament to the logic of community over individuality as a cultural norm (oral history interview, August 2022). The construction of the chariot follows traditional architectural design that is curated and taught by the Baraha community. This design is meant to model earthquake-resistant building design for the residents of the earthquake-prone city. Clearly there are mythological, ecological, architectural, sociological, geographic, and historical elements of the festival that warrant further study and analysis to develop insights into community resilience in the face of disasters.

Similarly, I am intrigued by elements embedded in a *Newa* story-telling tradition, the *Swasthani Bratakatha* that started in the 17th century or earlier, where the narrative of the story warns of oceans rising to engulf humanity if humans give in to greed and unrestrained extraction of oily substances from the ground.

These writings and teachings warrant close study, and it is my argument that it is a climate justice imperative to do so. A compelling area of future scholarship in climate justice, therefore, is to highlight the importance of moving beyond a victimhood status for Indigenous people and recognizing, acknowledging, and validating Indigenous climate epistemologies as a basis for climate action and decision-making.

In *Decolonizing Methodologies*, Smith (2012) presented compelling arguments for why the epistemologies – that is, ways of knowing – of Indigenous communities should be given due credit for their sophistication, rather than treating them as a cultural novelty or discarding them as irrational or strange. In environmental discourse, space is increasingly created for Traditional Ecological Knowledge (TEK), but there still is a tendency to tokenize it or judge its legitimacy through Western science. This tendency is due to the legacy of colonialism. Overcoming it requires careful, critical, and deliberate efforts to challenge it within academic, policy, and activist communities. There is an urgent need to build a research agenda that offers ways to move beyond tokenizing attempts to recognize Indigenous knowledge in climate discourse, and to offer a framework for taking Indigenous epistemologies seriously by conceptualizing linkages between ways of being, ways of knowing, and ways of governing.

Conclusion

In conclusion, I am arguing for a structural solution to the climate crisis and for the pursuit of transformative climate justice that entails reclaiming various commons by local communities. This involves modifying or creating institutional structures of governance to enhance the capabilities of Indigenous and other underrepresented communities to have access to or ownership of vital resources for livelihood by utilizing their own traditional practices and worldviews. Such actions increase resilience. By creating these transformative structural solutions, we also support distributive, participatory, and epistemic justice goals. We can see how they are all inter-related. A commons governance solution to the climate crisis therefore offers a framework that enables us to conceptualize and put into practice more expansive ways of pursuing distributive and participatory justice. Meanwhile, its success may depend on the extent to which epistemic justice is taken seriously.

References

Buckingham, S. and Kulcur, R. (2009) "Gendered Geographies of Environmental Justice," *Antipode* 41(4), pp. 659–83.

Chiro, G. (2008) "Living Environmentalisms: Coalition Politics, Social Reproduction, and Environmental Justice," *Environmental Politics* 17(2), pp. 276–98.

Climate Watch (2022) World Resources Institute. Washington, DC: Available online at: www.climate-watchdata.org

Crippa, M., Guizzardi, D., Solazzo, E., Muntean, M., Schaaf, E., Monforti-Ferrario, F., Banja, M., Olivier, J.G.J., Grassi, G., Rossi, S., Vignati, E. GHG emissions of all world countries – 2021 Report, EUR 30831 EN, Publications Office of the European Union, Luxembourg, 2021, doi:10.2760/173513, JRC126363

Dahal, S., Sanjay N., and Michael, S. (2013) "Examining Marginalized Communities and Local Conservation Institutions: The Case of Nepal's Annapurna Conservation Area," *Environmental Management* 53(1).

Executive Order 13914 of April 6, 2020. Encouraging International Support for the Recovery and Use of Space Resources. *Code of Federal Regulations* 20381–82, title 3.

Foster, B. (1999) *The Vulnerable Planet*. New York: NYU Press.

Goeminne, G. and Paredis, E. (2010) "The Concept of Ecological Debt: Some Steps Towards an Enriched Sustainability Paradigm," *Environment, Development, Sustainability* 12, pp. 691–712.

Joshi, S. (2021) *Local and Global Postcolonial Political Ecologies*. London: Routledge.

Hardin, G. (1968) "The Tragedy of the Commons," *Science* 162, pp. 1243–8.

Hopwood, B., Mellor, M., and O'Brien, G. (2005) "Sustainable Development: Mapping Different Approaches," *Sustainable Development* 13, pp. 38–52.

Low, N. and Gleeson, B. (1998) *Justice, Society, and Nature: An Exploration of Political Ecology*. London: Routledge.

Mohai, P., Pellow, D., and Roberts, J. (2009) "Environmental Justice," *The Annual Review of Environment and Resources* 34, pp. 405–30.

Ostrom, E. (1990) *Governing the Commons*. London: Cambridge University Press.

Pulido, L. (2016) "Flint, Environmental Racism, and Racial Capitalism," *Capitalism Nature Socialism* 27(3), pp. 1–16.

Roberts, J. and Parks, B. (2007) *A Climate of Injustice: Global Inequality, North-South Politics, and Climate Policy*. Cambridge, MA: MIT Press.

Satyal, P., Corbera, E., Dawson, N., Dhungana, H., and Maskey, G. (2019) "Representation and Participation in Formulating Nepal's REDD+ Approach," *Climate Policy* 19(1), pp. 8–22.

Schlosberg, D. (2007) *Defining Environmental Justice: Theories, Movements, and Nature*. London: Oxford University Press.

Smith, L. (2012) *Decolonizing Methodologies: Research and Indigenous Peoples*, 2nd ed. London: Zed Books.

Thwaites, R., Fisher, R., and Poudel, M. (2018) "Community Forestry in Nepal: Origins and Issues." In *Community Forestry in Nepal*, edited by Richard Thwaites, Robert Fisher, and Mohan Poudel. London: Routledge.

Walker, G. (2012) *Environmental Justice: Concepts, Evidence and Politics*. London: Routledge.

Part VI

Climate Governance

Introduction

Mark Vardy

It has been roughly 300 years since the invention of the steam engine that ushered in the Industrial Revolution. The transformation of Western Europe from pre-modern agricultural and feudal societies to what we now recognize as modernity involved complex processes that cannot be reduced to any simple assertion of cause and effect, but there is one parallel development that is worth noting here. Over roughly the same span of time that fossil fuels – in forms including coal, diesel, gasoline and jet fuel – were burned at rates that steadily increased the amount of carbon dioxide in the Earth's atmosphere to a point deemed dangerous by scientists, the political institutions associated with both colonialism and democratic governance also grew. That is, at the same time that scientists and engineers were designing ever more powerful ways of extracting fossil fuels and converting them to energy, the political institutions through which governance is enacted at national and international levels also took shape (Mitchell, 2013). If we look to the past several hundred years, then, we can say that we have inherited both the machinery that produces climate change and the political institutions that are now tasked to deal with it. The question is, do the political institutions that the world has inherited from the past need to be rethought or reformulated to meet the challenge of climate change, and if so, how?

A key characteristic of Science and Technology Studies (STS), which is exemplified by Hannah Knox in Chapter 16, is the detailed attention that it pays to how realities actually come to matter. Through spending several years closely observing the work of the city council in Manchester, UK, Knox shows how climate change challenges habitual ways of making the city a governable space. Consider how Manchester, which grew to prominence as a manufacturing center in the Industrial Revolution, wanted to reinvent itself as a low-carbon city. To work towards this goal, the city had to experiment with new ways of knowing and acting. For example, a lot of climate change data are produced for use by nations and are not actionable by cities. But even after it teamed up with a scientific organization to produce data at the city level, the municipal government encountered new challenges related to aligning historical divisions of responsibility with this new data. And then when it came to regulating homes for energy efficiency, climate change once again unsettled ideas of reality. Knox shows how eco-homes in Manchester became experimental objects that provided material for testing new ways of engaging with the realities of climate change.

Moving from the city to the national and international levels, Chapter 17 by Reiner Grundmann and Chapter 18 by Mark Vardy focus on the Intergovernmental Panel on Climate Change (IPCC). The IPCC is tasked with producing assessments of peer-reviewed climate science that are intended to be policy relevant but not prescriptive. The IPCC does not have a formal seat at the table when it comes to the international climate negotiations organized by the United Nations Framework Convention on Climate Change (UNFCCC), but they do inform the UNFCCC process. The 195 countries that comprise the IPCC are directly involved in establishing what

DOI: 10.4324/9781003409748-21

questions it should assess. But the actual work of assessing the literature and writing assessment reports is undertaken by volunteer scientists. The underlying assessment reports typically exceed 1,500 pages and are "accepted" by delegates from the member nation-states. But these governmental representatives "approve" each and every word in the much shorter Summary for Policymakers (SPMs), which typically are 30 to 40 pages.

The IPCC's relation to both science and politics, its status as a public authority on climate science, and its mandate to work through consensus are among the reasons why STS scholars find the IPCC such an interesting object of study (e.g. Beck and Mahony, 2021; De Pryck and Hulme, 2022). In Chapter 17, Grundmann considers issues that flow from the IPCC's history and its function as he explores the question, what kind of expertise does the IPCC actually embody? Grundmann lays out a compelling argument to consider the IPCC's expertise as a public commentator, which makes it all the more important to pay close attention to the IPCC's narratives and how it develops them. In Chapter 18, Vardy points to the different process of shaping knowledge in the underlying reports and the SPMs. He argues that while national self-interest can shape the SPMs, we should be attentive to concepts such as social transformation in the underlying reports.

The final chapter in Part VI expands the scale from the IPCC to historical relations between the Global South and North. The transition to modernity, centered in Western Europe, was financed in large part through colonialism, setting up an exploitative pattern of relations that continues to this day. In Chapter 19, Tiago Ribeiro Duarte discusses the ramifications of these histories of modernity and colonialism in terms of trust in expertise. More specifically, the growth of modernity was accompanied by the increasing social division of labor and greater specialization of occupations. Under such conditions, a great deal of trust must be routinely placed in the expertise of others. As Duarte shows, trust in experts is necessary for the computer models that are central to the IPCC to occupy the influential role that they do in framing climate change. But trust is fragile, and Duarte shows how it can be challenged on two fronts. One is the denialist countermovement of fossil fuel interests, and the other stems from ways that the Global North exerts power through scientific practices that reflect the interests of rich industrialized nations.

Taken together, the chapters in Part VI shed light on how STS approaches a range of issues related to expertise and the governance of climate change, from the most intimate issues of trust and what "home" means, to the relations between science, colonialism, and nationhood.

References

Beck, S. and Mahony, M. (2018) The IPCC and the new map of science and politics. *WIREs Climate Change.* 9:e547.

De Pryck, K. and Hulme, M. (2022) *A Critical Assessment of the Intergovernmental Panel on Climate Change.* New York: Cambridge University Press.

Mitchell, T. (2013) *Carbon Democracy: Political Power in the Age of Oil.* New York: Verso.

16 Climate Change as Ontological Unsettling

A View from the City

Hannah Knox

Climate change tends to be thought about in two ways: as either a global or individual problem. Think of the graphs of rising temperatures, international school strikes, or images of climate catastrophes from around the world, which frame climate change as a global problem, and all the calls to ride bikes, fly less, save energy, or buy organic, which address climate change as a problem on an individual level. However, in this chapter I want to consider what we can learn about the challenges of representing and enacting the climate if we shift scale, zooming in from the global and up from the individual to settle on the collective and yet grounded scale of the city.

In this chapter, I argue that at the city level, climate change not only requires social and economic change, but also a radical reconfiguration of the very material objects through which change is achieved. I call this a process of **ontological unsettling**. By this, I mean that climate change is not just something that demands change in the world, but it is also something that profoundly challenges people's understanding of the nature of things *in and of themselves*. As things morph in the face of climate change, they take on new significance and demand new lines of responsibility and novel forms of action. *New objects demand new ways of governing*, and so it should not be surprising that in the face of climate change we see the appearance of governmental experiments oriented to the question of how to confront and shape the future.

Cities are particularly good places to see this effect happening in practice. As sites of governance, cities enable us to look at the technologies and techniques through which climate change comes to matter. As places where people live and work, they offer us a window into the concrete impact of how climate change is known on people and things, allowing us to see how subjects and objects become unsettled in practice in their confrontation with climate knowledge. In what follows, I draw from my research to show how these issues became visible.

From Economics to Ecologics

The first example of unsettling I want to describe is the way climate change challenges established paradigms of economic growth and development. The research I draw on comes from an in-depth, long-term study centred mainly on one city, Manchester, in the United Kingdom (UK). I was not a new visitor to the city when I began this study of climate change. I have lived in the city since my undergraduate days and conducted my doctoral research on a past form of urban future in Manchester which considered the question of how economic development could revitalise post-industrial cities. I studied post-industrial regeneration in Manchester, looking at plans and activities to stimulate economic investment in what at the time was being called the 'new media' sector. In this case, the proposal was clear. The old industrial model of manufacturing as the economic base of the city was no longer viable. To revitalise the city, what would be needed was the creation of a creative service sector made up of people working in advertising,

DOI: 10.4324/9781003409748-22

media, education, and emerging digital industries (Florida, 2002). The measures of success were also clear. If countries measure their economic fortunes through the measure of Gross Domestic Product or GDP, cities were measuring their value against a metric of Gross Value Added (GVA). Essentially GVA is a measure of the value that a city adds to the national economy, calculated by measuring the value of goods and services that have been generated, and put against the cost of necessary materials to produce those goods.

By 2010 Manchester was doing well on the metric of GVA averaging growth of over 4% since the early 2000s. After a concerted effort of post-industrial regeneration and economic boosterism, Manchester was thriving (Peck and Ward, 2002). Although there were still areas of significant deprivation, the city centre had become a place for businesses to locate, tourists to visit, and property developers to invest. But now there was also the problem of a looming climate catastrophe (While, Jonas and Gibbs, 2004).

When I began my climate research in 2011 there had already been a concerted effort within the city to act more effectively on climate change. In 2009, a report had been written which had committed the city authorities to two targets: to reduce the city's carbon emissions by 41% by 2020 (from a 1990 baseline), and to bring about a change in culture to transform it into a low carbon city. It was clear that this potentially posed a significant challenge to the paradigm of economic growth with which the municipality was operating. Carbon emissions are produced by the very same things that are indicators of economic growth, including transportation, the use of energy for heating homes and businesses, and a vibrant airport. How would the city of Manchester be able to confront its climate commitments, without compromising its commitment to economic vitality?

These were the questions that many were asking when I started my research (Hickel, 2021; Raworth, 2017). I spent two years in the midst of these issues, interviewing people, shadowing policy officers, getting to know activists, attending events, and participating in public meetings. I followed the process of trying to put a climate plan in place, participated in activities to refresh the plan after its initial five years were up, and joined discussions with and about other cities in Europe who were trying to do the same thing in their municipal contexts. From this I gained an understanding of the specific challenges that were facing cities like Manchester, and also some of the answers that were being devised.

Through all my research, I learned that while the headline problem of squaring economic growth with environmental sustainability was probably the most widely recognised form of unsettling that climate change generates (think of calls for degrowth, and arguments that locate the causes of environmental harm in capitalist extraction), economic regeneration was only one of many objects and entities that were being undone and reworked as they came into confrontation with climate data. In this chapter, I will show this by focusing on the destabilisation of one object in particular – buildings. But before I discuss the specific issue of buildings, I first explain how climate change itself appeared in Manchester as an agent capable of generating the ontological unsettling that this chapter goes on to explore.

Knowing Climate

To understand why climate change seems so destabilising, it is important to understand the particular form through which climate change comes to appear as what STS scholar Bruno Latour (2004) called a **matter of concern**. A matter of concern is an issue that people are confronted with and find themselves having to engage with. Thinking about issues as matters of concern, rather than matters of fact, allows STS scholars to engage with the social and political qualities of issues that are often dominated by a frame rooted in the natural sciences. In the case of Manchester, the matters of concern regarding climate change were not floods, wildfires, or megastorms so much as climate metrics (Knox, 2014). The reason for this was that for the city of Manchester to be able to act on climate change it first needed to be able to know what actually it was acting on. That is, it was all very well talking about reducing emissions in general terms, but what would an appropriate level of emissions reductions be for a city like Manchester? What would be a proportionate response to the challenges of global climate change? How could what was happening globally be scaled down to a set of measures that could be acted upon locally? The specific ways that climate change is measured are closely related to the type of political action that is expected. And the climate data that Manchester was confronted with were all generated with the expectation that it would be countries, not cities, taking action. To help make climate data actionable at the city level, the council collaborated with a climate science unit at the University of Manchester called the Tyndall Centre for Climate Research. The aim was to come up with a set of targets that the council could work against which they could measure their interventions.

As I discuss in more detail in my book *Thinking Like a Climate: Governing a City in Times of Environmental Change* (2020), one of the challenges that the council faced, was the mismatch between information on the causes of climate change on one hand, and the organisational set-up of city council on the other. So even after they obtained data that was actionable at the city level, they still had to figure out how to best use it. That is, the various lines of reporting through which the council works were not established in a way that made it easy to incorporate data on the causes of climate change. This is not surprising because the city is far older than the recognition of climate change, but it presented a few novel challenges. First, climate was defined as a problem at a global scale caused by the aggregate of human activities over time. This was far from amenable to local intervention because of the mismatch described above. To resolve this, the council worked with data that had been created by the Committee on Climate Change (CCC) to help the UK meet its legally mandated carbon reduction targets. The CCC had identified the required reductions in carbon emissions from key sectors namely 'services, electricity, residential, industry and transportation'. Here models had been devised that showed what reductions could be possible under these categories, and how much they would cost. However, the problem for the city council was that these categories did not align with the structure of the council and the domains of responsibility within which people worked. To make the numbers fit their organisational context, council officers worked with local climate scientists to come up with new categories that were simultaneously geophysically and politically meaningful. As a consequence, the categories were reworked and eventually a set of targets was produced that related to the more amenable classifications of 'buildings, transportation, energy and waste'.

Object Lessons

When I started my research, because the category of buildings seemed to be getting the most attention, I decided to follow what was happening as buildings became designated as a focus for low carbon intervention. As mentioned earlier, building development in the city had for some years been primarily an indicator of economic growth and urban success. I was intrigued, and I wanted to find out what would happen when buildings became the focus of carbon reduction targets.

It was in buildings that I first observed the destabilising effects of climate metrics. This led me to start thinking of climate change through the lens of ontological instability, or ontological unsettling (Law, 2015). If ontology is the area of philosophy concerned with the nature of things, then ontological instability is a process whereby received understandings of the nature of things are challenged, reworked, and set into relief. One of the places I saw this happening was in relation to domestic houses.

Houses are important for cities because they provide places for people, and more importantly – from the perspective of city governance – a workforce to live. Poor quality or poorly designed housing can lead to health problems or crime, while higher quality forms of housing can increase the desirability of an area. However, when houses are rethought in terms of climate change, attention moves from their importance as homes to their role as contributors to carbon emissions. In Manchester this opened up a whole set of questions about what the energetic qualities of the existing housing stock actually were. It turned out that as no one had asked this question before, no one actually knew. One of the first tasks of acting on houses in the city then was to collect information on those houses, in order to build up a picture of the city as a terrain of carbon emissions. Rather than just being concerned about categories like tenure, occupancy, or house prices, the questions being posed now were whether homes had loft insulation, double glazing, cavity walls, and gas boilers, or other kinds of heating. While this may seem at first glance like a minor shift, it is actually a dramatic change in how cities have envisioned the task of governing.

Climate change thus had the effect of both confronting and unsettling established approaches for both understanding and governing houses. One thing that this destabilisation raised was new questions about who or what should actually be responsible for the city as a built environment. During the 20th century, many homes in the city had been built with public money. But starting in the 1980s, the ownership of public housing in the UK has changed significantly. Many people were encouraged to buy their council-owned properties in the 1980s in what was known as the 'right to buy' scheme, and most of those that were not purchased were gradually moved from council ownership to management by housing associations or Arms Length Management Organisations (Malpass and Murie, 1994). When I was doing my research, the responsibility that city councils had for the management of buildings was restricted to a) managing planning applications for new houses and b) stimulating the growth of housing through local area planning. This meant that even if the council were to gain the necessary knowledge about which homes were contributing most to climate change, they would not have much leverage in doing anything with this new knowledge.

What we see here then is that the challenge of climate change mitigation exists not only in relation to established economic growth paradigms but also to mundane understandings of the objects that cities are made up of. Certainly, local area planning that established the need for new housing was based in part on a growth model of urban development, but the reasons why it was difficult to intervene to make changes to houses to make them less polluting was more complicated than a simple economic calculation that says that growth cannot be 'decoupled' from carbon emissions increases (Hickel, 2021). Instead it rested on more fundamental assumptions about what the nature of an object like a building is, and what this meant for the way responsibility for such objects had become established over time.

If the ontological unsettling of buildings opened up new questions about who should be responsible for these reconfigured objects, it also revealed the limits of the council's current capacity to act on these new objects. Unable to act directly on buildings newly conceived as carbon emitters, council officers had to find novel ways of intervening. One option was to put their efforts into developing plans and strategies, with the hope that homeowners, landlords, and housing associations might be encouraged to take up the recommendations of these plans. But a second way in which those who were trying to govern the climate found to act on these novel objects was in the form of what Harriet Bulkeley and Vanessa Castán-Broto (2012) have called 'urban climate change experiments'.

In a recent volume on climate urbanism by Vanesa Castán Broto, Aidan While, and Enora Robin, several authors explore the emergence of the experiment as a powerful form of urban governance that is being deployed to deal with complex problems, and in particular climate change (Castán Broto, Robin, and While, 2020). Experiments in climate urbanism take a range of different forms. They might take the form of living labs where the city is re-conceptualised as a laboratory or test bed for trying out new solutions to urban problems (Karvonen and van Heur, 2014). They might take on the form of pilot projects, which operate as provisional or pro-totypical interventions that are not subject to normal regulatory requirements (Jiménez, 2014). In my work this experimental ethos manifested as a direct response to ontological unsettling of buildings, offering a way of creating new understanding of what houses were and their sig-nificance to the making of a future city. Concretely, this experimental ethos manifested in the creation of several experimental or model eco-homes that appeared around the city at the time of my research.

Buildings as Experimental Objects

Eco-homes are houses that have been retrofitted to be more environmentally friendly. In Man-chester, there were many eco-homes which had appeared around the time of my research. First there were those which were set up as demonstration spaces where the latest in green tech-nologies such as energy-saving light bulbs, green roofs, or diverse forms of insulation could be showcased. The city council had been one of the first to establish an eco-show-home in a former terraced house to the east of the city centre. Another demonstration home I visited had been put together by a housing association and was located again in a terraced house in the north of Greater Manchester. Both of these homes were directed more to policy makers, manufacturers, and housing providers than to citizens, although the council-run home was open for tours for anyone who was interested. For the housing association, the purpose of the demonstration home was to build relationships with suppliers of ecological technologies, establish themselves as a frontrunner in ecological projects in the housing association sector, reassure tenants of the ben-efits of energy-based retrofit, and demonstrate a business case for ecological renovations within the housing association itself.

As well as institutionally created ecological demonstration homes, there were also several houses that had been renovated by their owners and were opened up occasionally for the gen-eral public to visit. These homes had generally undergone significant work, with people ripping out old windows and floors and renovating the home from top to bottom. Starting from scratch allowed people to make major improvements to the energy efficiency of their houses as well as learn a lot about how houses worked in terms of their energy dynamics. Some of those who had done whole house retrofits and were opening their houses up to the public had subsequently set themselves up as consultants or incorporated their learning into their work as engineers or architects.

In addition to these two kinds of eco-houses, there was a third example of an eco-house which was perhaps the more clearly a form of urban 'experiment'. This was the Manchester En-ergy House. Another terraced house, it had been reconstructed brick-by-brick inside a weather chamber on the campus of Salford University. The Energy House had been created as a testing centre, where companies could come and do experiments to see how energy-saving technolo-gies would fare in a setting that approximated 'real-world' conditions. It also offered the pos-sibility of conducting experiments on energy behaviours, with the house decorated inside like a real home and filled not only with sensors but also with beds, sofas, tables and chairs, and appliances like a fridge, a freezer, and an oven.

Finally, one further eco-home, which was created more recently, was a scale model of a house that could be taken to events and workshops and used to demonstrate the thermal properties of homes and of insulating measures. This model house was created by an organisation called the Carbon Coop, and was fitted out with heating elements, movable walls, and insulated and non-insulated areas. One of the uses of the model house was to view it through a thermographic camera, so that people could sense viscerally the effects of heating, cooling, and insulation. In all four cases, the eco-homes allowed people and organisations to engage with the problem of climate change as a matter of concern, making climate change real as a specific configuration of social, political, and material issues.

Conclusion

In a sense we might say that all of these houses were not homes but models. As models they offered a way of responding to the ontological unsettling I described earlier. That is, climate change presents humanity with a new problem that disturbs habitual ways of making sense and acting in the world, and as such, demands new ways of understanding ourselves and the potential for what politics could be. As sites of investigation and study, and as experimental objects, the eco-homes acted as models that helped policy makers and activists to diagnose the complex problem that is climate change, and build up a new understanding of the physical, institutional, and social relationships that are necessary for climate change to be tackled. The provisionality of models, trials, pilots, and experiments offered a way of both diagnosing the nature of the problem being confronted – ranging from the fabric of buildings to institutional and regulatory relationships necessary for change – and at the same time acting in ways that had not yet been institutionalised, accepted, or at times even rendered legal. When building a model or running a trial people could set aside existing barriers to action, such as cost, political buy-in, and questions of responsibility, that had become established under prior ontological conditions. But at the same time, by acting and demonstrating that the trial was possible to pull off or the model possible to build, such experiments could also be ways of engaging with new kinds of objects in order to prefigure a future that was not yet here, but which might now materialise in the future.

If turning unsettled objects into models has been an important way of acting on climate change to date, then the question I am left with is whether this will be enough going forward? For many years, cities around the world have been leading initiatives to tackle climate change, whether in the form of spectacular eco-cities, or more mundane attempts to change aspects of city life to make cities operate in a way that is both less environmentally destructive and that will be able to deal with the climatic changes that we are likely to face (Luque-Ayala, Marvin and Bulkeley, 2018; Sze, 2015; Castán Broto, Robin and While, 2020). Experimentation continues to be an important part of this work. However, this raises the question: if one of the things that experiments promise is a prefiguration of relationships to come, then what will actually need to be done to bring those relationships into being? Can models or experiments move from being mere prototypes into being pervasive infrastructures of life in the city? Whose shoulders will it fall on to make this change? What role will universities, students, activists, and others play in bringing about the kind of change that will ultimately result in lowering our city's carbon emissions to zero? Who will be involved in this process and who will be excluded? All these are questions we will need to keep asking as we look to cities to understand the ongoing challenges of action on climate change.

Further Reading

Castán Broto, V., Robin, E., and While, A. (2020) *Climate Urbanism: Towards a Critical Research Agenda.* Cham: Springer International Publishing: Imprint: Palgrave Macmillan.

Knox, H. (2020) *Thinking like a Climate: Governing a City in Times of Environmental Change.* Durham, NH: Duke University Press.

References

Bulkeley, H. and Castán Broto, V. (2012) *Government by Experiment? Global Cities and the Governing of Climate Change.* Oxford: Blackwell.

Castán Broto, V., Robin, E., and While, A. (2020) *Climate Urbanism: Towards a Critical Research Agenda.* Cham: Springer International Publishing: Imprint: Palgrave Macmillan.

Florida, R. L. (2002) *The Rise of the Creative Class: And How It's Transforming Work, Leisure, Community and Everyday Life.* New York: Basic Books.

Hickel, J. (2021) *Less is More: How Degrowth Will Save the World.* London: Windmill.

Jiménez, A. C. (2014) 'The right to infrastructure: A prototype for open source urbanism', *Environment and Planning D: Society and Space*, 32(2), pp. 342–362.

Karvonen, A. and van Heur, B. (2014) 'Urban laboratories: Experiments in reworking cities'. *International Journal of Urban and Regional Research*, 38, pp. 379–392.

Knox, H. (2020) *Thinking like a Climate: Governing a City in Times of Environmental Change.* Durham, NH: Duke University Press.

Knox, H. (2014) 'Footprints in the city: Models, materiality, and the cultural politics of climate change'. *Anthropological Quarterly*, 87(2), pp. 405–430.

Latour, B. (2004) 'Why has critique run out of steam? From Matters of Fact to Matters of Concern'. *Critical Inquiry*, 20, pp. 225–248.

Law, J. (2015) 'What's wrong with a one-world world?' *Distinktion: Journal of Social Theory*, 16(1), pp. 126–139. https://doi.org/10.1080/1600910X.2015.1020066.

Luque-Ayala, A., Marvin, S., and Bulkeley, H. (2018) *Rethinking Urban Transitions: Politics in the Low Carbon City.* Abingdon, Oxon: Routledge.

Malpass, P. and Murie, A. (1994) *Housing Policy and Practice.* Basingstoke: Macmillan.

Peck, J., and Ward, K. (eds) (2002) *City of Revolution: Manchester*: Manchester University Press.

Raworth, K. (2017) *Doughnut Economics.* London: Chelsea Green Publishing.

Sze, J. (2015) *Fantasy Islands: Chinese Dreams and Ecological Fears in an Age of Climate Crisis.* Berkeley: University of California Press.

While, A., Jonas, A. E., and Gibbs, D. (2004) 'The environment and the entrepreneurial city: searching for the urban "sustainability-fix" in Manchester and Leeds'. *International Journal of Urban and Regional Research*, 28(3), pp. 549–569.

17   The IPCC as a Body of Expertise

Reiner Grundmann

It is often said that the IPCC is the authoritative voice in climate science and that it is the primary scientific authority for policymakers. In fact, its reports are widely regarded as the most important and reliable publications on a global scale. But what kind of knowledge is created by the IPCC and how does it relate to climate policy? In order to answer these questions, we need to unpack the notion of expertise and examine the role of the IPCC accordingly.

The IPCC (2023) defines its goals as follows: 'Created in 1988 … the objective of the IPCC is to provide governments at all levels with scientific information that they can use to develop climate policies. IPCC reports are also a key input into international climate change negotiations.' An obvious interpretation of this statement would be that the IPCC sees itself as a science advisory body, which recommends specific courses of action to address the problem of climate change. However, this is not the case. While the IPCC provides 'information', it does not advise. This may seem like hair-splitting but it is not. The definition above has been carefully crafted from the very start of its operations. We need to understand the reasons behind it and what this means for the nature of expertise it provides.

As I discuss in more detail in my book *Making Sense of Expertise* (2022), the IPCC should be understood as a public commentator, which is a form of expertise that is normally overlooked. This approach challenges taken-for-granted notions of expertise which are science-based and depict the IPCC as a science advisory body. While the IPCC does not advise governments, it is an intergovernmental body in which national governments play an active role. Governments nominate the authors for the report, and they approve the Summary for Policy-makers (SPM) document in a line-by-line procedure (see Vardy, Chapter 18, this volume). This makes the IPCC a very special organization within the world of policy advice. It is a body that marshals, assesses, and thereby produces knowledge about climate change.

Meanings of Expertise

From an STS perspective, research on the IPCC has focused on the science advisory aspect (see, for example, Beck and Mahony, 2018; Guston, 2001; Hoppe, Wesselink, and Cairns, 2013; Mahony, 2013; Pielke Jr., 2007; Ravetz, 2011; Turnpenny, Jones, and Lorenzoni, 2011). This research has produced important insights into the role of the IPCC at the science-policy interface. But it tends to use the term 'expertise' without interrogating it. Here I want to discuss the notion of expertise and focus on the way the IPCC delivers expertise on climate change.

As I will show, the IPCC is not an advisory body in the usual sense. My argument is based on a typology that distinguishes between different roles and functions of experts (Grundmann, 2022). My typology posits two different dimensions, namely (1) the nature of the problem and (2) the type of expertise (see Table 17.1). Problems can be well-defined or ill-defined, and

DOI: 10.4324/9781003409748-23

Table 17.1 Typology of expertise

	Cognitive orientation	*Practical orientation*
Well-defined problems	Scientist	Specialist
Ill-defined problems	?	Advisor, Commentator

expertise can be cognitive or practical. This leads to four possible combinations. Scientists deal with well-defined problems, with problems for which they think a solution exists. They tend to avoid ill-defined problems, or re-define them as well-defined so they can get to work and develop a solution.

If we look at the practical aspects, then we can identify experts dealing with well-defined and ill-defined problems. Specialists deal with well-defined problems. These are often professionals applying rules to problems. They are able to do this as a result of training, skill, and experience. But there are also experts who deal with ill-defined problems and cannot apply rules to problems. This is the category in which we find the IPCC, which is tasked with the practical work of assessing very complex, messy, and ill-defined problems for policymakers. Experts in this category need to evaluate diverging pieces of evidence, and address questions of norms and values. Controversies are a common feature in this situation. It should be noted that there are two types of experts in this category, advisors and commentators, who are both dealing with ill-defined problems.

Advisors stand in relation to a client who requests expertise. The client seeks advice from the expert when facing uncertainty and conflicting values. The client typically wants to know what the best course of action would be, given the circumstances. The client expects actionable knowledge. Commentators, on the other hand, do not advise clients. However, they may also pronounce on the question of what should be done.

While the roles of specialists and advisors are familiar to social scientists, the role of the commentator is less so. It may even be questioned that this role should be included in the list of expertise. I hope to convince the reader that this is, indeed, a useful way of adding an important dimension to the function of experts, and that this will help to better understand the role of the IPCC.

Specialists work by applying rules to problems in routine ways. They do so out of the public's view, and they are typically not questioned about their activities. By contrast, commentators operate in public, providing opinion to an audience. Commentators engage in public discourse and thus tend to get embroiled in controversy. This is something that could be welcomed, as Mike Hulme has been tirelessly pointing out. But the very purpose of the IPCC is to eclipse controversy and establish a monopoly of expertise that is science-based and aims at depoliticizing the issue.

In what follows I sketch some of the important aspects of the IPCC's expertise. In so doing I will look at the genesis and function of the IPCC, its use of models and narratives, and the criticism it faces before finally discussing its commentator role.

Genesis and Function of the IPCC

To understand the kind of expertise that the IPCC embodies, we need to understand its role in global environmental governance. And to understand that, we need to go back to its origins in 1988. At its 70th plenary meeting in December 1988, the United Nations (UN) General Assembly created the IPCC with the mandate to 'provide internationally coordinated assessments of the magnitude, timing and potential environmental and socio-economic impact of climate change and realistic response'. The UN also requested the World Meteorological Organization (WMO) and the United Nations Environment Programme (UNEP) to utilize the IPCC to undertake a 'comprehensive review and recommendations with respect to … possible response strategies to delay, limit or mitigate the impact of adverse climate change'.

The IPCC was established as three Working Groups (WGs) that were tasked with different areas of focus: Science (WGI), Impacts (WGII), and Response Strategies (WGIII). After 1992, WGIII was given a different mandate when it was merged with WGII. A new WGIII was set up to deal with socio-economic and other cross-cutting issues. In 1995, WGIII became Economic and Social Dimensions of Climate Change, and from 2001 onwards it became Mitigation of Climate Change. The term 'response strategies' had disappeared altogether which is to say that the post-1992 mandate of WGIII eliminated the element of policy advice.

The US was keen to play a leading role in setting up the IPCC and took the initial lead of WGIII. The US probably wanted to gain control over a potentially unruly public debate about political interventions. After all, in previous cases a pattern had emerged where environmentalist campaigning received support from scientists, something that the US government wanted to avoid. And for their part, climate scientists welcomed the prospect of speaking with one voice, which they hoped would make political responses timelier and more effective. Communicating a consensus message was assumed to be instrumental in this process.

An additional feature of the IPCC is its intergovernmental structure, which enticed developing countries to participate by addressing their mistrust of science. The issue here is that science developed by rich countries often reflects the perspectives and interests of the Global North and neglects the interests and perspectives of the Global South (see Duarte, Chapter 17, this volume). As Bert Bolin, the chair of the IPCC in its early years put it, 'right now, many countries, especially developing countries, simply don't trust assessments in which their scientists and policymakers have not participated. … Don't you think global credibility demands global representation?' (Cited by Schneider, 1991, p. 25).

Developing countries were also wary of the potential for the IPCC to issue unwelcome political messages. They wanted to ensure that the IPCC restricted its focus to science, in which they demanded more participation. They wanted to ensure that political messages were avoided, especially those that would call for curbing CO_2 emissions and thereby endangering their economic development.

The IPCC formula of 'providing policy-relevant but not policy-prescriptive information,' which it started using in 2003, makes perfect sense in this context. It means that the development of response strategies and political recommendations should be outside of its remit. Still, the IPCC does influence public opinion and politics through its Assessment Reports (AR), which are produced every five to seven years. The production of assessments is a key way that the IPCC's influence becomes evident. I will first discuss this, and then detail two more ways that IPCC exerts its influence, selecting studies to include in its assessments, and by acting as a commentator.

Narratives of Alarm and Reassurance

Assessing the published research means foregrounding specific studies, and putting other studies in the background, thereby defining and framing issues for attention. The IPCC's framing activities include problem definitions, causal analysis, and future projections. Above all, framing means developing narratives about the risks of climate change, and the need for intervention. As in other examples where future risks are prominent, the IPCC frames climate change through two opposing narratives: an alarming and a reassuring narrative.

As I explain in more detail in my book, *Making Sense of Expertise* (2022), metrics, models, and metaphors are central elements of both narratives. Metrics are anchored in physical science and relate to greenhouse gases, climate sensitivity, or temperature changes. Metrics are particularly important because the physical aspects of climate change form the **epistemic core** of the IPCC. This means that these metrics are seen as the most important way that knowledge about climate change is gained. These metrics relate to each other in a causal framework that tries to translate levels of GHG emissions to levels of GHG concentrations, to levels of temperature increase, and to impacts on physical, biological, or social systems. In this framework the central variable is GHG emissions. Everything else is subordinated. Models, mainly through computer simulations, and metaphors expressed through images and rhetoric, have been used to illustrate the risks of increased GHG concentrations, and these are widely used in public discourse.

Taken together, metrics, models and metaphors are used to frame narratives about physical and temporal limits in terms of carbon budgets, burning embers, guardrails, and tipping points. Such narratives are both alarming and reassuring. If the narrative only emphasized alarming or catastrophic impacts, it would be politically counter-productive because it would suggest that policy efforts are pointless. A reassuring narrative is thus embedded in an effort to spur action. It is like the IPCC is saying that catastrophic developments are in the cards but we could be OK if we do the right thing. The five-minutes-to-midnight metaphor (or 'we have 12 years left') allows the IPCC to combine both prospects. However, this metaphor has been used for decades, thereby potentially undermining its credibility.

Limits of the IPCC's Assessment Process

As discussed above, the IPCC has been charged with a bias towards the Global North; a second source of bias related to the dominance of natural science disciplines, which this section focuses upon. While uneven geographical participation potentially challenges the legitimacy of the IPCC process, the disciplinary biases restrict the kind of knowledge that is produced, the kind of questions that are asked, the kind of problem that is constructed, and the kind of solutions that are envisaged.

Turning to disciplinary hierarchies, WGI and the physical sciences it represents get more political and public reception compared to the reports from WGII and WGIII. The policy-relevant knowledge the IPCC delivers is global and very general, embedded in a bureaucratic and diplomatic framework, and tied to the epistemic core of WGI.

WGI publishes results from model calculations, emphasizing observed changes in physical parameters, the causes of observed change, and projections into the future. As Minx et al. (2017) point out, 'there is a crucial role for social science research to assess the "real-world" practicalities of climate solutions' (p. 255). This is indispensable to understand what climate change means to people, and which climate policies do and do not work under specific institutional arrangements. But this task is avoided in favour of a science-first, top-down approach. The epistemic core of the IPCC remains always on the level of general categories, carefully avoiding any political implications. It tends to describe global forces and mechanisms, based on geophysics. Individual countries are rarely, if ever, mentioned. The most fine-grained level of analysis refers to regions, such as Europe, Eastern Asia, Africa, and so on, or to industry sectors.

One of the few places where countries are mentioned is in its latest report (AR6), where the IPCC mentions 'at least 18 countries which have sustained GHG emission reductions for longer than 10 years' (2010–2019) (WGIII, SPM, p. 9). But the countries are implied by region rather than name, and the report does not go into any discussion about the reasons for this (relative) success, nor if this could provide lessons. The question of cross-national policy learning is not even posed. Evaluating policy options and developing recommendations is not under the remit of the IPCC.

Even the assessment of the published research is done in a very peculiar way and has attracted criticism. The growth of climate-related research is exponential and so rapid that clearly defined procedures need to be in place to assess its content unless the reviewers want to be accused of cherry-picking. To put the growth in knowledge into perspective, a doubling of climate change papers occurs every 5–6 years, while overall scientific publications double in 24 years. To date, more than 350,000 publications on climate change are on record, and half of these have appeared in the last five years.

The IPCC has thus been taken to task for a lack of any systematic review of the literature: 'A clear evolution in the findings over time is lacking, which supports our claim that knowledge has failed to accumulate; the findings do not reveal a systematic assessment of the literature; and the statements are typically overly generalised or purely descriptive in nature with little policy relevance' (Minx et al., 2017, p. 255).

The absence of self-reflection is noteworthy but not surprising. By this I mean that the IPCC does not thematize different forms of assessments, and therefore the issue of systematic reviews is not discussed. The IPCC does not have a government as a client, and must not recommend policies, so this blind spot can be expected. It nevertheless needs to be addressed because the IPCC's audience, which includes all of us, expects an answer to the question: What should be done about climate change? Here commentators step in to fill the gap. These commentators are authors of IPCC reports itself, or other climate scientists, but in principle everyone: journalists, activists, politicians, business leaders, or professional organizations. A prominent commentator interpreted the message from AR6 as humanity being on a 'highway to hell, with the foot on the accelerator'. While this is a questionable summary of the knowledge base, the question remains how a rapid decarbonization of the world economy can be accomplished, and who should pay for it.

The IPCC as a Public Commentator

The argument so far has provided evidence for the claim that the IPCC is officially prevented from developing response strategies to climate change risks but provides a mix of science arbitration and commentary which are communicated through scenario-based narratives. The IPCC is dominated by scientists from rich countries who have the power to define issues and to frame options. The role of science arbiter is performed in a non-systematic way, giving rise to claims that the IPCC often acts as a stealth advocate (Pielke Jr., 2007), using its alleged objectivity and policy-neutrality as a source for credibility and epistemic authority (Asayama, 2021).

Because IPCC reports have many layers and messages, commentators of all persuasions use it as a reservoir for evidence they can use to advance their issue. In the past the IPCC has been criticized for being too dramatic and activist, especially after the publication of the fourth report (Pielke Jr., 2010). This changed after the Fifth Assessment Report (AR5) when it was criticized for being too conservative, underestimating the risks posed by climate change (Oreskes et al., 2019). The pendulum seems to swing in the other direction again with the publication of AR6.

Reports comprising AR6 were published 2021–2023. Compared to AR5, the headline statements from AR6 WGI SPM are more outspoken, and more dramatic than those from AR5. Not only are the changes in the climate system, and the human causes 'unequivocal'. These changes are 'widespread and rapid'. The scale of the changes is 'unprecedented'. Furthermore, 1.5°C and 2°C warming 'will be exceeded during the 21st century unless deep reductions in carbon dioxide (CO_2) and other greenhouse gas emissions occur in the coming decades'. At the same time the report leaves a more optimistic taste compared to the previous one. While AR5 hinted at the fate that we are committed to global warming ('Most aspects of climate change will persist for many centuries even if emissions of CO_2 are stopped'), AR6 points out that we could make a difference to global temperatures in the coming 20 years or so. This would be possible if the world were to adopt the assumptions that go into low-emission scenarios.

This leaves us with two possible interpretations of the difference between AR5 and AR6. One is that the climate system is more responsive to our policy interventions than we thought, and aggressive and quick GHG reductions can make a discernible difference in the near future. The other is that such policy interventions are now seen as a real possibility, while only a few years ago AR5 assumed this to be unrealistic. The first interpretation would presuppose a new, radically different understanding of the physics of the climate system, which is extremely unlikely. No big changes in the scientific understanding have been claimed by the IPCC since AR5. If the second interpretation is correct, this would presuppose that there is a greater readiness in society to implement climate policies, which has led to a more optimistic evaluation. In fact, there are signs of this, most importantly perhaps the adoption of Net-zero commitments by several governments (Zero Tracker, 2023). The IPCC reflects these social and political developments, albeit tacitly.

Bearing in mind that the SPMs are approved by governments around the world, it is remarkable that they have agreed to the wording above, that is, that we could make a difference to global temperatures in the coming 20 years or so. The growing concern among the younger generation could explain this shift. After all, the IPCC is linked to social and cultural norms and expectations via the participating governments. These same governments meet at climate summits where they negotiate treaties. While the IPCC reports are not a formal input into the negotiations, the social forces that have informed it are represented in the negotiations, too. If this is correct, the progress in climate diplomacy and politics is hardly the result of IPCC reports but of social and political developments. Science and society are not separate entities, but closely interconnected, an old insight of STS.

It is worth mentioning that the 2015 Paris Agreement went beyond AR5 (2014), which discussed mitigation pathways aiming at 'net-zero' emissions. The central IPCC reference point for balancing GHG emissions and their removal, published in AR5 in 2014, in time for Paris, was the end of the century. The Paris Agreement introduced a mid-century deadline, in a way snubbing the IPCC. Again, this indicates that social and political forces are dominating climate policies. This should be a sobering insight for those who believe that the IPCC is setting ambitious targets based on 'the science' which are then watered down at international climate negotiations.

Conclusion

Returning to the initial questions, I have argued that the IPCC primarily focuses on generating knowledge based on physical parameters at the global scale. Other dimensions such as social, political, ethical, and economic considerations are given less priority, with issues of inequality and social justice often considered as secondary concerns. However, if we were to perceive climate change as a multifaceted issue encompassing social, cultural, and political aspects, a more comprehensive analysis would be necessary. This would involve examining economic interests, political institutions, businesses, citizens, and local communities, thereby shifting the emphasis away from purely physical and global perspectives. Adopting such an approach would facilitate cross-jurisdictional learning but would also quickly venture into the realm of providing policy advice, which goes against the traditional scope of the IPCC. Consequently, delivering actionable knowledge would entail moving beyond abstract global-level analyses and engaging with national, regional, and local levels of intervention.

How does the IPCC relate to climate policy? As we have seen, it does not provide policy advice but is influencing public opinion through its reports and comments. It combines the roles and functions of science arbiter and commentator, while the role of advisor does not belong to its remit. The IPCC is unlikely to fulfil one of its main tasks properly, the assessment of the scientific literature. Leaving the selection of relevant studies to expert reviewers will inevitably expose them to the charge of bias and cherry-picking. IPCC scientists (especially coordinating lead authors) have some power to create, select, and communicate data, narratives, and scenarios, thus focusing the debate on specific issues and solutions.

While countries of the Global North have created the problem of anthropogenic climate change through their historic emissions, they also dominate the efforts to frame the problem and its solutions in specific ways. Within the IPCC this is mainly done through the creation of a global knowledge infrastructure which they dominate. Within this infrastructure there is a significant physical science bias which has led to a very specific and arguably problematic framing of the climate problem, namely one of physical limits and tight deadlines, viewed in a top-down perspective (Grundmann and Rödder, 2019).

The public discourse is dominated by statements from WGI about the physical aspects of a changing climate in the future. Results from computer models and sometimes apocalyptic messages dominate in the media. This scientized approach has been aimed at depoliticizing the issue. This has been pointed out by many social scientists, and especially by STS scholars. As Brian Wynne (2010) put it, 'scientific knowledge should be received less as predictive truth machine and more as reality-based social and political heuristic' (p. 295).

IPCC reports have become a central reference point in global discussions about climate change, but they are only loosely coupled to international climate negotiations where politics, after all, comes to the fore. One conclusion we can draw from this is that the IPCC is politically toothless. Its main activity relates to the physical aspects of climate change and its environmental and social impacts. The evaluation of the scientific literature is contested, and value conflicts remain unresolved. The IPCC is one among several voices, a situation in which its epistemic authority is potentially jeopardized. Some critics demand more reassuring, others more dramatic narratives. While the IPCC tries to establish a consensus view, this will always be preliminary and fragile. As long as the intergovernmental structure is maintained, this task is made easier. Governments like to refer to reports that have been written with their own input. But the demand for independent assessments will never go away, and the unique character of the IPCC might not last.

Further Reading

Grundmann, R. (2022) *Making Sense of Expertise: Cases from Law, Medicine, Journalism, Covid-19, and Climate Change*. New York: Routledge.

References

Asayama, S. (2021) 'Threshold, budget and deadline: Beyond the discourse of climate scarcity and control', *Climatic Change*, 167(3–4), pp. 1–16.
Beck, S. and Mahony, M. (2018) 'The IPCC and the new map of science and politics', *Wiley Interdisciplinary Reviews: Climate Change*, 9(6), p. e547.

Grundmann, R. (2022) *Making Sense of Expertise: Cases from Law, Medicine, Journalism, Covid-19, and Climate Change.* New York: Routledge.

Grundmann, R. and Rödder, S. (2019) 'Sociological perspectives on earth system modeling', *Journal of Advances in Modeling Earth Systems*, 11(12), pp. 3878–92.

Guston, D. H. (2001) 'Boundary organizations in environmental policy and science: An introduction', *Science, Technology, & Human Values*, 26(4), pp. 399–408.

Hoppe, R., Wesselink, A., and Cairns, R. (2013) 'Lost in the problem: The role of boundary organisations in the governance of climate Change', *Wiley Interdisciplinary Reviews: Climate Change*, 4(4), pp. 283–300.

IPCC (2023) Intergovernmental Panel on Climate Change: About the IPCC. Available at: www.ipcc.ch/about/ (Accessed 23 June 2023).

Mahony, M. (2013) 'Boundary spaces: Science, politics and the epistemic geographies of climate change in Copenhagen, 2009', *Geoforum*, 49, pp. 29–39.

Minx, J.C., Callaghan, M., Lamb, W.F., Garard, J., and Edenhofer, O. (2017) 'Learning about climate change solutions in the IPCC and beyond', *Environmental Science and Policy*, 77, pp. 252–59.

Oreskes, N., Oppenheimer, M., and Jamieson, D. (2019) 'Scientists have been underestimating the pace of climate change', *Scientific American*, 19(08). Available at: https://blogs.scientificamerican.com/observations/scientists-have-been-underestimating-the-pace-of-climate-change/ Retrieved 27 June 2023.

Pielke Jr., R. (2007) *The Honest Broker*. Cambridge: Cambridge University Press.

Ravetz, J. (2011) '"Climategate" and the maturing of post-normal science', *Futures*, 43(2), pp. 149–57.

Schneider S.H. (1991) 'Three reports of the Intergovernmental Panel on Climate Change', *Environment* 33, pp. 25–30.

Turnpenny, J., Jones, M., and Lorenzoni, I. (2011) 'Where now for post-normal science? A critical review of its development, definitions, and uses', *Science, Technology, & Human Values*, 36(3), pp. 287–306.

Wynne, B. (2010) 'Strange weather, again: Climate science as political art', *Theory, Culture & Society*, 27(2–3), pp. 289–305.

Zero Tracker (2023) 'Zero tracker: Data explorer'. Available at: https://zerotracker.net/ (Accessed 7 June 2023).

18 Consensus, National Self-Interest, and the Shaping of Climate Knowledge in IPCC Assessment Processes

Mark Vardy

Introduction

For one week in the middle of March 2023, I sat in a windowless auditorium as an official observer to the Intergovernmental Panel on Climate Change (IPCC) 58th plenary session. The meeting was held in Switzerland, and tourists enjoyed the early spring weather outside the meeting venue where the snowy Alps filled the horizon. Inside, I watched and listened as governmental delegates from around the world and IPCC authors argued over, debated, and edited – sentence by sentence – the exact wording of the "Synthesis Report of the IPCC Sixth Assessment Report (AR6): Summary for Policymakers" (AR6 SYR). This short document is only 42 pages but it brings together and summarizes scientific and social-scientific findings from thousands and thousands of pages of underlying IPCC reports. It soon became clear that agreement was difficult to achieve, and additional sessions were scheduled (see Figures 18.1 and 18.2 for photographs of the meeting). The second and third day were extended to 10:30 pm, and on subsequent days the sessions went well past midnight. The final day went through the night to the next morning and then continued into the evening. Some delegates and authors napped on couches outside the auditorium. I spoke with several participants who said they only had about two hours of sleep in 48 hours. When the AR6 SYR was finally approved, two days behind schedule, most government delegates from developing nations, who were unable to extend their stay in Switzerland, had already left.

The delay in securing agreement on the AR6 SYR at the 58th Plenary meant that the media nearly didn't get their promised story. But when agreement was reached, news outlets around the world were notified. Many news organizations, including *Al Jazeera* (Al Jazeera, 2023) and the *Washington Post* (Kaplan, 2023), used the report to anchor articles about how the world has only a few years left to reduce greenhouse gas emissions enough to stay below 1.5°C of warming. The process of writing IPCC reports takes years. By that time the next report is released, it is possible that the world will have gone screaming past the point at which scenarios that keep us below 1.5°C warming are feasible.

In this chapter, I take a closer look at the production of IPCC reports. As I discuss in more detail below, a typical IPCC assessment cycle takes five to seven years and produces several "special reports", such as the *Special Report on Global Warming of 1.5°C* (2018), and an "assessment report" for each of the IPCC's three Working Groups (WGs). The assessment reports often exceed 1,500 pages and do not go through the line-by-line approval process with government delegates. That process is reserved for the much shorter Summary for Policymakers (SPM), which is written for each of the special reports and assessment reports.

The topic of how consensus is shaped in IPCC reports is of great interest to STS scholars (Hulme, 2022). Mike Hulme and Martin Mahony (2010) provide a foundational analysis. They

DOI: 10.4324/9781003409748-24

Figure 18.1 View of the IPCC Chair Hoesung Lee, IPCC Secretary, Abdallah Mokssit, and IPCC authors on the podium during the IPCC's 58th Session.

Source: Photo by IISD/ENB | Anastasia Rodopoulou: Available at: https://enb.iisd.org/58th-session-intergovernmental-panel-climate-change-ipcc-58-16Mar2023 Retrieved 24 March 2023.

Figure 18.2 View of governmental delegates during the IPCC's 58th Session.

Source: Photo by IISD/ENB | Anastasia Rodopoulou: Available at: https://enb.iisd.org/58th-.session-intergovernmental-panel-climate-change-ipcc-58-16Mar2023 Retrieved 24 March 2023.

discuss the tension found in the underlying assessment reports between unanimous expressions of scientific judgment, or consensus, on one hand, and reasoned scientific arguments for dissenting judgments, or dissensus, on the other. A key reason for this tension flows from the inherent uncertainty and indeterminacy of something as complex as the Earth under conditions of anthropogenic climate change. So while consensus can lend weight to the IPCC's findings, it can also obscure climatic dynamics that are important to pay attention to.

In addition to how consensus shapes the underlying reports, STS scholars are also interested in the approval process in which government delegates and IPCC authors agree on the exact wording of the SPMs (De Pryck, 2022). In these SPM approval plenaries, consensus means that countries from around the world and IPCC authors have agreed on each and every word in the document, which is one key reason why the IPCC has such significant social authority. But the process of agreeing line-by-line on the wording between IPCC authors and governmental delegates produces a different kind of consensus than that in the underlying reports. As Kari De Pryck (2021) says of the approval process, "consensus building is the artful result of a complex layering of compromises" (p. 125). The SPM approval process can obscure important social and political dynamics of climate change (De Pryck, 2021; Livingston et al., 2018).

In this chapter, I argue that it is important to pay attention to the difference between how consensus is shaped in the underlying assessment reports versus the SPMs. This difference is important to pay attention to because it helps to understand the IPCC as a politically situated actor. Whereas governmental delegates in approval plenaries sometimes suggest wording that appears to be motivated by national self-interest, IPCC authors are not beholden to this same logic of national self-interest when they are composing the underlying assessment reports. The underlying reports can thus more easily include progressive arguments about social transformation. To make this argument, I outline the processes through which the IPCC works.

Shaping Assessments in Chapter Writing Teams

The IPCC is frequently discussed by STS scholars, but relatively few have observed how IPCC authors actually work together while writing assessment reports. However, the IPCC granted Indiana University anthropologist Jessica O'Reilly permission to observe the production of AR6, and in February 2016, I joined her project as one of four researchers along with Kari De Pryck and Marcela da S. Feital Benedetti. The research plan centered around the Lead Author Meetings (LAMs) for each Working Group. Starting with WGI in 2018, each Working Group was scheduled to have four LAMs, about one every six months. I attended three for WGI *The Physical Science Basis*, and three for WGII *Impacts, Adaptation and Vulnerability*. Each LAM lasted about a week. The basic structure of the meetings alternated between plenary meetings that all authors attended and chapter meetings where authors gathered in their chapter writing teams. I conducted over 100 semi-structured interviews with about 40 authors during coffee and meal breaks, interviewing several authors several times over several years to understand how they experienced the assessment process over time. In this chapter, I point out some salient features of the assessment-writing process. In this section, I focus on the underlying assessment reports.

The IPCC was established in 1988 to provide nation-states' scientific assessments of climate change that are intended to be policy-relevant but not policy-prescriptive. The IPCC has since gone through six assessment cycles. Assessment cycles begin with a week-long plenary like the one described at the start of this chapter (also see Figures 18.1, 18.2, and 18.3). At the beginning of each assessment cycle, government delegates from nation-states agree on what topics the assessment should include. The actual work of assessing the literature and writing the reports is then turned over to hundreds of volunteer authors. The authors are selected anew for each new assessment cycle, and they are divided into three WGs: WGI *The Physical Science Basis*; WGII *Impacts, Adaptation and Vulnerability*; and, WGIII *Mitigation of Climate Change*. In the Sixth Assessment Report (AR6), which lasted from 2017 to 2023, WGI had 234 authors, WGII had 270 authors, and WGIII had 278 authors.

Authoring IPCC assessment reports is accomplished through processes of social coordination. Each Working Group has between 12 and 18 chapters. Each of those chapters has six to 15 Lead Authors (LAs). Each chapter also has two or three Coordinating Lead Authors (CLAs) who not only are in charge of their own chapter but also have to communicate with CLAs from other chapters to coordinate common topics and themes. To one degree or another, all of these people must work with one another in a collaborative manner. While the IPCC encourages its authors to work through consensus, the actual experience of group authorship can vary considerably from one chapter to the next.

IPCC authors volunteer their time to assess the literature and write the assessment reports, which they often have to fit around their already busy full-time occupations, such as researching and teaching as university professors and raising children. Issues related to gender, age, ethnicity, cultural background, career status, and familiarity with English (the working language of the IPCC) might influence how much authors feel that they can fully participate in the discussions in their chapter writing teams. Moreover, chapter writing teams do a significant amount of work between LAMs, and additional challenges come from managing a heavy workload and scheduling online meetings across widely divergent time zones. Authors based in the Global South can face additional barriers related to a lack of reliable Internet and access to databases that house peer-reviewed literature.

Different CLAs have different leadership styles, with some being more authoritative and "top-down" and others taking a more "flat" or "non-hierarchical" style of leadership. But by the same token, leadership styles are informed by social contexts and cultural backgrounds, and different LAs have different expectations of what leadership should look like. There are different opinions between and within chapters about how CLAs and LAs should perform their duties. While some authors feel that their chapter writing teams build meaningful connections and work well together, others might feel that aspects of their chapter are more dysfunctional. Many of the authors who I interviewed said, unprompted, that learning to work with others in a diverse team is a big part of what it means to be an IPCC author. Overall, there is a fair amount of flexibility in terms of how chapters go about doing their work, which has led some scholars to critique IPCC processes.

The social processes of learning to work with others from a range of backgrounds as a chapter writing team distinguishes how consensus or agreement is shaped in the longer assessment reports versus the shorter SPMs. Before I turn to examine the SPM process, I first want to go into the assessment reports themselves to show why it is important to pay attention to the way knowledge is shaped differently at different parts of the process.

Uncertainty and Social Transformation

The AR6 WGI *Physical Science Basis* includes a chapter on extreme weather events which contains the following excerpt:

> One factor making such [extreme weather] events hard to anticipate is the fact that we now live in a non-stationary climate, and that the framework of reference for adaptation is continuously moving. [...] As warming continues, the climate moves further away from

its historical state with which we are familiar, resulting in an increased likelihood of unprecedented events and surprises.

(Seneviratne et al., 2021, p. 1535)

What I find remarkable about this excerpt is that it makes it clear that the physical sciences of climate change raise questions of uncertainty that we are used to hearing from sociologists of risk and modernity, such as Ulrich Beck. Writing in the 1980s when the threat of nuclear disasters loomed large, Beck argued that the very scientific and technological developments that are key characteristics of modernity create new forms of risk and give rise to new forms of uncertainty. In the statement above, we can see a form of support for Beck's thesis insofar as the physical sciences of anthropogenic climate change are reporting that one of the qualities of climate change is its capacity to generate uncertainty. That is, in accordance with, but independent of, Beck's thesis of reflexive modernization, physical scientists are saying that the Earth's climate—having been impacted through the kinds of modern industrialized processes that Beck analyzed—is now creating conditions in which there is an "increased likelihood of unprecedented events and surprises."

The second example I consider comes from Chapter 6, "Cities, Settlements and Key Infrastructures," in AR6 WGII *Impacts, Adaptation and Vulnerability*. A section on "Equity and Justice" contains this statement:

Social, economic and cultural structures that marginalize people by race, class, ethnicity and gender all contribute in complex ways to climate injustices and need to be urgently surfaced in order for adaptation options to shift to benefit those most vulnerable rather than mainly benefitting the already privileged and maintaining the status quo.

(Dodman et al., 2002, p. 72)

In the above excerpt, and in many other places in the underlying assessments, we find evidence of IPCC authors grappling with climate change as a problem that challenges a status quo which enshrines the self-interest and private profit of a few corporations and individuals above the needs of the many.

Taking the above statements from both WGs together, we can see that there is an increased likelihood of extreme weather events occurring in a surprising or unprecedented manner and that a crucial way of preparing for such unanticipated events requires redistributing resources so that marginalized and vulnerable populations are better prepared. Through discussions of "social transformations" and related concepts, assessment reports provide evidence that acting now, on all levels including the local level, to transform social, economic, political, and cultural systems and structures that perpetuate the inequalities that make people vulnerable is an actionable and viable way to address climate change (Lidskog and Sundqvist, 2022). The question then becomes, what happens to these understandings when it comes to the shorter summaries for policymakers?

Shaping Assessments in Approval Plenaries

The specific dynamics of the IPCC, in which government delegates first approve the outline of the underlying assessment reports, and then approve the shorter reports that summarize their findings, makes the IPCC a prime candidate for study as what STS calls **boundary work** (Gieryn, 1999). That is, rather than maintaining the idea that science and politics are two readymade spheres that are distinct from one another by virtue of essential demarcation criteria, STS scholars show how the boundaries between science and non-science are actively constructed and maintained through rhetorical justifications. This approach does not mean that science cannot tell us valuable things, but it does mean that we must be attentive to the specific translations that occur as scientific knowledge is worked up from its many diverse points of origins into a narrative that travels beyond the local sites in which it is produced. STS scholars have used this approach to study the IPCC, and in so doing detailed how boundaries are drawn in different ways at different parts of the IPCC process (Hoppe, Wesselink, and Cairns, 2013). Boundaries need to be continually redrawn and shored up; they are always in a process of being constructed. One site where boundaries are drawn is the SPM approval plenaries as governmental delegates and IPCC authors agree on the wording. But it is important to note that the reason given for governmental involvement in this process is to ensure the report has the greatest policy relevance while not being prescriptive. To this end, all suggestions must be couched in scientific language.

There is a vitally important difference between the longer underlying assessment reports and the shorter summaries for policymakers. When they are completed, the longer assessment reports are "accepted" by IPCC government delegates. "Acceptance" means that the Panel acknowledges that IPCC authors have written the reports and that government delegates received them. Government delegates do not have the power to alter any of the wording of the underlying assessment reports. However, the much shorter summaries for policymakers are "approved" by government delegates, and that approval consists of a sentence-by-sentence process of reviewing, debating, and arguing over the precise wording.

The draft SYR is shown as a Word document on giant screens in the auditorium. The sentence that is under consideration is highlighted in yellow. The chair reads the sentence out loud and asks for approval. If no objections are voiced, the chair bangs their gavel, the sentence is highlighted in green to indicate that it has been approved, and they move on to the next sentence. However, if a government delegate wants to ask a question or raise an objection to the phrasing, then they may do so. The IPCC authors who drafted the report will confer amongst themselves and then suggest alternative wording that is consistent with the underlying reports. Sometimes government delegates suggest their own preferred wording. In this back-and-forth process, in which a single sentence can be argued over for hours, the authors must agree that any changes remain consistent with the underlying reports, and all of the government delegates must approve the wording. The opposition of any one country is sufficient for the sentence to remain open.

It often happens that no agreement can be reached on a sentence or cluster of sentences. In these cases, the chair may decide to "park" it and move on to other parts of the draft. Or the chair may decide to send the sentence or cluster of sentences to a formal "contact group" or a less formal "huddle" for IPCC authors and government delegates to try to find agreement (see Figure 18.3). Wording that is agreed upon in a contact group or huddle is highlighted in cyan (or turquoise). Text in cyan is returned to the plenary session where it is once again highlighted in yellow for the chair to once again ask for approval. The rules of procedure state that if a country participated in the contact group or huddle where text was agreed upon, then they cannot voice objections in the plenary.

In the process of going back and forth between science and politics during plenary sessions, each country will argue for wording that they say is better suited for policymakers. While it is true that some suggested edits are intended to clarify confusing sentences, many more are suggested to shape the sentence in ways that better reflect national self-interest. Let me give one example that I observed and that can also be found in the *Earth Negotiations Bulletin* (IISD, 2023, p. 12), which provides a public record of IPCC meetings. The example pertains to the form of geoengineering known as carbon dioxide removal (CDR). Assumptions about the capacity of humans to invent, develop, refine, manufacture, deploy, scale-up, manage, and effectively use

Figure 18.3 View of an informal huddle during the IPCC's 58th Session.

Source: Photo by IISD/ENB | Anastasia Rodopoulou: Available at: https://enb.iisd.org/58th-session-intergovernmental-panel-climate-change-ipcc-58-18Mar2023. Retrieved 27 March 2023.

CDR to remove carbon dioxide from the Earth's carbon cycle are baked into several of the scenarios that keep Earth below 1.5°C warming. But many of these assumptions are unproven, and there are good reasons to be wary of the assertion that CDR has the capacity to do in actual reality what it is assumed to do in several of the IPCC scenarios (see Part 10, this volume). For this reason, several countries wanted to introduce wording that pointed out the limitations of CDR so that policymakers and decision-makers would not perceive it as a simple panacea. But a country that appears to favor continued fossil fuel production argued against wording that expressed caution about CDR. To this end, they said that if wording was introduced to warn policymakers about the limitations of CDR, then they would introduce similar wording about the "feasibility" of renewable resources such as solar and wind power. In this tit-for-tat exchange, which is just one example of many, we can see how the "consensus" that is achieved in the plenary reflects a compromise position (De Pryck, 2021).

Conclusion

Given that the IPCC reports provide science advice for the international negotiations on climate change, it makes sense that governmental delegates will argue for wording that frames the science of climate change in ways that will advantage their national self-interest. This is not a cynical statement but rather a recognition that the IPCC is constrained by the structures that surround it, which include the international system of sovereign nation-states that emerged in Western Europe around the mid-1600s. Social constructivist scholars working in international relations and STS show how the historical development of science is inextricably linked with the development of forms of social order, including the sovereignty of the nation-state and the system of international states as that which defines the space of political action (Shapin and Shaffer, 1985; Walker, 1993). In other words, science and society are **co-produced**. This STS concept points to how science and society are mutually produced by human and non-human agents through processes of **co-production** (Miller and Wyborn, 2020). While boundary work is useful for analyzing specific instances or cases in which the lines between science and politics are argued over and negotiated, the idiom of co-production provides a way of examining more entrenched ideologies and practices that shape society at a fundamental level (Irwin, 2008). When looking at the SPM approval process, it becomes clear how science is co-produced with concepts of political agency tied to sovereign nation-states acting within the international system of states.

Climate change is now showing the limitations of particular ways of co-producing science and society. New ways of forming relations are vitally important. But the presence of national self-interest is not sufficient reason for the wholesale rejection of the IPCC. If you know where to look and how to interpret them, assessment reports and SPMs draw our attention to the need for social transformation. That is, they do show how reducing the marginalization and vulnerability that people are already experiencing, locally as well as internationally, is a viable approach to engaging productively with climate change. National self-interest restricts the scope of climate action that is conceived in the IPCC, but read with this caveat in mind, IPCC reports can provide support for the argument that we should reconfigure social relations for a more just and equitable world which will be better prepared for a changing climate.

Further Reading

De Pryck, K. and Hulme, M. (2022) *A Critical Assessment of the Intergovernmental Panel on Climate Change*. Cambridge: Cambridge University Press.

References

Al Jazeera (2023) "UN calls for rapid, ambitious action to tackle climate crisis." March 20, 2023. Available at: www.aljazeera.com/news/2023/3/20/world-can-tackle-climate-change-but-must-be-more-ambitious-ipcc (Accessed March 23, 2023).

De Pryck, K. (2021) "Intergovernmental expert consensus in the making: The case of the summary for policy makers of the IPCC 2014 Synthesis Report." *Global Environmental Politics*, *21*(1), pp. 108–129.

De Pryck, K. (2022) "Governmental assessments." In K. De Pryck and M. Hulme (eds.) *A Critical Assessment of the Intergovernmental Panel on Climate Change*. Cambridge: Cambridge University Press.

Dodman, D., B. Hayward, M. Pelling, V. Castan Broto, W. Chow, E. Chu, R. Dawson, L. Khirfan, T. McPhearson, A. Prakash, Y. Zheng, and G. Ziervogel (2022) "Cities, settlements and key infrastructure." In H.-O. Pörtner, D.C. Roberts, M. Tignor, E.S. Poloczanska, K. Mintenbeck, A. Alegría, M. Craig, S. Langsdorf, S. Löschke, V. Möller, A. Okem, B. Rama (eds.) *Climate Change 2022: Impacts, Adaptation and Vulnerability. Contribution of Working Group II to the Sixth Assessment Report of the Intergovernmental Panel on Climate Change*. Cambridge: Cambridge University Press, pp. 907–1040, doi:10.1017/9781009325844.008.

Gieryn, T.F. (1999) *Cultural Boundaries of Science: Credibility on the Line*. Chicago: University of Chicago Press.

Hoppe, R., Wesselink, A., and Cairns, R. (2013) "Lost in the problem: The role of boundary organisations in the governance of climate change," *Wiley Interdisciplinary Reviews: Climate Change*, 4(4), pp. 283–300.

Hulme, M. (2022) "Scientific consensus-seeking." In K. De Pryck and M. Hulme (eds.) *A Critical Assessment of the Intergovernmental Panel on Climate Change*. Cambridge: Cambridge University Press.

Hulme, M. and Mahony, M. (2010) "Climate change: What do we know about the IPCC?" *Progress in Physical Geography*, 34(5), pp. 705–718.

IISD (International Institute for Sustainable Development) (2023) "IPCC-58 Final." *Earth Negotiations Bulletin*, 12 (819). Available at https://enb.iisd.org/sites/default/files/2023-03/enb12819e_0.pdf (Accessed June 27, 2023).

IPCC (2023) "Synthesis Report of the IPCC Sixth Assessment Report (AR6): Summary for Policymakers." IPCC. Available at: https://report.ipcc.ch/ar6syr/pdf/IPCC_AR6_SYR_SPM.pdf (Accessed March 25, 2023).

Irwin, A. (2008) "STS Perspectives on scientific governance." In E.J. Hackett, O. Amsterdamska, M.E. Lynch, and J. Wajcman (eds.) *The Handbook of Science and Technology Studies*, Cambridge, MA: MIT Press.

Kaplan, S. (2023) "World is on brink of catastrophic warming, U.N. climate change report says," *Washington Post*. March 20, 2023. Available at: www.washingtonpost.com/climate-environment/2023/03/20/climate-change-ipcc-report-15/ (Accessed March 23, 2023).

Lidskog, R. and Sundqvist, G. (2022) "Political context." In K. De Pryck and M. Hulme (eds.) *A Critical Assessment of the Intergovernmental Panel on Climate Change*. Cambridge: Cambridge University Press.

Livingston, J.E., Lövbrand, E., and Olsson, J.A. (2018) "From climates multiple to climate singular: Maintaining policy-relevance in the IPCC synthesis report." *Environmental Science & Policy*, 90, pp. 83–90.

Miller, C.A. and Wyborn, C. (2020) "Co-production in global sustainability: Histories and theories." *Environmental Science & Policy*, 113, pp. 88–95.

Seneviratne, S.I., X. Zhang, M. Adnan, W. Badi, C. Dereczynski, A. Di Luca, S. Ghosh, I. Iskandar, J. Kossin, S. Lewis, F. Otto, I. Pinto, M. Satoh, S.M. Vicente-Serrano, M. Wehner, and B. Zhou (2021) "Weather and climate extreme events in a changing climate." In V. Masson-Delmotte, P. Zhai, A. Pirani, S.L. Connors, C. Péan, S. Berger, N. Caud, Y. Chen, L. Goldfarb, M.I. Gomis, M. Huang, K. Leitzell, E. Lonnoy, J.B.R. Matthews, T.K. Maycock, T. Waterfield, O. Yelekçi, R. Yu, and B. Zhou (eds.) *Climate Change 2021: The Physical Science Basis. Contribution of Working Group I to the Sixth Assessment Report of the Intergovernmental Panel on Climate Change*. Cambridge: Cambridge University Press, pp. 1513–1766, doi: 10.1017/9781009157896.013.

Shapin, S. and Schaffer, S. (1985) *Leviathan and the Air-Pump*. Princeton, NJ: Princeton University Press.

Walker, R.B. (1993) *Inside/outside: International Relations as Political Theory*. Cambridge: Cambridge University Press.

19 Trust at the Climate Science-Policy Interface

Tiago Ribeiro Duarte

Introduction

While climate change is taking place at an increasingly fast pace, policies to mitigate it are lagging behind. One might then ask why, if the science is increasingly clear about the reality and the risks of anthropogenic climate change, is political action so slow? In this chapter, I explore this question by focusing on the relation between trust and the climate science-policy interface, to which Science and Technology Studies (STS) scholars have dedicated significant attention. Examining the relation between trust and the nexus of science and policy provides insights into understanding the complexity of reaching a global consensus and implementing policies at a national and global level to deal with climate change. I examine how trust is connected to two phenomena: the geopolitical divide between the Global South and the Global North, and climate denialism.

Trust, Expertise and Climate Science as a Social Activity

Trust has been understood in different ways in the social sciences. In this chapter, I deploy Anthony Giddens' concept from his book *The Consequences of Modernity* (1991). Giddens points out that the nature of trust has changed with the emergence of modernity. In traditional societies trust is strongly linked with interpersonal relations. But in modern societies, the character of trust changes because our lives are influenced to an unprecedented level by a range of systems and structures that are operated by unknown others. An example is traffic lights. Very few people are involved with the programming and maintenance of these infrastructures and yet those living in urban centers have their lives regulated by them, which implies that our lives in traffic depend on the actions of these absent others. Even if we knew all people involved with keeping these infrastructures working, we would still not have the expertise to assess their competence. In this sense, trust, in modernity, comes into play in contexts where people lack access to complete information, having no other choice than trusting or distrusting complex systems.

Trust in science and technology therefore cannot be taken for granted. From the point of view of STS, trust has to be actively built and maintained through social processes, such as science education, science communication, interactions between experts and laypeople, and so on. In none of these processes can the act of presenting scientific information be taken as sufficient for people to trust them. Trust or distrust will be built depending on factors such as how information is transmitted, the language used to transmit it, the information source, and on whether or not the person or institution transmitting it is regarded as trustworthy. As a result the science-policy interface should not be understood through what has become known as the **linear model**, which says that scientists produce relevant knowledge, communicate it to policymakers, and the latter

DOI: 10.4324/9781003409748-25

develop rational policies based on the scientific knowledge that was communicated to them (see Zehr, Chapter 6, this volume).

The importance of trust at the nexus of science and policy is evident in the case of climate models. Climate models run simulations of the climate system to reconstruct past climate events, study particular climate phenomena, and simulate future climates. Such models are crucial tools in climate science for at least two reasons. First, the global climate cannot be observed directly; it is necessary to deploy tools such as models to bring together data from several local sources, such as temperature measurements, to come up with data that is global. Second, because of the global scale of climate change, there is no way to carry out large-scale experiments on it. Modeling is thus a tool that allows scientists to simulate how the climate system works and how it would work if particular changes occurred. Third, modeling enables scientists to make simulations about future climates, which observational and experimental sciences cannot do. However, while climate modelers have familiarity with the strengths and limitations of climate models, most policymakers – and even scientists from other areas of climate science – have no direct access to them. When they work on climate policies, policymakers might use information produced by climate modelers, but in most cases they have not been to modeling centers, met with teams of modelers, or watched them working on their computers. Policymakers thus work in a context in which potential policies are informed by work that they have not seen and produced by people they do not know. The world of climate modeling is therefore opaque to them, and this opaqueness can potentially bring about issues of trust.

There is another layer of complexity in this relation, which is the lack of expertise of most policymakers to assess the quality of climate models and the competence of groups of climate modelers. Climate modeling is, as all scientific fields, a very specialized area, and being able to write computer codes, run models, debug them, and interpret their results requires years of training with highly-skilled experts. Modelers themselves are reliant upon fields of science, mathematics, and computing they must trust in order to build and run climate models. As a result, even if policymakers went to modeling centers, they would not be able to assess the quality of the work done there. In this sense, climate modeling is opaque to policymakers, as well as to scientists from other research areas. Indeed, climate modeling is an activity that is not directly accessible to most people and, even if it were, the direct access would not make much difference as a high level of specialized expertise is needed to assess these complex machines and their outputs. As I will discuss below, the significant levels of trust associated with climate models can be exploited by individuals and groups who wish to discredit climate science. First, however, I want to discuss another area where a lack of trust can be found in the science-policy interface, namely between Global North and South countries and institutions.

Trust and the Divide Between Global South and Global North

The complex and important relations between trust and the science-policy interface can be illustrated by research that I started in the 2010s. Back then, inspired by the work of Myanna Lahsen who writes about her research in Chapter 2 of this volume, I was interested in the climate-policy interface in Brazil and geopolitical aspects, particularly those related to the divide between the Global South and Global North. This led me to research the Brazilian Panel on Climate Change (BPCC), a scientific panel that was founded in 2009 by the Brazilian Ministry of Science and Technology and the Ministry of Environment to review the relevant scientific literature on climate change in the country. The policymakers involved with its creation were prestigious Brazilian climate scientists, who, at that time, had key positions in the ministries responsible for developing climate policy in the country. These scientists were also members of the Intergovernmental Panel on Climate Change (IPCC).

The BPCC sought to mirror the IPCC in its structure, procedures, template, and topics. Similarly to the IPCC, the main official goal of the BPCC was to produce review reports that could inform climate policymaking in Brazil. However, during my research I found it had additional informal goals. The founders of the BPCC believed that their reports should be translated into English so that research that was originally published in Portuguese could be integrated into the IPCC reports. Furthermore, they thought that junior scholars who took part in the BPCC would acquire experience and skills that would enable them to become part of the IPCC in the future. In other words, the BPCC was founded by a group of scientists and policymakers who trusted the IPCC, believed it was a good model for a panel at the science-policy interface, and were willing to promote further integration between Brazilian science and the intergovernmental body.

However, not all policymakers in Brazil were so enthusiastic about the IPCC and the BPCC. Among my informants was an influential Brazilian climate policymaker who held a position in the Ministry of Environment and who had represented the country in international negotiations on climate change agreements. This policymaker, who I will refer to through the pseudonym Antônio, was very suspicious of the IPCC.

To understand this suspicion it is important to take into consideration that although the IPCC is an international panel composed of scientists from all continents, since its first report in the early 1990s there has been significant inequalities in the origin of its members. Most of them were working at institutions in the USA or Europe (Corbera et al., 2016). For Antônio this resulted in the IPCC being biased toward Global North interests. As a consequence, he argued the panel would focus on issues that were more pressing for Global North countries, but, more seriously, would have errors in the data presented in its reports. According to him, these errors were evident when it came to issues related to the Global South. In particular, he argued that the contribution of rainforest deforestation to climate change was inflated in IPCC reports. Antônio further argued that this was the case because most rainforests are in the Global South. The bias would therefore result in the attribution of a larger share of responsibility for climate change to countries in the Global South than they actually had.

In short, Antônio did not trust the IPCC. He believed that what was needed was not a Brazilian panel that would mirror the intergovernmental body at the national level. Rather, he had attempted, some years before the foundation of the BPCC, to create a counter-panel. He wanted to bring together top Brazilian climate scientists to review drafts of the IPCC reports and find errors in the way climate change in Brazil was being portrayed. By finding these errors, it would be possible to "react" to the IPCC and make a stronger case that Brazil had a smaller contribution to climate change than the international panel had implied. The counter-panel never came into reality, but the distrust in the intergovernmental body continued to be part of the worldview of this Brazilian policymaker for several years.

This could be regarded as an exception, a single policymaker with idiosyncratic views. In fact the accusation of biased data in the IPCC reports is not shared by mainstream Brazilian scientists. Antônio, however, was influential in climate policy in Brazil and had key positions in several different governments. His views of the divide between Global South and Global North were shared by other policymakers who worked closely with him as well as with Brazilian diplomats working on climate negotiations at the international level. He did not gain much traction within the scientific community, which was much more trusting of the IPCC and of Global North science. Yet, throughout the years he helped shape the views of other policymakers spreading distrust in the climate science-policy interface.

In sum, this case illustrates the role of trust at the nexus between science and policy. As science is opaque to those distant from their sites of production, policymakers are left to trust or distrust scientists and their work based on non-scientific judgments. Antônio, an experienced policymaker, had a very clear view of the geopolitical inequality between the Global South and the Global North. He assessed science based on these inequalities, or, in other words, on its origin and the interests he believed it represented. Whether or not he trusted particular theories or data was influenced by his perceptions of the trustworthiness of the people who produced them, which itself was influenced by his worldview.

Trust in Inauthentic Science

The issue of trust at the climate science-policy interface also emerges in relation to climate denialism, which is a significant risk given the degree to which denialists attract attention from the media, public, and politicians and can influence policy – or the lack of it. There is an important literature on climate denialism that sheds light on the strategies deployed by industry, particularly oil companies, to convince publics and policymakers of substantial doubt and insufficient scientific evidence for political action (see the sources listed in the Further Reading section below). These strategies were used in periods in which there was already significant consensus within the scientific community about the risks of anthropogenic climate change. Yet, in countries such as the USA they were influential enough to prevent this country, one of the greatest greenhouse gas emitters, from developing internal policies to mitigate climate change and become part of international climate agreements. An important part of these strategies was mobilizing individuals with scientific credentials and distinguished careers in science to spread doubt and support denialist narratives.

But, if some distinguished scientists are claiming that there is still doubt about the reality of a phenomenon, wouldn't that be a situation in which there is a genuine scientific controversy that should influence the development of policies? Shouldn't they therefore be *trusted* as reliable sources of information for policy making? Isn't skepticism a crucial value of science so that systematic criticism of scientific theories should be supported and be taken into consideration at the science-policy interface? Skepticism is indeed an important value of scientific activity. But there are situations in which the public perception of a significant scientific dissensus can be manufactured with the support of distinguished scholars in order to prevent or delay the implementation of policies. These situations need to be differentiated from genuine scientific controversies. There is an evolving literature on how inauthentic controversies are created and the tactics deployed to make them look genuine and trustworthy to publics and policymakers. A recent paper that I co-authored with an interdisciplinary group of Brazilian scientists, entitled "The risk of fake controversies for Brazilian environmental policies" (Rajão et al., 2022), developed a typology to classify these tactics with the following concepts: manufacturing uncertainty, manufacturing pseudo-facts, misusing scientific credentials, and disregarding scientific literature.

Manufacturing uncertainty and misusing scientific credentials can be illustrated with examples from the well-known book by Naomi Oreskes and Erik M. Conway, *Merchants of Doubt* (2010). This book documented how a group of physicists from the USA, who held distinguished careers during the Cold War and worked closely with the government in military projects, were involved in manufacturing doubt about a number of issues, ranging from the risks of smoking, through acid rain and the ozone hole, to climate change. For the purposes of this chapter, what is particularly important in Oreskes and Conway's study is that these physicists were casting doubt on issues on which they did not have the relevant expertise. They had made significant contributions to particular areas of physics during their careers. However, scientific fields are highly specialized and no one can claim to have expertise in all areas of science. For a scientist to become an expert in a particular field of climate science, it takes years of immersion in the relevant community under the guidance of experienced mentors. Yet these physicists cast doubt on the science of various scientific fields in which they did not have training or a relevant record or contributions. In other words, they *misused their scientific credentials* to give publics and policymakers the impression of a controversy in a field in which substantial consensus was already reached. They did so by claiming, for instance, that climate models were not reliable tools, that scientific institutions such as the IPCC were corrupt, that there was too much uncertainty surrounding climate change to justify political action, and so on. In other words, they *manufactured doubt* over well-accepted scientific tools, institutions, and theories so as to justify distrust in them. Their argument was enabled, in part, by the considerable amounts of trust that accompany the use of climate models, as discussed in the first section above.

Manufacturing pseudo-facts is closely related to the tactic of manufacturing doubt. Here, denialists spread alternative facts, which have no support in the scientific literature. Climate denialists have, for instance, claimed that climate change was not caused by human activity, but by solar phenomena, even though the scientific community had already reached a significant consensus around the anthropogenic nature of the rise in global temperature.

The final tactic, *disregarding scientific literature,* can also be illustrated with reference to denialists from the USA. Their efforts to spread doubt on climate science (and other fields of science with regulatory implications) did not take place in the scientific literature and did not engage with it. They mostly addressed policymakers and publics, rather than the scientific community itself. They did so through the publication of letters and interviews in the media or on the internet, and through booklets and books published by commercial publishers without a proper scientific review process. They also published in fringe or predatory journals, that is, journals that have the appearance of genuine scientific journals, but that publish papers that would not succeed if submitted to a serious peer-review process.

The fact that they were not directly engaging with their peers is a strong indicator of an inauthentic controversy. One might argue that science is a critical activity and that this should not be named denialism, but rather a healthy skepticism that should be encouraged so that knowledge could move forward. This argument misses the point that science is a collective enterprise.

Ever since at least the publication of Ludwik Fleck's *Genesis and Development of a Scientific Fact* (1935) and Thomas Kuhn's *The Structure of Scientific Revolutions* (1962) it is known that science is done collectively and that there are periods when there is open debate about an idea which are succeeded by others in which a consensus is formed and new debates come about. This is an important part of scientific progress since if all theories were continually reopened for debate, scientific progress would be immobilized. No basis for the future development of science would be stabilized and the scientific community would spend a significant amount of time dealing with endless controversies. In this sense, the peer-reviewed literature is an important mechanism. Whenever the scientific community reaches a substantial consensus over a particular phenomenon or theory, dissidents will no longer be able to publish in relevant scientific journals as reviewers will recommend the rejection of their work.

Science, then, is composed of both controversy and consensus. There are times when genuine controversy over certain theories and phenomena should be taken into consideration in the science-policy interface, but there are also periods when scientific dissensus is emulated and denialists become highly trusted by publics and policymakers, which can prevent the implementation of important policies or at least delay them. This brings about significant risks to the environment and human societies as phenomena such as climate change can be highly destructive.

Final Remarks

Trust is a key concept for understanding the dynamics of the climate science-policy interface. It helps us understand why certain scientific tools, institutions, or theories become influential in informing policy making in particular contexts while others do not. In the context of policy making, there are good reasons why those in the Global South may not trust scientific assessments that overemphasize perspectives grounded in the Global North. Further, it is a useful concept to reflect on climate denialism as denialists need to attract the trust of publics and policymakers if they are to succeed in preventing or delaying the implementation of climate policies. Trust in science therefore should not be taken for granted. Trust is a sociocultural product that needs to be cultivated and maintained through social processes. Denialists understand that and deploy a range of tactics to convince publics and policymakers that they and their ideas are more trustworthy than those of mainstream science. In certain contexts and among some social groups they have been very successful. Studying trust and how it can be developed in different contexts and among different social groups is therefore a crucial task if appropriate climate policies are to be put in place to prevent the disastrous consequences that climate change can bring about.

Further Reading

Dunlap, R.E. and McCright, A.M. (2015) "Challenging climate change," in Dunlap, R.E. and Brulle, R.J. (eds.) *Climate Change and Society: Sociological Perspectives*, New York: Oxford University Press, pp. 300–332.

Lahsen, M. (2004) "Transnational locals: Brazilian experiences of the climate regime," in Jasanoff, S. and Martello, M. (eds.) *Earthly Politics, Worldly Knowledge: Local and Global in Environmental Politics*. Cambridge: MIT Press. pp. 151–172.

Lahsen, M. (2008) "Experiences of modernity in the greenhouse: A cultural analysis of a physicist 'trio' supporting the backlash against global warming," *Global Environmental Change*, 18, pp. 204–219.

McCright, A. and Dunlap, R. (2000) "Challenging global warming as a social problem: An analysis of the conservative movement's counter-claims," *Social Problems*, 47(4), pp. 499–522.

Oreskes, N. (2004) "The scientific consensus on climate change," *Science*, 306, p. 1686.

References

Corbera, E., Calvet-Mir, L., Hughes, H., and Paterson, M. (2016) "Patterns of authorship in the IPCC Working Group III report," *Nature Climate Change*, 6(1), pp. 94–99.

Fleck, L. (1935[1981]) *Genesis and Development of a Scientific Fact*. Chicago: University of Chicago Press.

Giddens, A. (1991) *The Consequences of Modernity*. Stanford: Stanford University Press.

Kuhn, T. (1962) *The Structure of Scientific Revolutions*. Chicago: University of Chicago Press.

Oreskes, N. and Conway, E. (2010) *Merchants of Doubt: How a Handful of Scientists Obscured the Truth on Issues from Tobacco Smoke to Global Warming*. New York: Bloomsbury.

Rajão, R., Nobre, A., Cunha, E., Duarte, T., Marcolino, C., Filho, B., Sparovek, G., Rodrigues, R., Valera, C., Bustamante, M., Nobre, C, and Lima, L. (2022) "The risk of fake controversies for Brazilian environmental policies," *Biological Conservation*, 266, 109447.

Part VII

Energy, Sustainability, and Sociotechnical Transitions

Introduction

Stephen Zehr

Researchers actively engaged in climate change research regularly promote the idea that major societal and technological changes are necessary to reduce the severity of and adapt to climate change. The concept **sociotechnical transition** is used to describe the encompassing nature of necessary changes. Change from fossil fuels to renewable sources in electricity production is one example. Science and technology studies (STS) research on sociotechnical transition understands that change in major technological systems is complex, involves a heterogeneous set of actors, and is far from a linear and straightforwardly rational path. Change is neither economically nor technologically determined. While any sociotechnical transition may be of interest to STS researchers, much research attention focuses on transitions away from fossil fuel technologies toward more environmentally sustainable technologies. This research holds dual goals of understanding the sociotechnical transition process and potentially directing change toward sustainable development goals that include social equity and human well-being along with environmental protection.

STS researchers employ a basic model to account for sociotechnical transitions which has the following features. New "niche" developments often face resistance to adoption from existing sociotechnical "regimes" that hold historical momentum and enjoy a supportive social, political, cultural, and economic "landscape." For example, fossil fuel-driven electricity production may enjoy regime status with an array of supportive policies, structures, and ideologies that maintain its centrality long after it can be rationally defended on environmental, social, cultural, and even economic grounds. A coal-fired power plant in the US State of Montana, for instance, may be maintained for the coal mining jobs in some communities and cozy co-relationship with state and local politicians even though the electricity produced by it is more expensive than wind or solar, without even considering the cost of environmental impacts. Many new sociotechnical niches will fail due to the difficulty of penetrating and destabilizing the existing regime and landscape. However, occasionally the landscape may destabilize the regime (e.g., climate change becomes a widespread public and political concern), opening an "innovation space" for new sociotechnologies to enter. Early on these sociotechnical niche developments often need the support of public policies and social networks of committed users to stay afloat until they become large and economically efficient enough to replace the existing regime. For example, for many years solar photovoltaic panel technology benefited from public policies and funding and social networks of committed users who installed rooftop solar systems before it became economically efficient and culturally acceptable for large electricity providers.

Ongoing STS research focuses on several features of sociotechnical transitions relevant to climate change. Research investigates the resistance that new, environmentally efficient

DOI: 10.4324/9781003409748-26

sociotechnologies face from politically conservative actors, economic vested interests, or steeped cultural practices. Researchers have addressed the speed of change within sociotechnical regimes and the political forces and social movement activity that affect it. STS research emphasizes how dimensions of social inequality are integrated into or ignored in sociotechnical transitions. How, for instance, are working- and lower-class members integrated into the transition from gasoline to electric vehicles? STS research considers consumers' standpoints, emphasizing how everyday social and cultural practices around existing sociotechnical regimes affect the adoption of new technologies that are allegedly environmentally and economically more efficient. STS research also investigates and contributes to the actual design of new sociotechnologies that aim for sustainable development principles of environmental protection, human well-being, and social equity.

The chapters included in this part represent a sample of that research. David Hess addresses aspects of governance and social movement activity that resist or push us toward clean energy transitions. Hess recognizes the challenges of generating political support, especially in polarized political climates. Importantly, Hess identifies spaces where political opportunities exist in polarized political climates. Even in politically conservative regions, environmentally efficient energy sociotechnologies may be adopted and receive social movement support if represented in the right way. This chapter generates optimism that STS research might speed up energy transitions through fine-grained analysis of specific spaces and places where supportive political action is possible, even when political conditions would seem to dictate otherwise.

Marianne Ryghaug, Tomas Moe Skjølsvold, and Robert Næss address sociotechnical transitions from the consumption side. They note that researchers and technology developers have often treated consumers/users as mere rational economic actors in sociotechnical transitions readily pushed or pulled into consumer behaviors as economic and technical conditions dictate. An STS approach, on the other hand, emphasizes the hybrid roles and agency of consumers/users in sociotechnical transitions and their potential effect on social inequality.

Daniel Breslau draws upon STS research on economic transactions to show how market behavior involving energy transitions is not a simple matter of buyers and sellers coming together to transact well-defined goods (e.g., rooftop solar systems and smart home meters). Rather, drawing upon an STS *agencement* perspective, Breslau shows how elements of markets must be *constructed* before buying and selling can occur. As Breslau emphasizes, this shift in focal point is crucial for understanding and assisting energy sociotechnical transitions that require changes in current conditions for economic transactions in energy assets and commodities. Breslau illustrates this argument with analysis of emerging markets for energy storage (increasingly required in transitions to renewable energy) and market manipulation claims surrounding subsidies for renewable energy.

Clark Miller reflexively discusses necessary adjustments in STS research for it to become more proactive in facilitating energy transitions. Closely linked to Part XI in this primer, Miller emphasizes that STS researchers shift their attention toward the future by working with engineers, scientists, and publics to develop sociotechnical imaginaries to reach sustainable development objectives. Miller maintains that STS researchers already hold the conceptual and methodological tools for this purpose, but must be more proactive in creating organizational infrastructure that can further open research avenues where diverse sociotechnical imaginaries integrate views of equally diverse publics.

20 Energy Transitions in a World of Polarized Politics

David J. Hess

Global institutions such as the United Nations Framework Convention on Climate Change have helped to develop widespread recognition of the importance of decarbonizing our societies and technological systems. However, progress has been slow. Rather than solving the problem during the 1990s when it first became widely recognized, the world has consistently failed to limit greenhouse gas emissions at a level that can reduce the likelihood of severe effects of climate change. One of the great challenges of the social sciences is to understand how to increase the momentum for **energy transitions** (changes in industry, technology, and society to reduce global emissions) while also addressing pervasive problems of injustice and weak democratic institutions. This problem is complex, and there is opposition to many proposals. However, as a research problem, the analysis of energy transitions is a challenge worthy of a generation. This chapter reviews some of the political and governance dimensions of energy transitions, with attention to how Science and Technology Studies (STS) perspectives on technology can help to elucidate both existing problems and potential solutions.

Factors That Limit or Slow Energy Transitions

Prior to thinking about opportunities for solutions, it is necessary to understand the scope of the factors that make it so difficult for the world to wean itself off carbon-intensive energy systems. The most obvious factor is opposition from both industrialists and workers in the fossil-fuel sector. They may exert substantial influence on political parties and government leaders, and there is now a growing body of research on "**climate obstruction**," or the organized resistance to energy-transition policies that would address climate change. (See the Climate Social Science Network for more information.)

Governments are central to increasing the momentum for decarbonization policy, but there are many reasons why governments in most regions of the world have not enacted policies that lead to reduced carbon dioxide emissions. Many countries, especially those with a low or middle income per capita, face a multitude of other challenges. For example, they may be undergoing rapid growth or facing the need to electrify the country. Existing hydropower resources may be challenged by unstable weather or social justice and social movement concerns. Moreover, the transition to low-carbon alternatives can require costly investments and risks of higher prices for consumers. The potential income from the export of fossil-fuel resources may be essential for governments to balance their budgets and pay for debt.

Geopolitical factors also limit the capacity of governments to support energy transitions. As we have seen with the war in Ukraine and the wars in the Middle East in previous decades, control over energy resources is an important part of geopolitical rivalry, and a country's dependence on foreign energy can create political and economic risks. Where countries have high

DOI: 10.4324/9781003409748-27

levels of fossil-fuel resources, they may be reluctant to shift toward dependence on foreign energy, including low-carbon technologies such as solar, wind, and nuclear. Furthermore, to the extent that the dollar, pound, yen, and euro remain global currencies, much is linked to their role as the currency used in international oil production and sale.

As if these factors are not enough, the rise of political extremism increases the challenge for advocates of energy-transition policies. The causes of political polarization are complex, and they involve factors that are not always directly connected to energy and climate change, such as global migration, income inequality, the declining spending power of families, and general anger at the visible effects of globalization. Right-wing conservative political leaders often articulate support for domestic fossil fuels, but in some countries, such as Mexico, progressive populist leaders have also done the same. The message of opposition to decarbonization policy can involve support for domestic workers in the fossil-fuel sector or for consumers who believe that decarbonization will increase household expenditures. In some cases, there are also local opposition movements to renewable energy infrastructure and power lines that transmit electricity from renewable sources, just as there are opposition movements to fossil-fuel extraction and infrastructure.

Finding Political Opportunities

In this chapter, I point to the need to think about finding political openings or opportunities for change even where there is opposition to energy-transition policy. Approaches from the field of Science and Technology Studies (STS) can be a helpful part of the mix of research into finding solutions. For example, one approach draws on the STS approach to **technological systems** (e.g., the electricity, water, highway, or rail systems) as complex networks of natural resources, infrastructure, hardware, software, organizations, consumers, governing rules, and cultural values. This approach to thinking about energy transitions is often described as **sociotechnical**; that is, it attempts to view technological systems from a perspective that is not limited to engineering, economics, and reliability. Part of this approach involves analyzing the systems from a design perspective that includes techniques for clarifying the choices and values that are embedded in what appear to be merely technical decisions. These values can include technical goals of efficiency and reliability, but they can also include broader societal goals and concerns such as environmental sustainability, social fairness, privacy, safety, and security. One of the fundamental insights of STS thinking is to show how the technical is also the political. For example, some types of energy-transition policy (such as support for small-scale solar) appear to cross political divisions more easily than other types (support for large-scale solar or wind farms). There are even cases in the U.S. of conservative mobilizations in support of solar energy as a conservative, property rights issue. In other words, there are possibilities for gaining broad support for small-scale solar even if the utilities prefer large-scale solar farms, which they can control more easily.

More generally, some types of policies and associated frames for defending them may also enable bridges to be built across political divisions (e.g., Hess, Mai, and Brown, 2016). For example, policies that provide mandates (government requirements for action) may trigger opposition from small government conservatives if the mandates are directed toward households or businesses. However, if the policies are mandates that require energy efficiency for government buildings and operations (and therefore they can be communicated as policies that reduce government spending), the policies may gain acceptance among both conservatives and progressives. Likewise, incentives that can be helpful to businesses, homeowners, schools, or nonprofit organizations—but that work through market mechanisms of loans or small government mechanisms of tax credits—can also become the basis of broad agreement, especially if there is support from businesses that want access to the credits. One example is policies that enable loans from government bond funds to support energy-efficiency or renewable energy investments in homes and businesses.

When attempting to identify opportunities for change in political cultures such as the U.S., where much of the country is dominated by conservatives who are generally opposed to energy-transition policies, one should recognize that conservative political parties in general, both in the U.S. and in other democratic countries, are not a monolith. In Europe, the situation is more visible because there are often different parties that represent moderate and right-wing conservatives (Hess and Renner, 2019). However, in the U.S., the Republican Party also shows evidence of **divided conservatism** (a range of divergent views across conservative parties or within a single conservative party). An example of the politics of energy transitions is the division in the Republican Party over support for the right to build rooftop solar or for solar development companies to have access to financing. There are even some, albeit relatively rare, coalitions across political divisions, especially in support of solar energy, such as the "green tea" convergence of the conservative Tea Party and progressive Sierra Club (Hess and Brown, 2016). Moreover, there are also differences across generations, with younger conservatives often more willing to support decarbonization policies than older conservatives.

It is also important to pay attention to political institutions and policy processes. In the U.S., state governments are often the primary sites for both policy innovation and retrenchment on energy transitions. Even in conservative states, it is possible to have some favorable legislative support for renewable energy and energy efficiency (Brown and Hess, 2016). Having early negotiations with all stakeholders prior to the legislative process can lead to fruitful settlements where opposing parties (e.g., the utilities and solar industry with alignments to Republicans and Democrats) have in some cases worked out compromises. Sometimes reframing energy-efficiency laws as public health laws (such as air quality in the conservative state of Utah) can lead to agreements across political divides. It is also possible to shift more contentious negotiations from the legislature, where the politics can be highly polarized, to quasi-independent government agencies such as a public utilities commission. Doing so can enable the passage of legislation without having it stall amid negotiations on details.

Connecting energy-transition policies to job development is also an important pathway to depolarized agreements. For workers in the fossil-fuel sector in carbon-intensive regions such as Appalachia, it is important to have well-developed **just transition** policies (policies that create new opportunities in the places that people call home rather than require them to relocate to cities (see Hess, McKane, and Belletto, 2021)). High-quality Internet infrastructure for small towns and rural areas in these regions can be as important as economic development programs that help to create new businesses in the decarbonizing region. Overall, it is important for political leaders to underscore the number of jobs created in the clean-energy and energy-efficiency industries. The phrase "good, green jobs," which some unions embrace, also signals the need for these jobs to enable pathways out of poverty by including training programs and opportunities for promotion. Old industrial clusters in areas facing deindustrialization may provide elements for new **clean-tech clusters**, that is, a network of businesses, government agencies, worker training programs, and services that support industrial innovation. Again, the focus on business development can provide a common ground for building support across political divisions.

Attention to how policymakers, advocates, scientists, engineers, and other actors construct problems and resolve them at different levels of spatial scale is also crucial. It is often possible to find agreements at the community level, where there are common concerns involving issues such as air quality, traffic congestion, green spaces, local food, and quality of life. In the U.S., there has been a remarkable growth in support for a transition toward 100% renewable sources, and these efforts can cross political divisions. Likewise, the movement for community control of electricity has opened opportunities for local energy-transition projects (Hess, 2019). In some states, there is support for community choice (a type of energy contract negotiated by a local government unit for electricity customers) or community solar (access to the benefits of solar energy for those who do not have rooftop solar).

Another approach that can help to improve support for decarbonization involves working directly with the private sector and households. Large businesses increasingly want decarbonization policies, and they may even push conservative state governments to change policies to improve their access to low-carbon energy. The scale of private-sector changes can be significant, sometimes larger than the greenhouse gas emissions of smaller countries. Although private governance entails risks of greenwashing and requires careful monitoring, researchers now recognize that it is simplistic or even mistaken to assume that private governance exists in a trade-off relationship with public governance (government policy change). Instead, there can be significant positive spill-over effects from private governance to public governance reforms (Tzankova, 2020). Changes in household-level behavior can also be a source of significant emissions reductions (see Ryghaug, Skjølsvold and Næss, Chapter 22, this volume). Policies that support (but do not mandate) such changes can be configured to cross political divisions and must include trade-offs based on income and resilience exposure.

At the international level, it is possible that the war in Ukraine may provide opportunities for energy transitions. Certainly, European countries will be looking for new energy sources, and other countries will be examining vulnerabilities to imported energy supplies. The outcome favors domestic energy, but that preference can include domestic fossil-fuel resources in addition to renewable energy. The effects of the emergent new world order will likely be mixed. For example, the immediate dependence on oil and gas will likely weaken the motivation to decarbonize where national security or powerful domestic industries are at risk. The government of Poland has viewed its coal industry as a source of independence with respect to imported fossil fuels from Russia. Likewise, the U.S. can increase its global influence by exporting higher levels of natural gas. However, decarbonization can also be a pathway to achieving independence from foreign fossil-fuel resources.

Finally, it is also important to question the assumption that policies and programs supporting adaptation to environmental risks and hazards from climate change (e.g., floods, storms, or drought) involve a trade-off with policies that support the mitigation of greenhouse gas emissions and other environmentally damaging effects of modern industries. Under the trade-off scenario, a government or other organization has limited resources, and it can either invest the resources in adapting to climate change or preventing climate change. Again, an STS lens encourages us to think creatively about design that can involve adaptation *and* mitigation. This thinking also involves policy work to guide rebuilding after disasters, including managed retreat from areas that have become too hazardous for rebuilding.

Conclusion

This brief discussion raises a challenge to students in their roles as voters and future researchers, advocates, activists, business owners, and policymakers for the twenty-first century. Rather than give up in despair or write utopian, normative statements about what should be done, I suggest a middle path that attends to current political and industrial opportunities and finds new ways of constructing problems and solutions that may take root in often inhospitable soil. STS provides an invaluable sociotechnical perspective that encourages us to look carefully at the design of systems and technologies, and to understand the political valences of seemingly limited technical choices.

Further Reading

Gilligan, J. M. and Vandenbergh, M. P. (2020) "A framework for assessing the impact of private climate governance," *Energy Research & Social Science*, *60*, 101400.

Hess, D. J. (2009) *Localist Movements in a Global Economy*. Cambridge, MA: MIT Press.

Hess, D. J. (2016) *Good Green Jobs in a Global Economy*. Cambridge, MA: MIT Press.

Hess, D. J. and Gentry, H. (2019) "100% renewable energy policies in U.S. cities: Strategies, recommendations, and implementation challenges," *Sustainability: Science, Practice, Policy*, 15(1), pp. 45–61.

Hess, D. J. and Lee, D. (2020) "Energy decentralization in California and New York: Value conflicts in the politics of shared solar and community choice," *Renewable and Sustainable Energy Reviews*, 121: 109716. (April). doi.org/10.1016/j.rser.2020.109716

References

Brown, K. P. and Hess, D. J. (2016) "Pathways to policy: Partisanship and bipartisanship in renewable energy policy," *Environmental Politics*, 26, pp. 971–990. 10.1080/09644016.2016.1203523

Hess, D. J. (2019) "Coalitions, framing, and the politics of energy transitions: Local democracy and community choice in California," *Energy Research and Social Science*, 50, pp. 38–50. https://doi.org/10.1016/j.erss.2018.11.013

Hess, D. J. and Brown, K. P. (2016) "Green tea: Clean-energy conservatism as a countermovement," *Environmental Sociology*, 3(1), pp. 64–75. DOI 10.1080/23251042.2016.1227417

Hess, D. J., Mai, Q D., and Brown, K. P. (2016) "Red states, green laws: Ideology and renewable energy legislation in the United States," *Energy Research and Social Science*, 11, pp. 19–28.

Hess, D. J. and Renner, M. (2019) "Conservative political parties and energy transitions in Europe: Opposition to climate mitigation policies," *Renewable and Sustainable Energy Reviews*, 104, pp. 419–428. https://doi.org/10.1016/j.rser.2019.01.019

Hess, D. J., McKane, R. G., and Belletto, K. (2021) "Advocating a just transition in Appalachia: Civil society and industrial change in a carbon-intensive region," *Energy Research & Social Science*, 75 (May), 201004. https://doi.org/10.1016/j.erss.2021.102004

Tzankova, Z. (2020) "Public policy spillovers from private energy governance: New opportunities for the political acceleration of renewable energy transitions," *Energy Research & Social Science*, 67, 101504.

21 Configuring Markets and Transactions for Energy System Transition

A Role for STS Research

Daniel Breslau

The UN's Intergovernmental Panel on Climate Change stated in 2018, "Limiting global warming to 1.5°C will require rapid, far-reaching and unprecedented changes in all aspects of society" (IPCC 2018). One important and often neglected aspect of society that must change is the economic framework for commerce in energy – the way we buy and sell energy and how we invest in new energy systems. Economic arrangements can either hinder or facilitate the transition to an energy system that will limit or reverse climate change. Science and Technology Studies (STS) scholars have a distinctive approach to the central features of economic life, markets, and transactions. This chapter describes the features of this approach and shows how it makes an important contribution to understanding the way economic arrangements must change to achieve a transition to a sustainable energy system.

STS and the Economic World

The production and use of energy entails a myriad of economic transactions. Whenever energy is transferred from producers to middlemen, and on to consumers, economic transactions take place. Moreover, whenever any part of the energy system undergoes transformation, financial transactions between developers, operators, and investors are needed to mobilize resources that go into the energy system such as wires, pipes, software, labor, land, and machines.

We sometimes have the illusion, reinforced by economic analyses, that economic transactions and the markets where they occur are the spontaneous result of buyers, sellers, and goods coming together to satisfy human wants. But, as STS scholars have pointed out, the elements of these transactions – the buyers and sellers, and the goods that they transact – are not formed in advance, only to meet in the marketplace. Instead, for markets and economic transactions to take place, the agents themselves must be defined: who are they, what kinds of calculations and comparisons do they make? Also, the goods that are the objects of those transactions must be made so they can be detached from the production process. They must be put in a form that enables evaluation, pricing, and transferral to new settings for consumption and use. Even the buyers and sellers themselves don't come into the world ready to make market calculations. They are configured as calculating actors with the aid of technologies and information for measuring and comparing. STS scholar Michel Callon calls this process **agencement**, a French term loosely translated as "configuration" to refer to how elements of economic transactions are organized. These elements include economic ideas, measurements, technologies, information systems, among others. The *agencement* perspective is different from a traditional economics perspective. The latter considers buyers and sellers with preferences and utility functions meeting to buy and sell goods that are already well-defined. The *agencement* perspective studies the way all the elements of market transactions are shaped so that transactions can be made. While

DOI: 10.4324/9781003409748-28

economists ask how the price of a good, electricity for example is the result of supply and demand in a market, an *agencement* approach studies the way prices are shaped by technologies for metering electricity or the software buyers and sellers use to make their calculations.

There are three crucial features of *agencement* that make changes in economic markets and transactions an important and interesting focus of STS research. The first is the role of economic knowledge in the formation of markets. We often treat the discipline of economics as a way of studying economic life, distinct from STS, with its own strengths and limitations. But in studying market *agencement* it is more fruitful to see economic knowledge as an important *part* of the economic practices and institutions we are studying. Economic knowledge, including the theories of economists and techniques for gathering data and measuring economic phenomena, does not simply describe markets. It also is deeply involved in making and organizing markets. Economic analyses, for instance, constantly compare actual markets to the idealized competitive market of economic theory and recommend ways to induce participants in the market to behave as predicted by the theory. The economic analyses put forth market rules that will channel participants' actions into desired outcomes. Callon and others have referred to this role of economics as **performativity** – the markets are performances of economic theory. Economists are important actors in the *agencement* process.

Second, markets are not simply sets of rules and norms. They also depend on technological devices of many kinds. Many of these are information technologies. Think of the ticker-tape machines in the early 20th century that allowed people who were not physically present on the stock market floor to nonetheless track prices and participate in trading. Similarly, metering systems for electricity made it possible to measure consumption and develop pricing techniques. The way that electricity was priced depended on the features of the metering technology. However, information technologies are not the only technologies necessary for enabling market exchanges. Methods for mechanically cooling and dehydrating agricultural products such as tobacco standardized their water content, making it possible to compare lots of the product by weight and adopt a shared system of pricing. In another example, there is no global market for oil without a global system of tanker ships. With the development of liquified natural gas (LNG) that can be shipped around the world, we are beginning to see a global market for natural gas that was impossible when this commodity could travel only in point-to-point pipelines.

Third, these features of markets are political. By that we mean that the sociotechnical arrangements allowing economic transactions to occur are subject to struggle among actors with conflicting interests. Buyers and sellers in markets do not simply compete to find the lowest price or to outsell their competitors. They also compete over the way the market itself is set up including the rules of exchange or the way prices are set. It is as though players in a game are not only competing to win the game, but are competing over the definition of the rules, the arrangement of the playing field, the required equipment, how points will be allocated, even the question of what game is being played! Buyers and sellers are naturally in conflict, with the former seeking the lowest price and the latter the highest. But even within these groups there are important conflicts. For instance, sellers who own different kinds of production technologies may compete to define market rules that give them an advantage. Large producers may have different interests than their smaller competitors. Established incumbent producers in energy markets are likely to have interests that conflict with new entrants to those markets.

STS Research on Markets and Climate

How does all of this relate to climate? Two examples from my research show how STS research contributes to understanding the economic dimensions of energy transition through the investigation of these three elements of *agencement.*

Example 1: Configuring Markets for Electricity Storage

The transition to renewable energy leaves behind the great advantage of fossil fuels which led to their rise as the engine of industrial society. Coal, and then petroleum and gas, could be extracted, transported, stockpiled, to be used at any time or place where energy was needed. Andreas Malm has shown how, even when production processes using water power were no more expensive and no less efficient, industrialists adopted coal because it freed them from the constraints of time and location. Renewable energy returns us to sources, principally solar and wind power, that provide energy only at certain locations and times. These sources are out of sync and out of geographic alignment with our needs in industry, transportation, and household use. We produce more than we need during some hours and less than we need during others, and generally not in the locations where we need it. Thus, the adoption of renewable energy systems must also incorporate, in a massive way, technologies for matching the time and location of energy capture with the time and location of its use.

One answer is storage. Storage is our primary way of moving electricity in time. Like any kind of storage, electricity storage is a way of bridging different time periods by charging batteries when electricity is abundant and cheap and discharging them when it is scarce and expensive. Though we have several kinds of storage, the most promising one for rapid growth is chemical batteries, the large-scale version of the Lithium-ion batteries we are familiar with. Battery storage has grown rapidly in some national energy systems, but in no case is it being adopted at the rate needed to meet national emission-reduction goals. The fact that the market *agencement* around storage – the definition of what exactly is being invested in, bought and sold, how it is able to participate in markets, and how its value is calculated – has yet to be worked out is an impediment to the adoption of this necessary set of renewable energy technologies.

Means for calculating its value are necessary for storage to be smoothly integrated into the power system. Its value depends on the revenues it captures in markets. Investors must be able to calculate the expected return on investment and compare it to other possible destinations for their financial resources. These revenues depend on a very complex set of processes. How does the owner of a large battery, like the "utility scale" systems that can store hundreds of megawatt-hours of energy, decide when to charge and when to discharge the battery? It makes sense to charge when power is cheap and discharge when it is expensive. But to use big batteries efficiently they need to be discharged within four to six hours of being fully charged. It might not be worth it to wait longer for higher prices because too much energy would be lost by just waiting. At the same time, responding to every up and down in the price will cause the battery to degrade much faster. Algorithms for deciding when to charge and when to discharge are therefore an essential feature of storage. But they are also financial technologies, since they create the value of storage as something that can be calculated and predicted.

Another *agencement* problem around battery storage has to do with the multiple purposes it serves and the problem of turning each of those into goods, in this case services, that can be priced, bought, and sold. Batteries provide "ancillary services" to the grid, notably frequency regulation, which refers to the instantaneous provision of small adjustments in the flow of power to keep the power system operating at the right frequency. At other times batteries provide "load services," by taking excess energy off the grid when demand is not high enough and later providing it back to customers in the distribution system. The key question, however, is how do battery systems decide when to provide which service? Promoters of battery storage have advocated for something called **value stacking**, providing multiple services simultaneously, each with its own revenue stream.

Some participants in electricity markets are concerned that owners of batteries will be able to "double bill," selling retail electricity to customers while also billing for wholesale energy provided to the grid. This presents a technological problem in the design of metering and control systems, and a regulatory problem of creating rules for how battery owners are allowed to charge for their services. As one regulator commented in a recent regulatory debate, "it is questionable whether it is possible, either technically or as a matter of policy, for a single electric meter to sort between electrons, with some being used for wholesale purposes and some being used for retail purposes."

Developers of battery storage have argued that the market rules create barriers to their participation, affecting their ability to earn revenues and attract investors. But owners of other kinds of resources, especially large power plants that use fossil fuels or nuclear energy, oppose making provisions for the special characteristics of batteries. They have argued for preserving the principle of "technological neutrality," by which the market does not favor any particular technology. Appealing to economic theory, they hold that neutrality allows competition to select the best technology to provide a particular product.

This is hardly a complete discussion of all the elements of an *agencement* around battery storage. But it illustrates the ways that the economic arrangements that make transactions possible play a crucial role in the transition of the energy system. How is storage to be defined, measured, and valued as an object of investment? Who are the buyers of the services provided by storage and how are they charged? The way these questions are answered will have important consequences for how much storage of what kind gets built and where. The answers to these questions involve economic ideas, technological devices (e.g., metering systems), and political conflict. Located right at the intersection of these elements, STS studies of *agencement* help to identify the ways that market arrangements facilitate or block the needed energy transition.

Example 2: Are Subsidies for Renewable Energy a Case of Market Manipulation?

The US Federal Energy Regulatory Commission (FERC) is tasked with assuring that markets are fair in the US, so that prices are set according to law as "just and reasonable." Some states in the Eastern US started offering various kinds of subsidies to electricity producers for reasons of state policy. In some cases these subsidies were meant to keep large nuclear power plants in operation. Many states also have adopted subsidies to promote renewable energy development to make it more profitable. Many owners of fossil-fuel-burning generators objected to the subsidies arguing that renewable energy generators would have an unfair market advantage and suppress prices for everyone. They claimed the subsidies were a kind of market manipulation.

For many years, the FERC sided with fossil fuel power plants. The FERC agreed that subsidized generators would distort market prices by offering electricity into the market below its cost. The subsidy guaranteed them a profit. So the FERC required that energy markets have a rule and a set of economic "screens" to detect when a subsidized plant was taking advantage of its subsidy and offering electricity into the market below its cost. When such an exercise of market manipulation was detected, the offender's offer price was replaced by a higher offer representing what it would be predicted to offer if it was not receiving a subsidy. Here the offer predicted by economic theory replaces the "distorted" initial offer.

In many cases, this rule meant that subsidized plants would not be able to compete because their substituted price was too high. Electricity purchasing utilities would purchase lower-cost fossil-fuel-generated electricity. Subsidized plants would not gain revenues in the market. These generators and the states that subsidized them protested vigorously, arguing that the rule would cancel out the effect of state policies to promote renewables or other generators such as nuclear plants. They argued that subsidized generators should have a market advantage – that was the purpose of the subsidy!

From around 2006, when these rules were first adopted, until 2021, the politics of these markets slowly changed. With the growth of state subsidy policies, more and more companies, states, and politicians were invested in them and wanted them to work. At the same time, members of the FERC started to either change their minds or were replaced with new commissioners more sympathetic to state renewable energy policies. By 2021, the FERC had a majority of members in favor of accommodating state subsidies, and the rules were changed.

Did federal regulators decide that market manipulation through subsidies was acceptable? Well, no. The new approach to state subsidies evolved from a new framework for thinking about competition and the role of the market, where subsidies are now thought of as shaping the actual costs of resources offered into the market. They were no longer considered a market distortion. In this case, the economic theory *performed* by the market was revised. While regulators once thought of the market as the sole determinant of the price of electricity, it is now considered a way of setting prices *only after* all subsidies are taken into account.

It should be noted that even if state subsidies were not allowed to affect electricity markets, the markets would be no more "free" or "natural." Fossil fuel generation creates enormous social costs through the effects of greenhouse gas emissions on the climate. Yet those costs are not reflected in the price of electricity. In effect, those generators are receiving a large subsidy since it is the broader society, not them, who pays for those environmental and social effects. Their costs are *socialized* in the sense of being passed on to publics both nationally and globally. Market rules that do not add those costs to electricity generated by fossil fuels are therefore part of a market *agencement* that provides a competitive advantage to these polluters.

Conclusion

Markets are not what just automatically happens when we let producers and consumers or sellers and buyers loose to "truck and barter" (in Adam Smith's words) as they please. Rather, they are complex configurations of rules, technologies, information, and political interests among other elements without which market transactions do not happen. But the particular configuration of those elements, the market *agencement*, has important consequences for the way that needs are shaped and finance channeled. They can work to hinder an energy transition to renewable forms or facilitate it.

The social and technological organization of the buying and selling of energy, and of investment in the energy system, will be decisive in our ability to transition to a system that no longer churns out greenhouse gas emissions. To avoid the worst impacts of climate change, we need to be attentive to the changes in economic arrangements that will be needed to facilitate a rapid and just transition. By studying these arrangements as a market *agencement*, the configuration of technologies, knowledge, regulations, and market rules that structure transactions, STS provides powerful tools for examining this connection of energy transition to economic life.

While the configuration of the economic world around energy systems is crucial for a just transition to a sustainable system, the ability of people with different resources and degrees of power to influence the politics of markets is very uneven. Powerful players, such as utilities, generation owners, and even regulators, will try to shape those economic arrangements in ways that meet their own needs. Rather than conceptualizing markets and prices as independently shaped by buyers and sellers, STS research can inform public discussion of them through its focus on all the elements of market *agencement*. This research might enable people who rely on energy markets to have a voice in that transition.

Further Reading

Silvast, A. (2017) "Energy, economics, and performativity: Reviewing theoretical advances in social studies of markets and energy," *Energy Research & Social Science*, 34, pp. 4–12.

References

Intergovernmental Panel on Climate Change (2018) "Summary for Policymakers of IPCC Special Report on Global Warming of 1.5°C approved by governments." www.ipcc.ch/2018/10/08/summary-for-policymakers-of-ipcc-special-report-on-global-warming-of-1-5c-approved-by-governments/

Malm, A. (2016) *Fossil Capital: The Rise of Steam Power and the Roots of Global Warming*. London: Verso Books.

22 The Role of Users in the Energy Transition

Marianne Ryghaug, Tomas Moe Skjølsvold, and Robert Næss

Introduction

In this chapter, we show how concepts and insights from Science and Technology Studies (STS) challenge understandings of users in energy transitions. We zoom in on the demand-side of sustainability transitions to emphasize its importance, but not just as the site where new technologies are used or adopted. We also look at the way STS has contributed to a conceptual and practical broadening of the role that people play in transitions. Rather than being interpreted as relatively passive adopters of new technologies, STS emphasizes that people or technology users play important roles in innovation processes, in making sense of and giving meaning to the material world, and through these processes actively participating in and shaping future societies and sustainability transitions.

To reduce climate change and avoid further environmental deterioration, there is a strong push from policy makers and innovators toward a system-wide transition away from energy and transport systems based on fossil fuels. Researchers who seek to understand such transitions emphasize that they entail shifting systems consisting of technologies (e.g., by phasing in renewable energy production at the expense of fossil fuel power) and social elements (e.g., by implementing new ways of organizing, new actor roles, and institutions). Thus, such transitions are what researchers call **sociotechnical**.

Based on such insights, scholars have focused on the relationship between stability and change, with a particular focus on the drivers and barriers that enable or prevent the wide uptake of technologies such as photovoltaics, wind turbines, or electric vehicles, and the establishment of new sociotechnical systems around such technologies as smart grids and electric infrastructures. In such research there has been a bias in favor of technology-centered narratives about innovation journeys as well as framework conditions for the successful diffusion of such technologies, for example, through probing policies and policy mixes. Such narratives have tended to downplay the importance of what we can broadly call the demand- or consumption-side of transitions. The **demand-side** of energy transitions is often thought of as "end users" or just "users," but can also be denoted as consumers, publics, or just "people." Often, people are assumed to be rational economic actors, whose main role is to passively accept or reject new technologies without question. Through this, they either contribute to or impede diffusion of new energy technologies and solutions.

Through broadening the perspectives on users, consumers, or people, STS also sheds new light on innovation, policy-making, and justice implications of transitions. This is because how policy makers and innovators perceive or envision users is important for how they design technologies and policies – and in turn how technologies are used or perceived. This suggests that at the intersection of technology design and use, there are possibilities for the emergence of

DOI: 10.4324/9781003409748-29

new injustices. For instance, access to "green" technologies tends to be unevenly distributed in society, but sometimes biases and injustices are also built into technologies themselves, such as when "smart" energy technologies require a set of skills, personal interest, and the ownership of a large home with high energy consumption levels to be used as intended. Through a focus on users, or the consumption-side, STS can also provide insights to innovators and policy makers.

STS Approach

STS researchers have typically stressed the active contribution of users in energy transitions and the different or "hybrid" roles people can have in them (Oudshoorn and Pinch, 2003). This contrasts with other disciplinary perspectives, which assume users have limited roles such as being primarily market actors or consumers. Understanding people as consumers translates into seeing their designated role located at the end of the innovation line where activity is confined to buying and adopting products. As examples, psychology focuses on people's motivations for adopting certain products such as photovoltaic (PV) panels or electric cars, while economists are interested in finding the right economic incentives or price mechanisms to foster certain environmentally benign behaviors. In contrast, STS research is broader. STS probes the ways that a technology becomes embedded in society through cultural, political, institutional, and economical means and how the process of embedding unfolds (Sovacool et al., 2000; Ryghaug and Skjølsvold, 2021). Hence, the interest is broadly on the processes that make technology part of society.

Technology diffusion is not a natural or seamless process that simply happens once a technology becomes accepted or competitive. Instead, consumption and adoption entail substantial work on the consumer or user side and recognizes the active role of both people and things in such processes (Hyysalo, 2021). As an example, in places with success in fostering a large share of electric vehicles, their adoption was not only affected by the price of the car, as often claimed in the news media. Rather, it has typically resulted from a combination of policy measures (such as environmental policies, economic incentives, infrastructure development) and gradual changes in the routines and practices of drivers (Ryghaug and Skjølsvold, 2023). Drivers having positive user experiences and talking positively about EVs, gradual emergence of user-oriented interest organizations that promote EVs, and so forth are important for their diffusion.

It is also important to stress the **hybrid roles** people have in technology adoption. Users are not one homogenous group that adopts new technology in the same way. Different types of users *adapt* the technology for different reasons and for different types of uses. Using the example of EVs again: people do not passively adopt an electric vehicle but rather adapt it to their own lifestyle, everyday life, practice, and context. As described below, several concepts from STS have highlighted the active and hybrid agency of users in sociotechnical transitions.

Findings: The Active User

People often have different interpretations, understandings, and knowledge of technologies. They are subject to what STS scholars call **interpretive flexibility**, meaning that new technologies are ascribed different meanings and associated with different needs by different social groups. A classic example within STS research describes how different relevant social groups of users (such as female cyclists and sport cyclists) contributed to the construction of "what" a bicycle eventually came to be. The bicycle was developed in the late 19th century to respond to problems defined by different groups of users (Pinch and Bijker, 1989). In other words, users played an active role in the actual design of the bicycle. Prospective technology users often make context-dependent assessments when adopting new technologies, asking how the technology would fit into everyday life and how it would potentially conflict with values, beliefs, and daily routines, or just simply: "*how may the technology be adapted to fit my needs*"?

Building on such a perspective, prospective users, or people, might also interpret technologies in ways that make them oppose and even protest and impede developments. For example, some people were opposed to the early use of bicycles because they interpreted them as dangerous and felt that they should be outlawed (Pinch and Bijker, 1989). Taking users and their thoughts, actions, and experiences as a starting point when thinking about energy transitions is therefore key to understanding how political measures might work, and how the diffusion of climate-friendly technologies might unfold. A theoretical approach that captures the embedding of new knowledges, policies, and technologies in everyday life is **domestication theory**.

Domestication theory emphasizes three important dimensions: the practical, the cognitive, and the symbolic. Each should be considered when using this theory to understand how users embed technologies in their everyday lives. The cognitive dimension describes knowledge acquisition and the expertise required to use technology. The practical dimension concerns how to use the technology in everyday life and adapting it to existing practices. The symbolic dimension is about identity creation, norm formation, and value and meaning expression associated with the use of technology.

Domestication theory contrasts with perspectives that stress how new technologies are adopted when their price is considered competitive, or their utility is considered superior to existing technologies. The theory highlights that adopting new technologies is a more complex process. Domestication theory stresses the mental, practical, and cultural work that takes place when users are trying to fit and adapt the technology to everyday life and habits. Focusing on domestication helps us describe the process of embedding technologies in everyday life, and the ways technology and society are "mutually shaped." By this, we mean that technologies don't just shape society as they are adopted. Rather through the context of users' adaptation of technology the latter itself is being shaped.

Let us take an example: In many cities around the world, authorities want to reduce car usage and encourage people to walk, bike, or use public transportation. In this context, electric mobility technologies have been highlighted as means to reduce the use of fossil-fueled vehicles. Electric bicycles (e-bikes) are one such technology. Initially, e-bikes were interpreted as assistive technology for people with disabilities, as the battery makes them much easier to cycle compared to a mechanical bike. Thus, e-bikes were first used by people who saw them as an effective means of transportation for disadvantaged groups in society. However, as an increasing number of user groups started using e-bikes, different e-bike models were developed to cater for different user needs. Today, e-bikes comprise a whole range of different bikes used for different purposes such as for exercise and mountain-biking, transporting kids, or even for commercial purposes such as transporting goods.

In an e-bike trial study where two electric bicycles were rented out for free to investigate the domestication process, we found that e-bikes were domesticated *cognitively* as most users had little knowledge about electric bicycles, *practically* as users gained experience and were able to incorporate them into their daily lives during the trial period, and s*ymbolically* as users increasingly self-identified with the technology and liked the way riding an e-bike conveyed meaning (e.g. being environmentally oriented or independent of cars and other transport modes) to others. It was fascinating to observe how people domesticate different technologies in various ways to meet their diverse needs. Some participants had such positive experiences with using the bike that they wanted to purchase an e-bike after the trial period.

Through studying use, we could also observe how the trial contributed to re-domesticating the bike. Many of the participants had at some point stopped cycling, but while using the e-bike they re-discovered the advantages of cycling (and even more so when comparing it to the conventional bike). They re-discovered cycling as a fun activity as in their childhood days and the good feeling and convenience of easily and effectively getting around. The e-bike made them explore the city and their surroundings more and gave them more time to visit friends. A similar study of electric vehicle (EV) owners showed how EV driving altered user habits by making transportation needs more salient and raised both technological and energy consumption awareness of Norwegian users (Ryghaug and Toftaker, 2014).

Two concepts have been coined to refer to the active agency of and diverse ways that consumers are mobilized to participate in energy transitions. **Material participation** (Marres, 2016) stresses that objects may be used to create relationships to more abstract issues such as climate change. Noortje Marres used the example of an "augmented teapot" to illustrate the relationships formed between the issue of climate change and energy use through this object. The teapot exhibited a particular color when it was best to put on the kettle from an environmental point of view. In this way she highlighted how mundane objects and everyday actions may connect to larger political issues such as the climate change problem.

Building on this perspective, Ryghaug, Skjølsvold, and Heidenreich (2019) developed the concept of **energy citizenship** to demonstrate how acquiring, using, and embedding new energy-related technologies in everyday life may shape new practices and ways of thinking about climate, environment, and energy. For instance, their research showed that introducing and using emergent energy technologies such as EVs, smart meters, or installing PV panels not only directly altered everyday practices, but also sensitized people to the overall issue and prompted them to become better energy citizens. For instance, through using EVs people became more aware of energy consumption and more environmentally conscious and interested in new ways of using and producing energy. Similarly, the acquisition of PV panels prompted many users to become more interested in acquiring an EV because they wanted a way to use their own energy.

However, one challenging aspect of seeing objects as gateways to more participation and engagement is that not everybody has equal access. We live in an unequal world where the purchase of an EV, for example, is easily affordable for some and inaccessible to others. Thus, one needs to think about the injustices in user involvement in energy transitions.

Users can also be involved more actively in demand-side activities by taking part in what is typically called **grassroots innovation** activities. Grassroots innovation involves activities that usually are carried out "by networks of activists and organisations generating novel bottom-up solutions for sustainable development; solutions that respond to the local situation and the interests and values of the communities involved" (Seyfang and Smith, 2007, p. 584). By actively adapting and tinkering with low-carbon technologies (small-scale renewables, solar PV, heat-pumps, and renewable heating systems), people enabled the development and deployment of such technologies and services necessary for an energy transition to unfold (Hyysalo, 2021). Lately people have not only contributed to such "bottom-up" initiatives individually, but also through "energy communities" that focus on collective energy generation. Collectives of citizens organize to produce energy to benefit their community. As they do so they also become important drivers of a larger energy transition due to their energy production participatory engagement.

Findings: The Socially Constructed User

So far, this chapter has primarily focused on processes associated with technology use or demand in the classical sense. However, an STS perspective also studies users more broadly in the indirect ways they feed into innovation processes. One example is how innovators envision or imagine users. STS studies have often shown that technology producers have an envisioned user in mind when designing new technologies. There are many examples of this within new energy and transition-oriented technology. One is smart energy technologies made to help people reduce their consumption of energy. STS research has shown that developers of these technologies are strongly biased toward thinking that most people are technology enthusiasts, motivated by economic gains, and strongly interested in optimizing resources and energy consumption. As a result, STS has shown that these technologies are poorly matched to the messy everyday lives of many potential users. There also may be strong gender or age biases built into these technologies. Therefore, smart energy technologies may have little impact on energy consumption. A smart thermostat, for example, may be ignored by residents or simply not used due to lack of knowledge and interest in synching it to one's phone.

This line of reasoning has been particularly prominent in the past few years around flexible energy demand (see Breslau, chapter 21, this volume). Because of the implementation of new and variable renewable energy production technologies such as wind and solar power, there is not always a good match between energy production and energy demand. Trends such as transport electrification increase this challenge and lead to new peaks in electricity consumption. Therefore, there is a need for shifting electricity consumption to other times of the day or week, or for "flexible demand." This shift positions households as important sites for managing electricity consumption patterns. Policy makers and innovators have recognized this, and directed much work toward making technologies and incentives to steer household behavior. The digitalization of household electricity management through smart electricity meters, home management systems, and smartphone apps has to a large degree enabled new relations to be constructed between producers and consumers of electricity (Adams et al., 2021). Again, STS researchers have been central in pointing out the sociotechnical work that goes into delivering flexibility (e.g., Fjellså, Ryghaug, and Skjølsvold, 2021), as well as the potential social injustices for those less able to be flexible (because they cannot produce their own energy or have batteries for storage or appliances with timers) or those who may feel forced into being flexible and may consequently risk their health, comfort, and economic well-being (Fell, 2020).

Limitations

Research on the role of users or the consumption-side of sociotechnical energy and climate transitions has focused primarily on the uptake of new technologies or practices. There has not been an equal focus on technologies that we should stop or reduce usage. For example, non-sustainable practices such as car-based mobility or flying need to be radically reduced or eliminated. Such unsustainable practices are often locked into current sociotechnical systems around transport or energy use, and thus difficult to destabilize and break free from. (The other chapters in this part address some of these issues.)

Conclusion

This chapter illustrates how an STS perspective on the role of users, people, or the consumption or demand-side in energy and climate transitions significantly expands on the relevance of this group. By this, we mean that STS provides new ways to think about this group practically and politically and that it might open new avenues for research. Our discussion has significant policy implications. First, policymakers must address people more broadly than as mere rational consumers. While economic policy instruments targeting consumers might work under certain circumstances, they might also risk alienating people with other political or practical orientations. Users of technology also contribute to processes of innovation. Their role in technology diffusion goes beyond accepting or rejecting new technologies. Recognizing users' agency in technological innovation opens the door to new policy avenues. A key question then becomes: How can one cultivate a more actively engaged citizenry through new policies and through technological development?

Further, it is important that policy makers actively probe the unintended and often noxious effects of demand-side policies. Such effects might include the strengthening of existing lock-ins or the creation of new lock-ins. An example can be found in the results of Norwegian EV policies. These policies triggered a massive uptake of electric vehicles, which has helped in decarbonizing the transport sector. But they also have entrenched the dominance of privately owned automobiles, subsidized the consumerism of wealthy people, and made more radical alternatives such as implementing shared mobility solutions seem less attractive.

Above all, our chapter illustrates that through the eyes of STS scholars, energy and climate transitions can come across as messy and heterogenous. Beyond what we have discussed so far, we want to end by noting that the household demand-side in energy transitions should be understood as part of a broader social and political context. In this context there are a range of actors that actively attempt to shape and influence "what" the households are and how they are supposed to act in a transition. This can be observed in pilot and demonstration projects that test new energy technologies, such as smart energy technologies intended to promote flexible demand. In such projects, the role of households becomes carefully orchestrated and curated through the actions of others. In our research (Skjølsvold et al., 2018) we have explored how this orchestration unfolds amongst policy makers and regulators, researchers, and innovators. These actors are engaged in producing visions, creating networks, and scripting technologies in ways that are all intended to affect the actions unfolding at the demand-side of a transition.

It is important to note how these visions on the demand-side are highly technology-centric. Most orchestration is geared toward the adoption and use of technologies such as smart homes, solar panels, or electric vehicles. Moving forward, however, it is essential to shift this focus. Given the dire messages of the Intergovernmental Panel on Climate Change (IPCC) with respect to climate change, we believe that people and societies should be orchestrated in ways where energy consumption is reduced, mobility demand is lowered or changed, and high-tech solutions are considered less desirable. The challenge is not only one of achieving the domestication of new technologies, but also de-domesticating or phasing out old ones, while reducing the demand for new ones. We believe STS has a strong role to play in advancing an understanding of how this could unfold.

Further Reading

Hartmann, M. (ed.) (2023) *The Routledge Handbook of Media and Technology Domestication.* New York: Routledge.

Hyysalo, S. (2021) *Citizen Activities in Energy Transition: User Innovation, New Communities, and the Shaping of a Sustainable Future* (p. 190). New York: Taylor & Francis.

Ryghaug, M. and Skjølsvold, T. M. (2021) *Pilot Society and the Energy Transition: The Co-shaping of Innovation, Participation and Politics* (p. 130). Cham: Springer Nature.

Suboticki, I., Heidenreich, S., Ryghaug, M., and Skjølsvold, T. M. (2023) "Fostering justice through engagement: A literature review of public engagement in energy transitions," *Energy Research & Social Science*, 99, 103053.

References

Adams, S., Kuch, D., Diamond, L., Fröhlich, P., Henriksen, I. M., Katzeff, C.,... and Yilmaz, S. (2021) "Social license to automate: A critical review of emerging approaches to electricity demand management," *Energy Research & Social Science*, 80, 102210.

Fell, M. J. (2020) "Just flexibility?," *Nature Energy*, 5(1), pp. 6–7.

Fjellså, I. F., Ryghaug, M., and Skjølsvold, T. M. (2021) "Flexibility poverty: 'Locked-in' flexibility practices and electricity use among students," *Energy Sources, Part B: Economics, Planning, and Policy*, 16(11–12), pp. 1076–1093.

Hyysalo, S. (2021) *Citizen Activities in Energy Transition: User Innovation, New Communities, and the Shaping of a Sustainable Future* (p. 190). New York: Taylor & Francis.

Marres, N. (2016) *Material Participation: Technology, The Environment and Everyday Publics*. New York: Springer.

Pinch, T. and Bijker, W. (1989) "The social construction of facts and artifacts: Or how the sociology of Science and Technology and the sociology of technology might benefit each other." In Bijker, W., Hughes, T., and Pinch, T. (eds) *The Social Construction of Technological Systems*. Cambridge, MA: MIT Press.

Oudshoorn, N. and Pinch, T. (2003) *How Users Matter: The Co-construction of Users and Technology (Inside Technology)*. Cambridge, MA: MIT Press.

Ryghaug, M. and Skjølsvold, T. (2023, forthcoming) "Electric vehicle diffusion and sustainability transitions: Learning about policies, actors and wider effects from the Norwegian case," *Proceedings of the National Academy of Sciences*.

Ryghaug, M. and Skjølsvold, T. M. (2021) *Pilot Society and the Energy Transition: The Co-shaping of Innovation, Participation and Politics* (p. 130). Cham: Springer Nature.

Ryghaug, M., Skjølsvold, T. M., and Heidenreich, S. (2018) "Creating energy citizenship through material participation," *Social Studies of Science*, 48(2), pp. 283–303.

Ryghaug, M. and Toftaker, M. (2014) "A transformative practice? Meaning, competence, and material aspects of driving electric cars in Norway," *Nature and Culture*, 9(2), pp. 146–163.

Seyfang, G. and Smith, A. (2007) "Grassroots innovations for sustainable development: Towards a new research and policy agenda," *Environmental Politics*, 16(4), pp. 584–603.

Sovacool, B. K., Hess, D. J., Amir, S., Geels, F. W., Hirsh, R., Medina, L. R.,... and Yearley, S. (2020) "Sociotechnical agendas: Reviewing future directions for energy and climate research," *Energy Research & Social Science*, 70, 101617.

Skjølsvold, T. M., Throndsen, W., Ryghaug, M., Fjellså, I. F., and Koksvik, G. H. (2018) "Orchestrating households as collectives of participation in the distributed energy transition: New empirical and conceptual insights," *Energy Research & Social Science*, 46, pp. 252–261.

23 STS and the Design of Futures

Clark A. Miller

Introduction

We live in an era permeated by the idea that by creating new technologies we can build better futures. It's a powerful idea. Each year, worldwide, it justifies extensive government and private-sector investments in the development and deployment of technology. This idea leads hundreds of thousands of college students to train as engineers. And, it contributes to the valorization of technologists, tech companies, and places like Silicon Valley in the media and popular culture.

Yet, we also live in an era in which the consequences of poorly designed technological societies stare us in the face every day. Traffic and air pollution clog our cities. Racial and ethnic discrimination is pervasive in data, information, and artificial intelligence systems. Social media undermines democracy, public trust, and community cohesion. To make matters worse, carbon-intensive energy systems are driving dangerous changes to the Earth's climate and planetary health in ways that exacerbate global inequality and insecurity.

This chapter argues that humanity needs to go beyond the idea that just inventing new technologies will make the future better. We need to get much smarter about how we *use* technologies, new and old, in fashioning where, how, and to what ends people live, work, play, and forge relationships with each other and with wider communities. It's one thing to be good at making technology. *It's a very different thing to be good at designing, building, operating, maintaining, repairing, and inhabiting technological societies.*

The field of Science and Technology Studies (STS) is uniquely positioned to help bridge this knowledge gap. At the heart of technological societies lie **sociotechnological systems**: hybrid systems that intertwine and integrate technologies with human ideas, values, behavior, relationships, and institutions. Leveraging the field's rich and heterogeneous insights into sociotechnological systems can help create new ideas, approaches, and tools that can help humanity escape the technological fix that it has created for itself and design and create better human futures.

Constructing the Future

The basic orientation of STS is **constructivist**. Broadly, constructivism entails two commitments. First, *STS considers the world around us to be the outcome, the product, or the construction of human work and decisions.* Humans are habitual makers. We make all kinds of things. We make our built environments: houses, office buildings, roads. We make systems that provide us with technological services: water systems, food systems, electricity grids, oil pipelines. We also make more abstract products of human ingenuity, creativity, and imagination: knowledge, ideas, paradigms, artworks, cultures. Arguably, we even make nature, although this may be more controversial to some readers. We certainly sculpt the world around us through things

DOI: 10.4324/9781003409748-30

like the pollutants we emit, the forests we cut down, the fields we plant, the areas we protect (or don't protect) as parks, and the species we exterminate, as well as the conceptual and mathematical models we develop to describe nature and the classificatory systems we invent to make sense of it, both of which inform the choices and policies we make. Finally, we make ourselves: our bodies, minds, identities, institutions, communities, economies, and societies. All are products of individual and collective choices and actions, taken together, whether coordinated or in aggregate, to construct what it means to be human.

The second aspect of constructivist thinking follows from the first: *the world need not be the way that it is*. We could have made it otherwise in the past, and we could make it otherwise in the future. This property is referred to as **interpretive flexibility**, which means that technologies can mean many things, be designed in many ways, and be used for many purposes. There are, for example, many ways to make a bicycle – and many ways and reasons to ride one. Some are for racing. Some for streets. Some for mountains. Some are designed for men. Others for women. Still others, with training wheels, for toddlers. Any particular bicycle, therefore, isn't just a bicycle. It's an integral component of a more encompassing set of arrangements that link together social, technical, and even natural elements, tethering particular technological designs to particular human purposes, values, practices, and know-how, not to mention business models, forms of work and labor, markets, and political economies – even gender and cultural identities. STS sometimes refers to these as assemblages, defined as constellations of diverse kinds of things, some human, some natural, some technological, that are assembled together to create a working whole.

One of the powerful capabilities of STS is excavating and unraveling the complex entanglements of social and technical elements that make up existing sociotechnical systems or assemblages. The purpose of such analyses is often to reveal and critique the consequences of how such systems have been designed and built. For example, considerable STS research has critically assessed the human dimensions of both existing carbon-based energy systems, which have proven extremely dangerous and damaging to both people and the planet, as well as the proposed low-carbon energy systems intended to replace them. *Building on such work, STS now can build a forward-looking project to assess the design of alternative potential future sociotechnical arrangements and the ways that those will entangle future values and relationships, ways of living and working, risks and benefits, and other facets of future societies built on future energy systems that people might care about.* Such work would provide valuable insights into classic STS questions such as "for whom?" and "in what ways?" concepts such as "better" and "more just" are defined, measured, and assessed – thus helping to inform the design of new energy systems that serve humanity better and help advance energy transitions that are not only carbon-neutral but also deliver more just and equitable futures.

Traditional approaches to technology development have failed to engage in the kind of forward-looking, human-centered design work that is informed by STS. This strikes me as a profound mistake. Such forward-looking work could be carried out in collaboration with or counterpoint to those involved in energy or other technology transitions. Over the past two centuries, humanity has remade virtually every aspect of human existence, from our behaviors to our organizations to our politics and economies, around key technological systems: automobile-based transportation, electrification of households and businesses, heat-driven industrial processes, digital computing and communication, large-scale agricultural and water systems, healthcare, and so on. Today, these systems are central to the critical challenges of our time, from climate change and human insecurity to inequality and threats to democracy.

Humanity is now tasked with remaking these systems. Along the way, that remaking will also remake what it means to be human, as well as the organization of human economies and societies. What is at stake is nothing less than the future of what some have called **the techno-sphere**: the layer of technological habitats that humans have created for themselves at the interface between the geosphere, the biosphere, and the atmosphere. STS has the potential to create powerful and valuable tools to illuminate how that transformation will impact and be implicated by the future of tomorrow's technological cities, societies, and systems. New STS-informed theories and insights can provide comparative frameworks, tools, and ideas to help assess diverse potential future sociotechnological system designs and pathways for achieving them. Absent such tools, we cannot hope to fully understand how the transition to clean energy will fully transform the technosphere or inform that process to make it more just and sustainable – either for the planet or for ourselves.

Designing Future Technological Societies

Developing new STS tools for designing the future is grounded in better understanding how technologies get used to make the worlds that people end up inhabiting. Today, our world is dominated by a paradigm of thinking focused remarkably narrowly on how technology is designed and made. Each year, universities around the globe teach hundreds of thousands of young engineers to imagine and create new technologies and, increasingly, to fashion startup companies to commercialize them. In the United States alone, the federal government spends over $100 billion annually on research and development aimed at advancing new technologies. Apple reports 1.8 billion active devices of its making worldwide, with new model iPhones and Macs appearing every year. Globally, 1.5 billion vehicles drive on the world's roads, with 80 million new cars sold each year.

By contrast, remarkably little attention is given to how people use technologies to fashion individual and family life, broader systems of organization, or whole societies and economies. We have thousands of engineering schools, but where are the schools in which people are trained in the design, operation, and maintenance of sociotechnological systems and assemblages? Where do people learn to guide the development of technological societies in beneficial directions? In the absence of such schools, is it surprising that, while new technologies are almost always imagined and marketed as tools of innovation and progress, the global integration of new technologies into human economy and life often goes awry?

The car – an engine of dramatically enhanced mobility – has wrought concrete, traffic-locked, smog-ridden cities. The car's dependence on global oil supply chains has, in turn, wrought extraction, inequality, and injustice around the globe, not to mention the planetary climate crisis. Coal-fired power plants, which have lit much of the world and provided the electrical foundations of industrial and now data-driven economies, have likewise accelerated climate change as well as air pollution, ill health, and extractive mining industries. Atomic bombs brought global fear, arms races, and a half-century of cold war. Looking forward, technologies like artificial intelligence and human genetic engineering promise similarly complicated and thorny legacies, if we do not learn to better fashion human lives, organizations, and societies and economies out of the technologies we invent.

What would it mean to attempt to make technological futures differently? In 1900, for example, at the onset of the automobile age, few people imagined what the future of the automobile might be, let alone how powerfully the automobile and oil industries might shape the societies, economies, and environments of the 21st century. Had we more ambitiously grappled with the design possibilities of automobile-centered futures along the way, could we have ended up, today, without suburban sprawl, global systems for extracting, refining, and transporting carbon-based fuels, or the climate crisis? Over the past century, who made the decisions that led to the present, guided by what visions of the future? What would have needed to be different, when, and in what settings, to at least enable people to see and to perhaps open up the possibility of arriving at a different technological future? What might we learn from contemplating how we arrived at today's automobile-dominated present for how to structure decision-making differently, going forward, for future technologies such as electric vehicles, solar panels, artificial intelligence, gene drives, and more? Going through such an exercise would valuably inform the search for alternative technological pathways into the future, providing critical insights into the kinds of alternative configurations of sociotechnological assemblages that might be possible, when, with what benefits and risks, and for whom.

I suggest in the rest of this chapter that STS offers a suite of tools and ideas that can contribute to the sociotechnological redesign of the technosphere now underway across many spaces of rapid technological innovation. Here I present three examples of STS tools and approaches to the design of futures, which is by no means a comprehensive list: STS approaches to the future, STS approaches to sociotechnical redesign, and STS approaches to democracy and justice. There are many more – and many more yet to be developed by innovative STS analysts and their allies across many fields of theoretical, professional, and practical work. Perhaps most important of all, I suggest in the conclusion, is the need for systematic investment in new schools of the design of futures. Such schools should be guided not by engineers who design technologies but by those whose focus is on how people use technologies to make technological societies. Such making includes the habitats, systems, places, and social, economic, and political arrangements within which people live. At the same time, it also includes the people themselves, as inhabitants of strange new worlds, with new livelihoods, new ways and forms of living and being, and new forms of individual and collective meaning, identity, and relationship to each other and to the other inhabitants of the Earth and planetary systems.

Sociotechnological Imagination of the Future

Over the past decade, one of the most important new trends in STS writing has focused on imagination – or more specifically imaginaries – of the future. If imagination is the capacity to envision things that aren't present, imaginaries are the product of the capacity to imagine. They are the things we imagine, especially when we consider the ways in which our imaginings are shared and collective and not just the product of an individual's fanciful thoughts. STS work on imaginaries has particularly emphasized imagination of the future – and even more specifically how to create better futures – linked to science and technology.

Today, almost all cultures in the world tell themselves stories about how science and technology will make people's lives better in the future. STS uses the phrase **sociotechnical imaginaries** to refer to the shared themes common to such stories in any given culture. The field has developed robust techniques for inquiring into who tells such stories, how their content varies across communities and cultures, whose stories (and what kinds of stories) attract or motivate larger societal and economic investments, and, by contrast, whose stories are neglected, dismissed, or explicitly rejected or suppressed. Like most STS scholarship, studies of imaginaries emphasize understanding and critically evaluating today's (or yesterday's) societies: how such societies (and powerful and not so powerful actors within them) think about and envision the future, how such visions guide the design, development, and deployment of new technologies, such as clean energy, and get embedded in the resulting sociotechnical systems, and how the construction of new sociotechnological futures reinforces or undermines the exercise of power in society more broadly.

A new strand of STS seeks to take the study of the imagination in new directions that we might call alternative imaginations of the future. This work starts with the proposition that the future will consist of ways of weaving technology into social, economic, and cultural fabrics – but that those future weavings remain interpretively flexible, capable of being sculpted in multiple potential ways. It then asks: what are the consequences of designing the future along those different possible pathways? This question can be asked, for example, of different technologies: how would a fossil fuel future differ from a nuclear or renewable-powered future? But the approach is even more powerful when it begins to leverage the commitment of STS analysis to focus on the details of sociotechnical assemblages and systems and illuminates how different ways of approaching even the same technology can result in vastly divergent human futures.

This approach to envisioning the future is quite different from contemporary sociotechnical imaginaries. Think, for example, of World's Fairs, consumer electronics shows, places like Disney's Epcot Center or Dubai's Museum of the Future, the rhetoric and hype that often accompanies either innovators' depictions of "the revolutionary potential" of new technologies, or large government investments in science and technology, like the recent CHIPS Act to promote the manufacturing of semiconductors in the US. Existing approaches to developing technology and envisioning technological futures generally view technology as an independent driver of social change, rather than as an outcome of human choices and something that will evolve in relationship to changes in societies. They also often depict singular, generalized, and idealized visions of the future, mostly utopian but sometimes dystopian, rather than more realistic attempts to discern how different future evolutions of sociotechnical arrangements might play out for different groups of people. Finally, they are often informed predominantly by the perspectives of engineers and technologists, rather than more diverse constellations of people with heterogeneous ways of imagining the future.

STS, instead, tries to take the opposite approach: depictions of multiple, realistic variants of sociotechnical futures, envisioned collaboratively by participants from many different technical and non-technical disciplines, professions, creative fields, and community groups. The goal is to explore potential future sociotechnological designs and the societal, environmental, and technological trade-offs they entail. For example, a series of recent studies in this vein has begun to open up the question of how solar-powered futures might vary across different ways of integrating solar technologies into human and natural ecosystems. The studies began with deliberate exercises – termed narrative hackathons – that brought together STS researchers in teams with solar energy engineers, artists, writers, and experts in speculative imagination to discover what it might be like to inhabit future societies built out of solar energy systems. The goal was not to understand what solar-dominant futures *will* look like but rather to explore the many different possible paths that solar-dominant futures *might* take. To inject diversity into imaginings of the future of solar societies – something few existing approaches to technology are inclined to do – participants were given initial starting points that encouraged their imaginings to go in very different ways. And to ensure that the resulting imaginative exercises emphasized sociotechnical variation (and not just different ways of doing technology), the teams were given the task of writing science fiction stories focused on the lives of people who inhabit the imagined technological environments of the future. The resulting stories were collected in books – *The Weight of Light* and *Cities of Light* – alongside essays and artwork from team members that further illuminate the divergent futures. Most importantly for our purposes in this chapter, the stories offer starkly divergent visions of the worlds that solar energy might bequeath us in the next few decades.

Building on this foundation of divergent solar futures, subsequent STS work has begun to excavate more detailed insights into how alternative future designs of the solar industry would lead to very different human outcomes. This work has explored various scales of human-technology interactions from individual *social practices* of energy (the routine behaviors through which we use energy to accomplish everyday tasks) to how *ownership* of energy technologies is structured (today, energy generation is largely owned by large companies, while energy use technologies, like a TV or a car, are owned individually; tomorrow could be very different) to *global political economy* (how wealth and power are distributed among the world's peoples and institutions). In turn, this work has informed the development of a framework for "intentional and responsible energy transitions" that emphasizes recognizing the importance of design choices in energy transitions – divergent ways of structuring solar and other low-carbon sources of energy and their role(s) in human affairs – as a key element in navigating toward carbon-neutral futures.

Sociotechnological Redesign

STS work on solar and other clean energy futures also builds on another important area of STS scholarship focused on the challenge of sociotechnical redesign, or the possibilities of unpacking how existing sociotechnological systems contribute to inequality, injustice, unsustainability, or other problematic outcomes and reconfiguring systems in ways that reverse those outcomes. Some of this work emphasizes the concept of **responsible innovation** – the integration of societal and ethical concerns early ("upstream") in the process of inventing new technologies so that the kinds of sociotechnological assemblages brought into being from the outset are more equitable or more just, bringing substantive benefits to a wider diversity of potential users. Other work, by contrast, has focused further "downstream" on inhabitants of technological environments that are particularly destructive (e.g., via environmental injustices) and on the possibility of re-designing their relationships with technologies in ways that improve their lives.

Again, solar energy is one of the technologies that has centered this work, grounded in the wide diversity of ways of pursuing decentralized or distributed solar systems through individual or community ownership. Of course, distributed solar systems are widespread around the globe, in remote villages, off-grid settlements, suburban rooftops, community-shared solar installations, and many other locations. Where STS has entered the conversation is via detailed analyses of the potentially substantial implications of even subtle variations in solar technology and industry design, perhaps most fully examined in the book *Solar Power*. Such analyses show that the distribution of benefits and risks – as well as the power and voice of diverse participants to shape solar imaginings, constructions, and futures – can be very different and should be an integral part of our analyses of solar projects and policies as we build out a low-carbon future. (In case you aren't aware, in 2022, the world built more solar energy than all other forms of energy combined, for a total of 268 GW of new solar power capacity. That means we're now building 1 TW of solar energy every four years. Humanity only uses about 15 TW of total energy of all forms.)

Taking advantage of these insights from STS analyses of solar-powered sociotechnical systems, STS scholars are now exploring the potential for re-designing energy systems in more equitable ways. In Puerto Rico, for example, STS scholars are working with low-income communities and engineering researchers to help community members access and deploy solar energy in ways that significantly enhance the benefits they receive from the energy they use and reduce vulnerabilities to hurricanes, electricity outages, and other risks. Potentially more than anywhere else on the planet, the crisis of existing energy systems has prompted wholesale bottom-up solar innovation, leading to dozens of different models for organizing solar energy into a resource for households and communities struggling to forge more affordable, more secure, and more sustainable ways of powering their day-to-day lives in the wake of escalating climate threats. In another example, in the sprawling suburban and exurban landscapes of metropolitan centers in the US Southwest, STS researchers recently launched a series of conversations with diverse stakeholders about potential future alternative sociotechnical assemblages for linking solar energy to electric vehicle charging in areas where solar energy is likely to dominate future energy generation. The goal of the conversations is to stimulate deliberation of how divergent future scenarios would work for different groups of people (across different incomes, ages, types of housing, modes of work, etc.), the various technological transformations and investments each scenario would entail, and how to ensure that important aspects of long-term sociotechnical futures are appropriately factored into the decision-making of communities, cities, electric utilities, and electric vehicle manufacturers over the next few decades.

Democratization of Sociotechnological Systems

STS has long considered the relationship between technology and democracy to be central to the field's concerns. For the purposes of this chapter, several key questions from this work are significant. How do we understand the distributed processes of decision-making that orchestrate the dynamic evolution of technological societies (ranging from the choices that individuals and families make about which technologies to buy and how to use them in fashioning their day-to-day lives and livelihoods to the collective legal and governance decisions made about technology by legislative, executive, judicial, policy, and regulatory institutions)? Who exercises voice and power in decision processes? What does it mean for those processes to be considered democratic and just? And how can they be made more so?

Energy systems and energy transitions epitomize the complexity of these questions. In the US, for example, the governance of electricity systems takes place across a complex web of institutional sites of decision-making that range from the individuals, families, and businesses that use the large majority of electricity to thousands of electricity service providers to local, state, and federal regulatory, policy, legislative, and executive agencies. This complexity – alongside the technological complexity of electricity systems – has largely made energy decision-making a place where members of democratic publics have been officially invited but at best tolerated, generally neglected, ignored, and often unofficially excluded.

As energy transitions have begun to take hold, however, publics around the world have begun to take notice and, in many cases, insist on being granted a greater voice. In some cases, this insistence on the democratization of decision-making reflects a sense that energy systems are acting too slowly in the face of catastrophic climate risks. In other cases, it stems from the very real consequences of energy system changes for people's lives and livelihoods and environmental outcomes that were discussed above. Energy systems will entail dramatic reconfigurations of many natural, social, cultural, economic, and technological features of today's world to which diverse groups of people attach importance, significance, value, and meaning. The less that significance is attended to in decision-making, the more people will rise in resistance to new ways of doing energy – and the harder it will be to achieve the rapid and deep changes needed to create sustainable and just energy futures.

STS offers diverse resources that could potentially help energy governance institutions do a better job of engaging publics in energy transitions. One of the most interesting recent examples is the introduction of a deliberative systems perspective into analyses of public participation in energy. Studies of deliberative systems highlight the importance of understanding and mapping the diverse locations distributed throughout society (e.g., coffee shops, dinner tables, legislatures, court rooms, regulatory proceedings, etc.) where people actually discuss and deliberate important issues, as well as how those locations communicate and exchange information. STS researchers have begun to map such ecosystems of public engagement in energy deliberations, to develop ways of assessing and evaluating the sum of engagement across such ecosystems, and to suggest pathways for enhancing overall engagement in energy deliberations by introducing multiple, additive reforms across different locations.

Final Thoughts

We are at one of the most important moments in the technological history of contemporary societies and economies. Energy systems are foundational to everything from transportation and industry to food and information. If the 20th century was the time of building energy systems, the 21st is shaping up to be the time of re-building them. Because of their deep integration of the social and the technological, how we rebuild energy in the next few decades will have transformational consequences for what it means to be human for centuries to come.

That fact should guide energy system redesign. But it can only do so if we find ways to make explicit what the design of divergent future energy systems means for divergent human futures – and then open those insights up to broad public deliberation and engagement in deciding how to proceed with energy transitions. More than any other field, I believe, STS is developing unique and powerful tools and perspectives for forward-looking sociotechnological design and evaluation of alternative potential futures that can help build these kinds of capacities. Now is the time for universities to recognize the importance of those tools. They are relevant for the reform of existing schools of engineering, helping young engineers recognize that they are not just designing technologies, they are designing fundamental elements of future societies. Even more important, however, they are relevant to the development of new schools for the design of the future oriented toward preparing professionals who are positioned – alongside engineers – to help understand and guide the use of technology to create future societies and economies.

Further Reading

Bijker, W. E. (1997) *Of Bicycles, Bakelites, and Bulbs: Toward a Theory of Sociotechnical Change.* Cambridge, MA: MIT Press.

Echevarria, A., Rivera-Matos, Y., Irshad, N., Gregory, C., Castro-Sitiriche, M., King, R. R., and Miller, C. A. (2022) "Unleashing sociotechnical imaginaries to advance just and sustainable energy transitions: The case of solar energy in Puerto Rico," *IEEE Transactions on Technology and Society.* doi.org/10.1109/TTS.2022.3191542

Eschrich, J. and Miller, C. (2021) *The Weight of Light: A Collection of Solar Futures* (Tempe: Center for Science and the Imagination, 2019) and *Cities of Light: A Collection of Solar Futures* (Tempe: Center for Science and the Imagination). Free to download at: https://csi.asu.edu/category/projects/solar-tomorrows/

Jasanoff, S. (2016) *The Ethics of Invention: Technology and the Human Future.* New York: W. W. Norton & Company.

Keeler, L. W., Bernstein, M. J., and Selin, C. (2019) "Intervening through futures for sustainable presents: Scenarios, sustainability, and responsible research and innovation," in *Socio-technical Futures Shaping the Present: Empirical Examples and Analytical Challenges*, edited by A. Lösch, A. Grunwald, M. Meister, and I. Schulz-Schaeffer. New York: Springer VS, pp. 255–282.

Miller, C. A., Rivera-Matos, Y., Echevarria, A., and Dirks, G. (2022) "Intentional and responsible energy transitions: Integrating design choices in the pursuit of carbon-neutral futures," in *Routledge Handbook of Energy Transitions*, edited by K. Araújo. London: Routledge, pp. 353–370.

Mulvaney, D. (2019) *Solar Power: Innovation, Sustainability, and Environmental Justice.* Berkeley: UC Press.

Winner, L. (1986) *The Whale and the Reactor: A Search for Limits in an Age of High Technology.* Chicago: University of Chicago Press.

Part VIII

Climate Change Adaptation and Resilience

Introduction

Tamar Law

Chances are in the last few years you have already seen headlines about the urgency of climate adaptation and resilience, such as the NYT's "Climate Change is Harming the Planet Faster Than We Can Adapt, U.N. Warns" (Plumer and Zhong, 2022) or the Guardian's "Adapt or Die: Resilience to Climate Change Needed, Says Environmental Agency (PA Media, 2021)." Climate adaptation and resilience are overlapping responses to the wide-ranging effects of climate change. **Climate adaptation** refers to adjustments in ecological, social, and economic systems to reduce vulnerability to climate impacts and enhance resilience. **Resilience** refers to the capacity of these systems to withstand such impacts. Efforts toward climate adaptation often include anticipating and responding to climate risks, whereas resiliency efforts aim to build capacity to cope with such changes.

For example, let's take the climate impact of sea-level rise, which threatens coastal communities in cities like Jakarta in Indonesia or New Orleans in the United States with flooding and storm surges. Climate adaptation efforts might include building coastal infrastructure such as seawalls and flood barriers, or even the restoration of natural coastal ecosystems such as mangrove forests and wetlands. For example, in Jakarta, the Indonesian government has invested heavily in constructing a large sea wall and has even gone as far as announcing a new capital location. On the other hand, resiliency efforts might focus on community education and engagement, such as community workshops on flood awareness and emergency preparedness. In New Orleans, the city has done exactly that and, moreover, implemented a resilient housing and infrastructure program, and retrofitted existing buildings to withstand stronger storms and flooding events.

These examples orient one toward what initiatives bolstering climate adaptation and resilience might look like in application. Now, take a moment and think of what forms of climate adaptation are occurring near where you live. What climate risks or impacts are these initiatives addressing? How are they understood? What evidence is used to inform this intervention? Who is spearheading and implementing the changes? What is the time span of the initiative? How might the program impact your own life? What does climate resilience look like for these initiatives? What might climate resilience look like for your community?

These questions push one to begin unpacking climate adaptation and resilience, akin to the research questions that Science and Technology Studies (STS) scholars pose. The headlines we read in the news about climate change make it seem like we need to adapt and build resiliency, right away, right now! The readings in this Part, however, suggest that climate change interventions need to be slowed down to understand and respond to their complicated sociopolitical and institutional arrangements.

DOI: 10.4324/9781003409748-31

Scholarship in STS examines the complex processes by which climate adaptation and resilience interventions are understood and produced through scientific and technological knowledge, implemented through political channels, and negotiated within social dynamics. Work in STS has revealed that climate adaptation and resilience are not simply fixed concepts but rather dynamic and contingent **sociotechnical** processes. This refers to the inseparability of social and technical aspects. To say it differently, climate adaptation and resilience are brought about through a range of human and non-human actors associated together in networks. Moreover, STS scholarship focuses on the sociopolitical dimensions of knowledge production for climate adaptation and resilience. STS reveals the role of expertise in adaptation and resilience building decision-making. As the chapters in this Part demonstrate, an STS approach elucidates both the possibilities and limitations of climate adaptation and resilience toward creating more just climate interventions.

In Chapter 24, geographer Tim Forsyth outlines the role of risk within climate adaptation and resilience, demonstrating how these terms are relational and self-enforcing. He begins by describing how risk is co-produced with ideas of social identity and agency. He then illustrates how formal approaches that are being propagated and implemented for adaptation and resilience, that are intended to reduce risk, may also co-produce certain cultural values and norms. In this way, these formal approaches to reduce risk may have the unintended result of causing **maladaptation**, actually increasing people's vulnerability to climate shocks, and failing to address underlying structural causes of vulnerability. Ultimately, he shows that an STS approach to adaptation and resilience must consider how ideas of risk and social response are produced together.

In Chapter 25, sociologist Marcus Taylor begins by deconstructing the conventional understanding of climate, unpacking the sociotechnical processes that imagine climate as "average weather." Then, Taylor illustrates an alternative understanding of climate informed by both social and natural processes and the political implications of this approach. Drawing from his research on rural resilience in India, he argues that such an understanding allows us to move beyond envisioning humans as adapting to an external climate. Rather, Taylor shows how humans are actively entangled within meteorological forces and the power relations that sustain them.

Just as climate is co-produced, adaptation is also shaped by social and political forces. Forms of effective climate adaptation are diverse across geographies and economic and social activities. In Chapter 26, geographer Tamar Law demonstrates that adaptation to climate change varies among agricultural actors in New York State. She examines how imaginaries of adaptation contrast across a state-led initiative toward Climate Smart Farming and small-scale alternative farmers. Through a tripartite typology of adaptation, Law reflects on the potential political implications and possibilities of alternative forms of agricultural adaptation.

Finally, ensuring effective climate change adaptation requires interdisciplinary collaboration across the humanities and sciences. In Chapter 27, anthropologist Sarah Vaughn's chapter addresses how collaboration might unfold through an in-depth case study in Guyana. Vaughn's research on mangrove restoration as a form of climate adaptation in Guyana illustrates how engineers, geoscientists, and beekeepers in Guyana develop shared knowledge that overcomes deficiencies in any one of these knowledge producers.

After reading these chapters, we invite you to think back to the original example you brainstormed of climate adaptation or resilience at your own local level. What other questions might these researchers ask to better understand these interventions? What other questions might you ask after reading these chapters? How has your own perception of this example shifted after reading these chapters?

References

The Guardian (2021). Adapt or die: resilience to climate change needed, says Environment Agency. *The Guardian*. [Online] 12 October. Available at: www.theguardian.com/environment/2021/oct/12/adapt-or-die-resilience-to-climate-change-needed-says-environment-agency (Accessed: June 20, 2023).

Plumer, B. and Zhang, R. (2022). Climate Change Threatens "Irreversible" Effects, IPCC Report Finds. *The New York Times*. [Online] 28 February. Available at: https://www.nytimes.com/2022/02/28/climate/climate-change-ipcc-report.html (Accessed: June 20, 2023).

24 Climate Change Adaptation and Resilience

Sociotechnical and Knowledge Dimensions

Tim Forsyth

International efforts to reduce and sequester carbon dioxide and other greenhouse gases are not yet slowing the rate of anthropogenic climate change. The lack of progress has encouraged scientists and policymakers to consider ways of reducing impacts, especially in poorer countries with the least capacity to cope. Analysts now discuss the terms adaptation and resilience as ways to reduce the impacts of climate change alongside climate change mitigation, or the reduction of atmospheric greenhouse gas concentrations.

Adaptation describes actions that can make the physical effects of climate change less damaging. It is sometimes also referred to as adjustments in natural or human systems in response to the impacts of climate change. **Resilience** refers to the properties of people, places, or objects, such as infrastructure, to withstand climate risks. Resilience overlaps with adaptation because it includes the ability to undertake adaptation. But unlike adaptation, resilience is a characteristic rather than an action. Accordingly, policymakers now undertake activities to build adaptation and resilience within other social and political interventions, such as international development, capacity building, and planning, rather than through short-term activities to manage extreme weather alone.

Scholars in Science and Technology Studies (STS) seek to make adaptation and resilience more effective and socially inclusive by examining the tacit (or unacknowledged) frameworks guiding approaches to risk and societal response. By definition, adaptation and resilience imply responding to risks, which are often defined in terms of physical climatic events such as storms, floods, and droughts. However, the extent to which these events present risks often carry assumptions about how these events are hazardous. These assumptions might also include tacit beliefs about how different social actors might respond to these risks. For example, events such as Hurricane Katrina in the USA showed that not every social group experiences the risks linked to storms in the same way. Focusing on the physical impacts of the storm alone overlooks the deeper social structures that might make different people more or less vulnerable. Indeed, some scholars now worry that seeking to implement adaptation and resilience based on existing understandings of risk might result in maladaptation—or difficult situations when interventions to reduce risk make people more vulnerable or displace risks onto others. Accordingly, STS scholars ask how ideas of adaptation and resilience reflect visions of social order reproduced in the supposedly neutral language of risk management and adaptive capacity. STS research on adaptation and resilience seeks to make these terms more sensitive to these tacit framings, and to increase the social inclusivity of understandings of risk and social responses.

In this chapter, I discuss the development of ideas about adaptation and resilience within climate change policy. I demonstrate the value of STS scholarship in three key areas: the constitutional role of unexamined models of risk, the co-production of risk and social identities and agency, and tacit worldviews underlying approaches to resilience.

DOI: 10.4324/9781003409748-32

The Constitutional Role of Risk

Risk can be interpreted in different ways. For many environmental scientists, risk is an indication of the statistical chance that events such as damaging storms or floods might occur. In this sense, "risk" is similar to "likelihood."

For many social and environmental analysts, risk also refers to the nature of hazards. Climate-related risks, for example, could include physical events arising from climate change such as floods, storms, and droughts. But this understanding of risk can also pose challenges. Events such as floods might not be equally hazardous to everyone. Moreover, the risk associated with floods might not arise from the physical event itself, but rather with the outcomes of these events such as displacement, loss of livelihoods, or death. Accordingly, there is a need to consider how far the terms used to indicate risk might actually represent where, how, and for whom risks are experienced.

STS scholars investigate these questions. For many STS scholars, "risk" refers to the frameworks that scientists or policymakers use to understand potential dangers. These frameworks of risk perform a constitutional role in defining how climate change might be hazardous and why people are at risk. The objective of STS research is to analyze how these frameworks might represent physical hazards or social responses in reduced ways, to ask what can be done to make each more diverse and inclusive.

Adaptation and resilience are examples of risk frameworks because they refer to responses to the risks posed by climate change. But discussions about climate risks, and hence adaptation and resilience, have undergone some important transitions over time. For some decades, research in anthropology and development economics discussed how vulnerable people might adapt to resource scarcity or other forms of hazards. These works tended to emphasize structural and political drivers of risk, such as long-term marginalization of people, and their ability to respond to risks through making institutions that regulated resource use (Watts, 2015). In the late 1980s, however, debates about adaptation focused increasingly on climate change and the influence of global systemic atmospheric change. The Intergovernmental Panel on Climate Change's (IPCC) first and second assessment reports in 1990 and 1996 initially defined climate risks in terms of additional units of atmospheric greenhouse gas concentrations because these can be linked to immediate impacts such as larger and more unpredictable storms, flooding, and drought. Accordingly, the IPCC Second Assessment Report defined adaptation and adaptability as:

> the degree to which adjustments are possible in practices, processes, or structures of systems to projected or actual changes of climate.
>
> (Watson, Zinyomera, and Moss, 1996, p. 5)

Some analysts have also argued that the IPCC represented risk in this way because it was consistent with its intention, especially during the 1990s, to represent scientific findings with one voice in a depoliticized manner. The IPCC makes it clear that its role is to summarize legitimate scientific research on climate change in ways that are policy-relevant, but never policy-prescriptive. This approach was justified to ensure that the knowledge generated by the IPCC can be trusted at a time when many climate change deniers claim that the science is politically motivated or reflects national interests. Referring to the atmospheric drivers of impacts therefore allowed the IPCC to represent risk in terms of global biophysical change, rather than in terms of social or economic structures (Beck and Forsyth, 2015).

This approach to risk, however, has limitations because it frames adaptation to climate change as a response to projected climate changes, rather than how these changes are experienced as problematic by vulnerable people. According to this view, the risks posed by climate change arise from additional atmospheric greenhouse gas concentrations, rather than vulnerability to those concentrations. For example, adaptation to these definitions of risk focused on strengthening sea walls, roads, and bridges, or providing storm shelters. These actions can save lives and allow economies to function in the face of extreme weather. But they focus on the immediate physical impacts of climate change rather than removing social barriers that cause certain people to be more vulnerable than others (Nightingale et al., 2020).

Moreover, these early systemic definitions of adaptation tended to see adaptation as an objective only if mitigation policies fail. Indeed, some scholars have argued that the focus on global systemic change has led to an implicit bias against adaptation policies in the IPCC reports and, by extension, in policy discussions. In 1998, for example, one UNFCCC-related meeting in Tokyo witnessed the Chinese delegation accuse the USA of insufficiently helping poorer countries respond to climate change, such as through technological development. The representative of the USA replied by saying "let me remind delegates that we are discussing a climate change convention, not a convention about development, and so we should only refer to atmospheric greenhouse gas concentrations" (personal observation).

Indeed, later debates about funding for adaptation have led to lines being drawn between budgets allocated to "climate change policy" and "aid and development" to indicate the impact of atmospheric greenhouse gases rather than other driving forces of risk (Klein and Mohner, 2011).

Over time, however, approaches to climate risk, adaptation, and resilience have diversified. There is now a greater acknowledgment of socio-economic drivers of vulnerability and of the greater role of context in risk (Ayers, 2011). For example, the Working Group II of the IPCC Fifth Assessment Report (2014) noted:

> The rational-linear process that identifies potential risks then evaluates management responses ... has been challenged on the grounds that it does not adequately address the diverse contexts within which climate decisions are being made, often neglects existing decision-making processes, and overlooks many cultural and behavioral aspects of decision-making.
>
> (Jones et al., 2014, p. 199)

In a document prepared for the Sixth Assessment Report, authors also acknowledged that risk should recognize the diversity of values and objectives associated with human and ecological systems. It emphasized that "the concept of risk should *not* be used to describe outcomes within physical systems only" (Reisinger et al., 2020, p. 6). Moreover, it stated that ideas of climate risk can refer to both impacts of, and responses to, climate change.

These changes in conceptualizations of risk allow a wider range of interpretations of adaptation and resilience. There is growing discussion of risk and loss in more flexible, personal, and emotional terms (Nightingale et al., 2021).

Co-producing Risk, Identity, and Agency

Research in STS also considers how frameworks of risk are co-produced with ideas of social identity and agency. For STS, **co-production** is a framework that acknowledges that the ways in which we know and represent the world (both nature and society) are inseparable from the ways in which we choose to live in it (Jasanoff, 2004). Co-production is therefore a way to analyze how knowledge about adaptation and resilience can reflect tacit social structures, values, and unseen codes of conduct. These factors can influence which knowledge is seen to be relevant, how it is collected, and how it is presented as authoritative. This information can relate to the representation of physical risks such as floods, or human responses to these risks.

Yet, in recent years, many environmental scientists and authors of IPCC reports have used the term co-production in different ways. These analysts have instead represented co-production as a cognitive process of co-owning research practice between scientists and relevant stakeholders through consulting with local people, or different users of research, while undertaking the research. Indeed, this kind of consultation has been called the gold standard of engaged science (Lemos et al., 2018).

Yet, the STS interpretation of co-production can illustrate ways in which formal approaches to adaptation and resilience can imply social responses to risk in ways that can be illusionary. One example is the distinction made under the IPCC's Fourth Assessment Report between "planned" and "autonomous adaptation" (IPCC, 2007). Planned adaptation referred to deliberate interventions to anticipate anthropogenic climate change. Autonomous (or spontaneous) adaptation included unconscious responses to climatic stimuli triggered by ecological changes in natural systems and/or market or welfare changes in human systems. Autonomous adaptation might include altering agricultural inputs, introducing water-managing technologies, altering cropping cycles, or diversifying economic activities.

This distinction between planned and autonomous adaptation, however, implies that people's changes in risk management only occur in response to climate change, rather than because of other numerous, long-term socio-economic transitions that might predate climate change. For example, much research on problems such as dryland degradation or soil erosion in developing countries has argued that adaptive responses at the local level are not simply driven by environmental changes *per se*, but in how environmental changes present hazards for vulnerable people's livelihoods and assets (Ribot, 2010). Adaptation, therefore, need not only include practices that lessen impacts of environmental change on currently productive resources, but also comprise forms of livelihood diversification that *de facto* make the impacts of these changes on existing resources less threatening. But also, some forms of so-called planned adaptation might impede pathways to this broader form of autonomous adaptation. For example, research in Thailand showed that the national government tried to encourage "adaptation" to climate risks by using tree plantations to stabilize slopes in agricultural zones to address risks such as landslides. But planting trees on agricultural land also reduced the possibility for farmers to continue or diversify agricultural livelihoods (Forsyth and Evans, 2013).

The point is that models of risk pre-shape expectations of how people might respond. But using fixed frameworks of risk can hide how people actually experience environmental changes as hazardous. They might also worsen those experiences of risk. This unfortunate outcome might also be a form of maladaptation.

As mentioned above, maladaptation occurs when attempts to build adaptation to climate risks end up increasing people's vulnerability or displacing risks onto other people (Schipper, 2020). Much early discussion of maladaptation highlighted technological interventions that were insufficient for addressing climate change: for example, coastal defenses that displaced wave action onto other locations, or high-energy air conditioners to counter heat waves. More recent research has defined maladaptation more in terms of interventions that fail to acknowledge deeper social and economic structures. For example, some adaptation interventions in Africa have assumed that stable labor markets are a pathway to resilience, but long-standing research in Africa has shown that flexible and dynamic markets offer a more diverse range of livelihoods for less advantaged workers. Adaptation defined in terms of stable labor markets can therefore exacerbate social inequality. A common problem here is how so-called "success" in adaptation projects might be defined by dominant development agendas rather than from a participatory and inclusive form of knowledge co-production (Eriksen et al., 2021).

For many STS scholars, however, maladaptation also arises from the application of risk frameworks that simultaneously simplify experiences of climate change and people's responses to them. For STS scholars, these challenges cannot be addressed by short-term consultation with stakeholders. Rather, there is a need to consider deeper, and less cognitive forms of consultation and social justice. This interpretation of co-production does not focus only on how to allocate climate policies more successfully to specific groups, but also considers how underlying beliefs and worldviews simultaneously shape both "what" is being allocated and to "whom."

The example of community-based adaptation (CBA) to climate change shows the challenges of co-producing risk and social agency in this way. CBA has been discussed within climate change policy since the early 2000s to allow local and vulnerable people opportunities to shape adaptation interventions (Ayers et al., 2014). It reflects long-term thinking about the benefits of community-based natural resource management and participatory development. It aims to achieve outcomes that are better attuned to the social drivers of vulnerability to climate change, including the social, economic, and political contexts of poverty. This work aims to diversify the understanding of risks posed by climate change and provides a more socially sensitive form of risk assessment that relies on predictions of physical changes such as floods and droughts alone. For example, community-based forms of adaptation can highlight the challenges for different social groups within localities to access freshwater or ensure that adaptation interventions such as new wells and water tanks are accessible to everyone.

There are many good reasons for local people to shape adaptation interventions. But critics have also questioned many assumptions underlying CBA. STS scholars have asked how far ideas of "community" and "climate risk" might be co-produced in ways that simplify and reduce scrutiny of both terms. For example, various scholars in international development have argued that the term "community" is simplistic and potentially damaging because it implies that local people act as a homogeneous unit. Many villagers or local groupings of people contain internal divisions and cases of exploitation, or even oppression, which can be a significant cause of poverty and vulnerability. Customary practices within communities, for example, can replicate exclusionary practices, such as forbidding women to own land or prescribing social roles that can be considered disempowering. Moreover, communities rarely can challenge deeper political, social, and economic driving forces of poverty, such as the state's role in controlling investment and access to services or the reliance of poorer countries and regions on international trade flows. Indeed, some development analysts have expressed concern that community-based adaptation and resilience approaches can be captured by elite interests or only succeed under restricted circumstances (Forsyth, 2013).

There are also concerns that CBA might focus on earlier understandings of global climate risk rather than the local and contextual drivers of vulnerability to risk. For example, some discussions of CBA might consider the location of physical infrastructure for risk avoidance, such as storm shelters and freshwater tanks. Such infrastructure can be crucial in saving lives during storms or tidal surges in locations such as Bangladesh. But this kind of discussion does not consider longer-term transitions in livelihoods and political rights that might drive a more holistic ability to withstand environmental hazards.

CBA can also represent climate risks and local communities in reduced ways. One example is floating gardens in northern Bangladesh. In this location, development organizations have highlighted how women have developed ways to use floating water hyacinth plants (a common weed) to make platforms to hold soil and grow vegetables at times when other land is flooded. Development agencies have used this example of successful CBA. Yet, critics have also pointed out that this colorful story also presents a supposed image of success that also presents women standing deep in floodwater in ways that bypass broader questions of gender inequality. Indeed, it is important to see how climate change policy approaches might become "adaptation regimes" based on how proposed solutions engage with various deep-set structural causes of vulnerability (Masud-All-Kamal and Nursey-Bray, 2022).

Some approaches to community-based adaptation therefore rely on representations of climate risk and community that can be reductionist and at times even romantic. Analysts therefore increasingly stress the importance of broader processes of social transformation, which are more powerful than narrow representations of community action alone.

Resilience and Normative Values

STS also considers worldviews and values in supposedly neutral science. This is now a theme for research on resilience to climate change. STS scholars therefore ask, how do tacit values shape what policymakers mean by resilience? What assumptions about risk and social agency do they carry? And is it possible to gain lessons for building resilience that are transferable between different contexts?

Conceptualizations of resilience have also changed over time. During the 1970s the term usually referred to physical properties of infrastructure or ecosystems, such as the ability of systems to withstand shocks (Holling, 1973). An increasing number of analysts, however, argue that resilience not only refers to physical properties of infrastructure or ecosystems, but also to socio-economic factors such as people's ability to access diverse livelihoods, or avoid long-term drivers of social vulnerability (Béné et al., 2014). For example, some development agencies have defined resilience in terms of the "3As" framework to combine anticipatory, absorptive, and adaptation capacities. These capacities focus on knowing in advance about risks, dealing with them when they occur, and then adopting long-term acts of adaptation that can reduce vulnerability.

But various analysts both inside and outside of STS have argued that these technical frameworks of climate resilience reproduce social and economic orders (Brown, 2016). In particular, some analysts have claimed that the term "resilience" has now become co-opted into neo-liberal thinking that has dominated economic and development policies since the 1980s based on ideas of market dominance, a small business-friendly state, and opposition to trade unions (Chandler and Reid, 2016). For example, strategies such as livelihood diversification or more formal development policies such as Sustainable Livelihoods Approaches have been criticized for enabling a discourse that national governments no longer have to work to reduce poverty or provide safety nets. According to this neo-liberal perspective, "successful" resilience is when individuals are free to serve markets, and states do not need to intervene.

Moreover, in these circumstances, various definitions and pathways to resilience might exist at the same time or fail to address deeper drivers of vulnerability. For example, research in Myanmar showed that different development organizations claimed to achieve resilience by diverse activities with different levels of engagement with social and political structures. In one region, a non-governmental organization (NGO) claimed to build resilience to climate shocks by sending flood warnings to people's mobile telephones. This activity added to "anticipatory capacity" by informing people of a significant hazard. But it failed to acknowledge social inequalities in land ownership or labor markets. For example, in one region of Myanmar where the warning was applied, land ownership might be concentrated among some 20 percent of households while the remaining 80 percent would work in casual labor, often for the landowners. Advance warning of flooding might therefore allow landowners to protect their livelihoods by harvesting crops, but it will do little to protect the livelihoods of casual laborers who might find a shortage of employment after the floods (Forsyth, 2018). Similar findings have been recorded in Bangladesh (Paprocki, 2018).

In this case, the definition of resilience reflected the underlying philosophy of development organizations. For example, some organizations focus on technological interventions such as providing wells or building sandbanks on riverbanks. Other organizations seek to build social rights and empowerment such as by choosing and hiring village champions (often women) who can then instigate local change. Each organization will define and measure resilience in different ways. However, all organizations also need to follow other constraints. For example, many development organizations work in authoritarian countries where governments suppress political rights and might expel organizations seen to be challenging their political order. Researchers and policymakers can adopt insights from STS by refusing to conflate short-term objectives of resilience (such as delivering text messages) with longer-term shifts in vulnerability. Moreover, definitions of resilience (similar to definitions of risk) should include social drivers of vulnerability, including political constraints on rights, but these are difficult to measure and monitor for methodological, institutional, and political reasons (Nightingale et al., 2018).

Conclusion

STS has contributed to the analysis of adaptation and resilience by highlighting the contingent and co-produced nature of risk, response, social values, and social identities. A core objective of this work has been to show that notions of risk are not fixed or separate from social context but reflect and can even shape inequality and values within society.

Yet, while this work has sought more flexibility and social inclusion in how we understand adaptation, many policy initiatives still seek to define adaptation in more universal and less contextual ways. The Global Stocktake, for example, is an assessment that considers how different countries have made progress in achieving long-term goals on mitigation and adaptation after the Paris Agreement. It is important not to dismiss progress on adaptation and resilience so far, as these can protect against climate change. But at the same time, it is crucial to consider how far these existing parameters of adaptation might still contain blind spots in how they define risk, why people are vulnerable, or how frameworks of risk and response might even reproduce existing social inequalities.

Moreover, much discussion within environmental policy acknowledges the need for greater social participation. These discussions include terms used in STS such as co-production, but the use of these terms in environmental science can be very different from the meanings adopted within STS. STS adds to these approaches by examining how assumed facts or scientific bases of policy (such as frameworks of climate risk, or the expected social responses of communities) are themselves held in place by larger social structures. Using STS allows a deeper level of democratization of adaptation and resilience by showing how these terms exist in association with different worldviews, values, or historic experiences. Listening to this research can allow researchers and policymakers to make approaches to adaptation and resilience more varied and socially inclusive, but also avoid potentially unhelpful outcomes.

Applying STS to adaptation and resilience, therefore, starts with asking how ideas of risk and social response are made together, and what social forces keep both stable and unquestioned. Diversifying ideas of risk and response can help make climate change science and policies more socially inclusive and effective for a wider set of challenges arising from climate change. It can also demonstrate and overcome the shortcomings of approaches to adaptation and resilience based on limited understandings of risk and social agency.

Further Reading

Beck, S. and Forsyth, T. (2015) "Co-production and Democratizing Global Environmental Expertise: the IPCC and adaptation to climate change." In R. Hagendijk, S. Hilgartner, and C. Miller (Eds.), *Science and Democracy: Making Knowledge and Making Power in the Biosciences and Beyond* (pp. 113–132). Abingdon: Routledge.
Nightingale, A. J., Eriksen, S., Taylor, M., Forsyth, T., Pelling, M., Newsham, A., . . . Whitfield, S. (2020) "Beyond Technical Fixes: Climate Solutions and the Great Derangement," *Climate and Development*, *12*(4), pp. 343–352.

References

Ayers, J. (2011) "Resolving the Adaptation Paradox: Exploring the Potential for Deliberative Adaptation Policy-Making in Bangladesh," *Global Environmental Politics*, *11*(1), pp. 62–88.
Ayers, J., Reid, H., Huq, S., Schipper, L., and Rahman, A. (Eds.) (2014). *Community-based Adaptation to Climate Change: Scaling It Up*. Abingdon: Routledge.
Beck, S. and Forsyth, T. (2015) "Co-production and Democratizing Global Environmental Expertise: the IPCC and adaptation to climate change." In R. Hagendijk, S. Hilgartner, and C. Miller (Eds.), *Science and Democracy: Making Knowledge and Making Power in the Biosciences and Beyond* (pp. 113–132). Abingdon: Routledge.
Béné, C., Newsham, A., Davies, M., Ulrichs, M., and Godfrey-Wood, R. (2014) "Resilience, Poverty and Development," *Journal of International Development*, *26*(5), pp. 598–623.
Brown, K. (2016) *Resilience, Development and Global Change*. Abingdon, New York: Routledge.
Chandler, D. and Reid, J. (2016) *The Neoliberal Subject: Resilience, Adaptation and Vulnerability*. London: Rowman and Littlefield.
Eriksen, S., Schipper, E. L. F., Scoville-Simonds, M., Vincent, K., Adam, H. N., Brooks, N., . . . West, J. J. (2021) "Adaptation Interventions and Their Effect on Vulnerability in Developing Countries: Help, Hindrance or Irrelevance?," *World Development*, *141*, p. 105383.
Forsyth, T. (2013) "Community-based Adaptation to Climate Change: A Review of Past and Future Challenges," *Wiley Interdisciplinary Reviews: Climate Change*, *4*(5), pp. 439–446.
Forsyth, T. (2018) "Is Resilience to Climate Change Socially Inclusive? Investigating Theories of Change Processes in Myanmar", *World Development*, *11*, pp. 13–26.
Forsyth, T. and Evans, N. (2013) "What Is Autonomous Adaptation? Resource Scarcity and Smallholder Agency in Thailand," *World Development*, *43*, pp. 56–66.
Holling, C. S. (1973) *Resilience and Stability of Ecological Systems*. Austria: International Institute for Applied Systems Analysis.
IPCC. (2007). Online Glossary, Climate Change 2007: Working Group II: Impacts, Adaptation and Vulnerability. from IPCC www.ipcc.ch/publications_and_data/ar4/wg2/en/annexessglossary-a-d.html
Jasanoff, S. (2004) "Ordering Knowledge, Ordering Society." In S. Jasanoff (Ed.), *States of Knowledge: The Coproduction of Science and Social Order* (pp. 13–45). London: Routledge.
Jones, R. N., Patwardhan, A., Cohen, S. J., Dessai, S., Lammel, A., Lempert, R. J., . . . von Storch, H. (2014) "Foundations for Decision Making." In C. B. Field, V. R. Barros, D. J. Dokken, K. J. Mach, M. D. Mastrandrea, T. E. Bilir, M. Chatterjee, K. L. Ebi, Y. O. Estrada, R. C. Genova, B. Girma, E. S. Kissel, A. N. Levy, S. MacCracken, P. R. Mastrandrea, and L. L. White (Eds.), *Climate Change 2014: Impacts, Adaptation, and Vulnerability. Part A: Global and Sectoral Aspects. Contribution of Working Group II to the Fifth Assessment Report of the Intergovernmental Panel on Climate Change* (pp. 195–228). Cambridge, UK: Cambridge University Press.
Klein, R. and Mohner, A. (2011) "The Political Dimension of Vulnerability: Implications for the Green Climate Fund," *IDS Bulletin*, *42*(3), pp. 15–22.
Lemos, M. C., Arnott, J. C., Ardoin, N. M., Baja, K., Bednarek, A. T., Dewulf, A., . . . Wyborn, C. (2018) "To Co-produce or Not to Co-produce," *Nature Sustainability*, *1*(12), pp. 722–724.

Masud-All-Kamal, M. and Nursey-Bray, M. (2022) "Best Intentions and Local Realities: Unseating Assumptions About Implementing Planned Community-based Adaptation in Bangladesh," *Climate and Development*, *14*(9), pp. 1–10.

Nightingale, A. J., Eriksen, S., Taylor, M., Forsyth, T., Pelling, M., Newsham, A., . . . Whitfield, S. (2020) "Beyond Technical Fixes: Climate Solutions and the Great Derangement," *Climate and Development*, *12*(4), pp. 343–352.

Nightingale, A. J., Gonda, N., and Eriksen, S. H. (2021) "Affective Adaptation = Effective Transformation? Shifting the Politics of Climate Change Adaptation and Transformation from the Status Quo," *WIREs Climate Change*, *13*(1), e740.

Olsson, L., Jerneck, A., Thoren, H., Persson, J., and O'Byrne, D. (2015) "Why Resilience Is Unappealing to Social Science: Theoretical and Empirical Investigations of the Scientific Use of Resilience," *Science Advances*, *1*(4), p. e1400217.

Paprocki, K. (2018) "Threatening Dystopias: Development and Adaptation Regimes in Bangladesh," *Annals of the American Association of Geographers*, *108*(4), pp. 955–973.

Reisinger, A., Howden Mark, Vera, C., Garschagen, M., Hurlbert, M., Kreibiehl, S., . . . Ranasinghe, R. (2020) "The Concept of Risk in the IPCC Sixth Assessment Report: A Summary of cross Working Group Discussions: Guidance for IPCC authors." Bonn: IPCC.

Ribot, J. (2010) "Vulnerability Does Not Just Come from the Sky: Framing Grounded Pro-poor Cross-scale Climate Policy." In R. Mearns and A. Norton (Eds.), *Social Dimensions of Climate Change: Equity and Vulnerability in a Warming World* (pp. 47–74). Washington, DC: World Bank.

Schipper, E. L. F. (2020) "Maladaptation: When Adaptation to Climate Change Goes Very Wrong," *One Earth*, *3*(4), pp. 409–414.

Watson, R. T., Zinyowera, M. C., and Moss, R. H. (Eds.) (1996). *Climate Change 1995: Impacts, Adaptations and Mitigation of Climate Change: Scientific-technical Analyses*. Cambridge: Cambridge University Press.

Watts, M. (2015) "Now and Then: The Origins of Political Ecology and the Rebirth of Adaptation as a Form of Thought." In T. Perreault, G. Bridge, and J. McCarthy (Eds.), *The Routledge Handbook of Political Ecology* (pp. 19–50). Abingdon: Routledge.

25 Rethinking Climate Change Adaptation

Marcus Taylor

Climate is what we expect,
Weather is what we get.

Saying attributed to Mark Twain

Climate change, we are correctly informed, is one of the greatest threats to human wellbeing in the present and coming decades. The increasing incidence of weather extremes and the resulting shocks and stresses upon human society require that we take urgent steps to adapt. For many policymakers, the big question is how adaptation should take place and what it should involve. In the following chapter, I show how an approach routed in science and technology studies (STS) provides interesting ways to reframe the issue of climate change adaptation by upending common assumptions and posing key issues in new ways. To do so, I argue, helps us reenvisage the parameters of what climate change adaptation involves, therein raising important questions around the direction of climate change policy.

There are three steps to this chapter's analytical process. First, the chapter deconstructs the conventional understanding of climate that underpins the majority of climate change scholarship. In the latter, climate is understood through socio-technical processes that create a statistical representation of the 'average weather' in any given locale. While this is one legitimate way of understanding climate that is essential for climate modelling, it is not the only way of understanding climate. On the contrary, conceiving of climate as an external and natural system limits the way we think and act on climate change with important implications for social justice. Second, the chapter then demonstrates an alternative understanding of climate, one that focuses on the coproduction of climates through intertwined social and natural processes. This enables us to move beyond thinking of humans as adapting *to* an external climate and rather examine how humans actively work meteorological forces into their daily lives. Finally, we turn to the political implications of this alternative approach to climate, emphasising how it facilitates a closer understanding of the power relations through which humans individually and collectively experience meteorological processes. I draw from my research on climate resilience projects in rural India to elaborate on these themes and drive home political implications.

Producing the Concept of Climate

The first element of an STS approach to climate change research is to unpack the core concept of 'climate'. Very often, the concept of climate is held to be a self-evident category. The World Meteorological Organization, for example, provides a short definition that states climate is simply the 'average weather' or, more expansively, 'the statistical description in terms of the

DOI: 10.4324/9781003409748-33

mean and variability of relevant quantities over a period of time ranging from months to thousands or millions of years' (IPCC, 2007, p. 869). This idea of climate is produced by measuring assorted meteorological processes, such as variations in temperature, humidity, atmospheric pressure, precipitation, wind, and atmospheric particle count to create statistical representations of what the weather 'ought to be' at any given time in any specific place. As the climate researcher Mike Hulme notes, to produce climate in this way requires an elaborate circuitry of technologies and institutions to measure and process long-term weather trends and turn them into this idea of 'climate' as the average weather. Through this process, local weather is captured in a set of quantitative measurements, circulated upwards through meteorological bureaucracies, and then aggregated into regional and global indicators of climate. Once sanitised in this way, they are transported back down to their initial localities as predicted trends for the future (Hulme, 2010).

The first thing to note about this way of thinking about climate is that it is divorced from the social world of experience and interactions between humans and weather. Indeed, by going to great lengths to set up measuring stations where – as best possible – they cannot be influenced by human processes, meteorologists seek to ensure that their measurements are not influenced by social factors. On the contrary, meteorologists are keen to model climate 'as it really is', prior to and external from subjective human interpretation and activities.

Second, this creation of a concept of climate as aggregated empirical data is indispensable for the purposes of climate science and its emphasis on modelling climate at both regional and global levels. Having produced a statistical reading of local climate by aggregating chosen weather indicators, this data is then processed through centralised meteorological stations to provide an account of weather trends and fluctuations at a wider level, including both short-term changes and long-term tendencies. Ultimately, such regional climates are then positioned within an overarching model of the global climate system that – through sophisticated computational techniques – seeks to capture the complex interactions that occur across atmospheric levels, the Earth's surface, and the biosphere. The global climate system is seen to evolve over time through either internal shifts or what are termed 'external forcings' that include volcanic eruptions, solar variations, or human modifications including the release of anthropogenic emissions of greenhouse gases. Anthropogenic dynamics are therefore represented as an important factor, yet they are seen in terms of an *outside* influence upon an otherwise coherent natural system (Hulme et al., 2008).

There are, however, contrasting models of climate systems, each of which is framed by different indices, parameters, baselines, timeframes, and spatial boundaries that are culturally and politically influenced. For instance, some models use a 30-year baseline to create the average of statistical meteorological readings that we then call climate while others use shorter or longer timeframes. The precise representation of climate as 'the average weather' therefore varies according to the professional conventions that scientists use to govern what is measured and over what temporal periods and spatial zones. This variability underscores some of the uncertainties about the robustness of the predictions generated by climate models (Edwards, 2010).

Notwithstanding this variation, modelling climate as a global system provides essential tools for representing and understanding changing biophysical relations, patterns and processes. It serves as an indispensable tool for developing a coherent analysis of the relationships between the Earth's atmosphere, landmass and oceans including changes in solar and terrestrial radiation, ocean temperatures, precipitation, atmospheric and other processes without its evolving techniques. Climate science is therefore an essential means through which we can better understand environmental change on a global scale.

Unsettling the Concept of Climate

What is less noted, however, is that producing this idea of climate as an independent and external physical realm of global atmospheric processes does not closely accord with the lived experience of climate, in which meteorological processes are bound up with social dynamics and experiences. For all its undoubted scientific complexity, the **socio-technical processes** through which modelling operates deliberately extract climate from its human and cultural settings. As Hulme (2008, 7) cautions:

> A rainstorm which offers an African farmer the visceral experience of wind, dust, thunder, lightning, rain – and all the ensuing social, cultural and economic signifiers of these phenomena – is reduced to a number, say 17.8 mm. This number is propagated into the globalised and universalising machinery of meteorological and scientific institutions and assessments where it loses its identity.

One might think that climate has always been understood in this fashion as 'the average weather'. It is notable, however, that this concept of climate is a relatively recent phenomenon that follows the rise of scientific rationalism in the nineteenth century. Earlier representations of climate instead centred upon the relationships between humans, meteorological phenomena and other living creatures that form part of our environments. The environmental historian Vladimir Jankovic, for example, notes how early modern naturalists considered climate as a spatial frame of reference used to categorise and evaluate local features of both weather and society in combination (Jankovic, 2010). Rather than denoting a long-term statistical average, climate in this sense was understood as a series of active and evolving relationships that encompassed local weather, topography, biological life and humans. Put simply, this earlier concept of climate focused not on statistical climate but rather on how humans live and work with meteorological trends and events. Climate in this earlier sense was a lived process: the way in which humans interacted with the weather in distinct environments that had inextricably natural and social components.

So, why should we care that climate science reduces climate to a statistical artefact that is extracted and held separate from human experiences and social dynamics? The problem is that this approach to climate is strongly engrained in how we address climate change in ways that create responses that are both limited and limiting. Importantly, having fixed climate as something biophysical and that exists in separation from society, climate science can only bring the two back together as external influences. Humans are seen as an externality to climate, much as climate is seen as external to society. Each is seen as part of a bounded system that corresponds to their own structuring processes to which the other exerts an external influence in the form of impacts, constraints or shocks.

For example, the periodic IPCC reports that draw together the cutting-edge of climate research follow a common mode of representing climate change in terms of these external relations. First, climate change is presented in terms of changes to global climatic systems captured by modelling. These are then seen to impact downwards in the form of external shocks to social and environmental systems. The latter must subsequently adapt as best they can. The causality is clear. First, the climate changes, this then impacts prevailing environmental conditions, ultimately leaving societies needing to adapt. This process in which climate change is represented in terms of external shocks and stresses leads us towards policy approaches to climate change adaptation that are highly technical and managerial: humans must adapt to a changing external environment through new technologies and practices.

But what if there was a different way of looking at the relationship between climate and society that didn't counterpose them as external systems? When we look at daily human activities, we can see clearly that climate is not external to us. Rather we continually and actively build meteorological forces into our **lived environment**s, both rural and urban. Let's consider two examples drawing on both rural and urban cases. First, consider how farmers consciously strive to produce landscapes that work with meteorological processes to produce amenable local climates suitable to their agricultural goals. To do so, they purposely employ numerous techniques such as planting or removing trees to influence localised temperatures, wind velocity, evaporation and exposure to sunlight. Similarly, they seek to shape precipitation by building infrastructures that condition how rainfall flows onto and within their environments. Such practices are often intimately connected to locally generated forms of knowledge. Michael Dove, for example, notes how farmers in rain-fed north-western Pakistan have complex understandings of the interactions between trees and crops that focused on how trees shaded crops, regulating temperature and altering the soil moisture content, with both good and bad consequences for yields and which necessitated localised management strategies (Dove 2005). In short, farmers do not simply adapt to climate, they actively seek to produce it.

Similarly, urban environments are also co-productive of climate. Although many urban dwellers might consider the weather as an outside background, our towns and cities have been built to shape how weather manifests itself, both in positive and negative ways. On a hot summer day, you only have to walk down a street planted with trees and grassed areas to know that it is far cooler than a purely concrete area. Indeed, heat islands in inner cities create microclimates by raising temperatures by several degrees compared to tree-lined, affluent suburbs. At the same time, cities are built to shape flows of water while buildings often provide their own microclimates, with air-conditioning that cools the inside albeit with the unintentional impact of furthering the outside heat island effect through the inexorable churn of AC generators. For example, across a 25-city study of urban heat islands, Chakraborty et al. (2019) find that poorer neighbourhoods experience elevated heat exposure owing to an excess of exposed concrete and an absence of green space. Put simply, social marginalisation leads to a very different tangible climate in those neighbourhoods compared to their more affluent neighbours, even though technically they all share the same abstract climate zone.

This makes clear that the local climate of a city or a rural area is not purely natural: it is co-produced by the **social infrastructure**s and built environments through which climate in its tangible form is experienced. Anthropologist Tim Ingold seeks to capture this expanded notion of climate through his concept of the inhabited '**weather-world**' in which life actively reproduces itself by binding the weather into substantial, living forms (Ingold 2007). For Ingold, these weather-worlds are indivisibly natural and social and take shape at the intersection of social organisation, meteorological processes and the role of other biological actors (trees, crops, animals, etc.). In this framing, climate does not pre-exist as a natural environmental system that provides external stimuli to human lives. Rather, it takes form in a 'hybrid realm' produced through the interactions of meteorological forces, social practices and built infrastructures, including the role of other, non-human agents including trees, plants and animals.

From Concepts to Policies

This may all seem like academic hot air – two contested ideas of climate written out in the pages of scholarly texts. In a world marked by climate change, the big question is what does this alternative understanding of climate mean for policy and practice? To clarify, here is an example that contrasts both conceptualisations of climate in a policy setting. In my research on climate change in rural India, I note how policymakers have increasingly identified climate change as the primary threat to agrarian livelihoods. Put simply, climate change is argued to present itself in increased risks from weather extremes, with increased likelihood of drought or extreme rainfall and other meteorological phenomena that threaten crops and livestock. This in turn has led to repeated public policies and planning aimed at helping smallholder farmers to adapt. Smallholder farmers, it is typically argued, are highly vulnerable to climate change and adaptation policies are therefore necessary to help build their resilience. Working according to the climate-society dichotomy, this framework represents adaptation as a process of planned adjustments to protect against a changing, external climate. Typical policies include introduction of drought-tolerant varieties of staple crops, better ways of collecting and maintaining rainwater, and new approaches to cultivating crops that use water more efficiently (Taylor and Bhasme, 2019).

All of this seems eminently sensible. Until, that is, you talk to smallholder farmers and listen to their very different experience and conceptualisation of the dynamics that produce their lived environment. For smallholder farmers, the concept of climate does not refer to the 'average weather'. Rather, they draw attention to how particular meteorological processes such as rain, humidity and heat are entangled with the social relations through which they practice agriculture. In short, when describing their own lived experiences, they implicitly use the second understanding of climate highlighted above. They are not simply adapting to climate change, they emphasise, but are simultaneously entangled in relationships of indebtedness in which they must borrow money from local landlords or merchants at high interest rates to buy the inputs for the season ahead. They also note how their vulnerability is predicated upon the lopsided contracts that middlemen bind them to for the sale of products. Or to the pests that have become tolerant to the expensive chemicals they spray on their lands. Indeed, it is the way that all these elements work together as part of the broader environment that is key.

Without doubt, such farmers readily acknowledge that increased incidences of extreme weather pose new problems. However, they insist that weather extremes are only one element of a broader lived environment that is indivisibly social and natural. A withered crop manifests itself in increased indebtedness that results in an inability to upgrade irrigation infrastructures which in turn means that farmers must pay one third of their harvest to more affluent farmers who can pump in crop sustaining waters (Taylor and Bhasme, 2021). In short, far from being vulnerable to climate change, smallholder farmers typically indicate that they are vulnerable to how weather processes interact with social dynamics. Together, these power relations structure their day-to-day lives, creating opportunities and constraints. While policymakers are rigidly focused on farmer exposure to weather extremes in the fields, farmers themselves stress that their relationship to weather is shaped in-and-through relationships of debt, hierarchy and dependency. For them, it is impossible to separate their experience of climate from this lived environment.

The policy implications are notable. What policymakers typically see as a technical problem of farmers *versus* an external changing climate is far more complicated. Farmers engage with climate in-and-through unequal social contexts. Reframed in this manner, the political space for addressing climate change expands. While climate change adaptation policies often advocate for development of climate-resilient crops as a technical fix for farmers to address an external problem, placing climate as part of a lived environment opens the possibility that reshaping the debt relations that disempower farmers from working productively with meteorological forces could be a more appropriate and socially just approach.

Conclusion

The political scientist Timothy Mitchell once noted that the more natural an object appears, the less obvious its discursive manufacture will be (Mitchell 2002, 210). Unpacking seemingly natural concepts – such as climate – and examining what they include and exclude is an important political act because it can help unmask alternative understandings of the processes at hand that are marginalised in official discourses. Building upon this sentiment, the current chapter started by questioning the conventional understanding of climate as 'the average weather' that can only be understood through complex socio-technical circuitries that measure meteorological trends and then model them through advanced computational processes. While such a conceptualisation of climate is vital to the important process of modelling past, present and future trends, it provides an insufficient basis for understanding the nature and scope of how societies must grapple with climate change. This is because humans do not simply encounter climate as an abstract, external system. Rather, they actively work with it by building meteorological processes into our lived environments. Climate as tangibly experienced by humans is not simply 'the average weather'. It is the interrelationship between meteorological processes, physical infrastructures, social relations and cultural values.

One strong advantage of adopting this expanded understanding of climate is that it enables us to better conceptualise how local climates are produced through complex power relations and social hierarchies. This insight is very important when we try to think through the climate justice issues intrinsic to adaptation and mitigation. As we saw in the final section, varied actors seek to shape how meteorological forces are harnessed within the lived environment in ways that provide benefits and security to them while externalising the risks and costs onto others. In short, from the rural to the urban, local climates are produced in typically inequitable ways in which risks and benefits are unevenly distributed.

Reframing climate in this way changes the fundamental questions we face when considering climate change policies. If humans do not simply *adapt to* climate change but rather *co-produce* the climates we inhabit, then there is a pressing need to move away from technical and managerial solutions that seek to protect society from climate change. Instead, we must reject the commonplace idea of adapting to climate change; and instead embrace the idea of transforming the way we produce societies and climates alike (Nightingale et al. 2000). This is a big statement, but one that is increasingly becoming recognised as marking an important shift in how we address the challenges posed by climate change. To learn to live with climate change requires us to engage and foreground how people experience climate in a tangible, experiential manner and to tease out the social inequities that produce very different climates for distinct social groups. To do so requires that we seek to empower climate policies from the bottom up, ones that build upon the lived realities of climate rather than the ideas of climate change that circulate within dominant institutions. This is the essential task before us for which new methodologies of research and practice are now emerging (Mehta et al., 2021).

Key Term Definitions

Lived environments: this concept emphasises that humans do not simply occupy an environment, we help create it. As such, our lived environments are constantly changing through the coupled social and biophysical forces that actively rework them over time.

Socio-technical processes: this term refers to the way that many driving forces of change in our world involve simultaneously social and technical components that are bundled together in ways that make them hard to tease apart. For example, to create the concept of climate as an abstract system of interacting biophysical forces on a global level requires both social processes (choices and conventions about what to measure, where and how to do so) with technical processes (advanced measurement systems and enhanced computer programming) that reinforce one another.

Social infrastructures: the varied institutions, practices and discourses that together shape the social services, including health, education, employment and security of a population.

Further Reading

Ingold T. (2011) *Being Alive: Essays on Movement, Knowledge and Description*, London: Routledge.

Mehta, L., Adam, H. and Srivastava, S. (2021) *The Politics of Climate Change and Uncertainty in India*, London: Routledge.

Taylor, M. (2015) *The Political Ecology of Climate Change Adaptation: Livelihoods, Agrarian Change and the Conflicts of Development*, London: Routledge.

References

Chakraborty, T., Hsu, A., Manya, D. and Sheriff, G. (2019) 'Disproportionately higher exposure to urban heat in lower-income neighborhoods: A multi-city perspective', *Environmental Research Letters*, 14(10).

Dove, M. (2005) 'Shade: Throwing light on politics and ecology in contemporary Pakistan', pp. 217–55 in *Political Ecology across Spaces, Scales, and Social Groups*, edited by Susan Paulson and Lisa Gezon. New Jersey: Rutgers Press.

Edwards, P.N. (2010) *A Vast Machine: Computer Models, Climate Data, and the Politics of Global Warming*. Cambridge, MA: MIT Press.

Fleming, J.R. and Jankovic, V. (2011) 'Revisiting Klima'. *Osiris* 26(1): 1–15.

Hulme, M. (2008) 'Geographical work at the boundaries of climate change'. *Transactions of the Institute of British Geographers* 33(1): 5–11.

Hulme, M. (2010) 'Cosmopolitan climates: Hybridity, foresight and meaning'. *Theory, Culture and Society* 27(2–3): 267–76.

Hulme, M., Dessai, S. Lorenzoni, I. and Nelson, D. (2008) 'Unstable climates: Exploring the statistical and social constructions of "normal" climate'. *Geoforum* (40): 197–206.

Ingold, T. (2007) 'Earth, sky, wind, and weather'. *Journal of the Royal Anthropological Institute* 13(S1): S19–S38.

IPCC (2007) *Climate Change 2007: The Scientific Basis:. WG I Contribution to IPCC 4th Assessment Report*. Cambridge: Cambridge University Press.

Jankovic, V. (2010) 'Climates as commodities: Jean Pierre Purry and the modelling of the best climate on Earth'. *Studies in History and Philosophy of Modern Physics* 41(2): 201–7.

Mitchell, T. (2002) *Rule of Experts: Egypt, Techno-politics, Modernity*. Berkeley: University of California Press.

Mehta, L. Srivastava, S., Movik, S., Adam, H., D'Souza, R., Parthasarathy, D., Naess, L.O. and N. Ohte (2021) 'Transformation as praxis: Responding to climate change uncertainties in marginal environments in South Asia', *Current Opinion in Environmental Sustainability*, 49, April 2021.

Moore, J. (2013) 'From Object to *Oikeios*: Environment-Making in the Capitalist World-Ecology'. In *Capitalism in the Web of Life: Ecology and the Accumulation of Capital*, edited by J. Moore, London: Verso Books.

Nightingale, A.J., Eriksen, S., Taylor, M. et al. (2020) 'Beyond technical fixes: Climate solutions and the great derangement'. *Climate and Development*, 12: 343–52.

Taylor, M. and Bhasme, S. (2021) 'Between deficit rains and surplus populations: The political ecology of a climate-resilient village in South India'. *Geoforum* 126: 431–40.

26 Farming in Climate Crisis

Agricultural Adaptation(s) in Central New York State

Tamar Law

Introduction

Seated across from me on a wooden slatted bench at a well-worn kitchen table, Yanna, an organic vegetable farmer, proudly points to her T-shirt, with the words "Farming in Climate Crisis" emblazoned across her chest. Her husband sits beside her in a matching shirt. She repeats a sentiment that many farmers shared over my eight months of fieldwork in the Finger Lakes region of Central New York (CNY)—climate change was beginning to impact their livelihood. The two farmers had deliberately worn this T-shirt for our interview, knowing that I was particularly interested in understanding how farmers in the region were experiencing more frequent extreme weather events. We had connected because another farmer told me that Yanna's story was a cautionary tale of the increasingly erratic and challenging weather.

Yanna's sun-spotted, slightly weathered hands gestured animatedly as she recounted their experience of an extreme weather event in 2018 that devastated her farm. Her husband nodded gravely along. In mid-August, their county experienced 11.53 inches of rain over two days, more than a three-month average. Most rainfall occurred overnight in a span of eight hours. Governor Andrew Cuomo called for a state of emergency in their county. They described waking up and looking out the window and seeing her fields washing away, the road awash in soil. Rushing outside, Yanna frantically put on her muck boots, which quickly filled with water, making it hard to walk. Miming moving through quicksand, she recounts:

> And I went like that … and I was wading through the water. And it was insane. Anyway, you know, my cucumbers, zucchinis, tomatoes came, you know, kind of floating by, floating by. And I'm like, this is bad. This is really bad. You know, it was like a big soup pot.

As Yanna watched her fall crops, sardonically referred to as her "mortgage payments," and her topsoil float away, this event solidified her commitment to adapting her farming practices to a changing climate. Yanna's story is just one example of how climate change is impacting New York State (NYS) farmers. My aim in this chapter is to describe how agricultural actors, including small-scale alternative farmers, extension agents, and policymakers, are conceptualizing and engaging agricultural adaptation, demonstrating that agricultural adaptation looks and is executed differently by different actors. This chapter situates the concept of agricultural adaptation by tracing a genealogy of "climate adaptation" more broadly, outlining some of its benefits and criticisms, and providing an overview of agricultural adaptation. I then describe the agricultural landscape of the Finger Lakes, offering a brief history of the land, and the impact of climate change on agriculture in the region. Through empirical examples from my fieldwork, I describe three different forms of adaptation (adjustment adaptation, everyday adaptation, and

DOI: 10.4324/9781003409748-34

transformative adaptation) to show that agricultural adaptation in the same place can hold different visions of climate adaptation, and serve different political, economic, and social purposes.

Situating Agricultural Adaptation: A Brief Genealogy of the Adaptation Concept and Critiques

Climate Adaptation

Adaptation has become a foundational approach for responding to climate change threats. The logic of adaptation has a long history that is worth recalling. The concept of adaptation emerged in the 1960s from intellectual currents in the fields of cultural ecology, ecological anthropology, and systems theory. In these fields, **adaptation** refers to the process by which communities adjust their behaviors and practices in response to changes in their environment, including for example changes in natural resources, weather patterns, and socio-economic conditions. Since its inception, adaptation quickly became a popular approach to both studying and managing environmental change, as it provides a way to understand and reduce vulnerability, improve resilience, and produce positive environmental and health co-benefits. **Climate adaptation**, as it has been applied in the context of global warming, is the practice of adapting and modifying human behaviors, infrastructure, and ecosystems to respond to the impacts of climate change. The primary goal of climate adaptation is to mitigate the vulnerability of societies and ecosystems to climate-related hazards while enhancing their ability to cope with evolving climate conditions.

Agricultural Adaptation

Since the early 2000s, with increasing anxiety about challenges and threats to food systems posed by shifting seasonality and extreme weather events, there has been growing interest in agricultural adaptation to climate change. Agricultural soil is intimately linked to climate change, not only in regulating greenhouse gases and storing carbon but also as a leading source of emissions, with some estimates attributing 25 percent of emissions to agriculture (Stockmann et al., 2013). Soils overall maintain around 2,000 gigatons of organic carbon and provide a crucial role in cycling greenhouse gases in the atmosphere (ibid.). However, this role has been undermined, due to large-scale deforestation and land conversion for agricultural purposes, and industrial agricultural practices, such as excessive tillage and monocultural production, which impair soil's structural integrity and microbial diversity.

If agriculture is problematically contributing to climate change, it also faces significant impacts from climate change. As global temperatures rise, there has been an increase in the frequency and intensity of extreme weather events, posing significant challenges to agricultural production systems worldwide. As deforestation, land conversion, and agricultural practices disrupt soil structure, topsoil becomes increasingly vulnerable to the impacts of heightened droughts and flooding. Such challenges, in turn, must be dealt with. **Agricultural adaptation** refers to the process of adjusting agricultural practices, technologies, and policies in response to the impacts of climate change. Examples include the development of drought-resistant crop varieties, changes in planting dates and crop rotations, the adoption of conservation agriculture practices to improve soil health and water management, and the use of early warning systems to anticipate and respond to weather extremes. Agricultural adaptation seeks to increase the resilience of agricultural systems and to maintain or enhance agricultural productivity and food security in the face of climate variability.

Studying Adaptation and its Critiques

As climate adaptation has increasingly been used as a policy paradigm from international to local scales, scholarship in Science and Technology Studies (STS), anthropology, and human geography have unpacked the concept of adaptation and its application in climate interventions. This work understands that climate adaptation is a complex, socially constructed, situated, and at times contested process. It examines adaptation's social, cultural, political, and economic dimensions. STS scholarship emphasizes the role of science, technology, and knowledge production in shaping adaptive practices, policies, and interventions. Evaluating adaptation as a sociotechnical process, such scholarship reveals how adaptation is embedded in societal and institutional contexts, values, and interests—exploring how different actors, including scientists, policymakers, communities, and organizations, envision and enact adaptation strategies in response to climate change. These visions of adaptation are **sociotechnical imaginaries**, the wider cultural and social narratives that shape views, anticipations, expectations, and attitudes regarding technology. Ultimately, such work elucidates the entanglements of science, politics, and society that shape adaptive responses to climate change to inform more transformative and just forms of adaptation.

Such critical work has questioned if "adaptation" actually reduces vulnerability, whether or not it serves communities over the long term, and if it can be implemented equitably. Let us outline these general critiques, which we can then take up in the case of the agricultural sector. Socially, scholars caution that adaptation can overlook entangled socio-environmental and socio-political relations which could lead to **maladaptation**, ultimately exacerbating social and environmental vulnerabilities. Politically, adaptation has been critiqued for addressing the impacts of climate change in the short term to maintain a status quo, while ignoring the long-term root causes of greenhouse emissions. Climate adaptation is critiqued for functioning as an **anti-politics**, de-politicizing and concealing the socio-political causes of the climate crisis and its impacts, rather than enabling transformational change. Economically, some have argued that the cost and benefits of adaptation are not equally felt. Vulnerable and front-line communities may bear a disproportionate burden of the costs, while wealthier populations may derive more of the benefits. Yet, despite numerous critiques, adaptation remains a pillar in responding to climate change, including in agriculture. For the remainder of this chapter, I follow an STS approach to evaluate the values, conceptions, and plurality of agricultural adaptation in CNY.

Agriculture and Climate Change in CNY

As adaptation is situated in particular geographies and social histories, this section outlines the bucolic agricultural landscape of the Finger Lakes Region of CNY. This region has a powerful intellectual and practical history of agriculture and soil science. Cornell University was founded in 1865, in large part to serve the surrounding agricultural communities through a robust, statewide cooperative extension network. Widely known for its wine production and terroir, the region boasts the largest expanse of prime farmland in the Northeast and the most productive soil for field crops due to its limestone components. As one cooperative extension agent explained to me, this area was agriculturally bountiful, because "the soil can take a 'lickin' and keep on 'tickin'." The homeland of the Haudenosaunee (Iroquois) people, the history of the land is marked by violent claims over access to this fertile soil. According to the earliest soil survey in this region, when George Washington sent John Sullivan and his soldiers in the genocidal Sullivan-Clinton scorched earth campaign of 1779 to forcibly remove the native people from their land, the soldiers were struck by the quality of crops. In time, many of the soldiers returned as settlers, starting their own farms on stolen land (Bonsteel, Fippin, and Carter, 1905).

Now the region is marked by rural poverty. Once defined by small dairies and family farms that dotted the rural expanse in the early 1900s, the spread of large-scale conventional agriculture led by the former Secretary of Agriculture Earl Butz's doctrine of "get big or get out" changed the character of agriculture in the region during the 1970s. As many small farms shut their doors, unable to maintain economic viability and ways of life, local economies faltered, and the historical agricultural landscape changed. Driving on the backroads, decaying farmhouses and barns are as familiar as hundreds of acres of corn and soy planted in rows. These fields were described to me by one land access advocate as a false mirage. While people driving by may be comforted by the illusion of agricultural abundance, these fields in fact represent a damaged monocultural landscape. However, over the past 30 years, this large-scale conventional farmscape has slowly been changing.

Beginning in the mid-1980s, the back-to-the-land movement brought a wave of new farming efforts to the region. Many first-generation farmers arrived, enticed by loamy soil and dirt-cheap land prices with a vision of an alternative form of agriculture. These farmers began what some farmers termed an agricultural renaissance, the process of re-agrarianification with a renewed emphasis on creating local, alternative, and small-scale agriculture, an emphasis that has only continued to grow. Alternative agriculture, broadly defined as agricultural practices that do not follow conventional methods, encompasses a wide variety of farming practices, including but not limited to organic, sustainable, restorative, permacultural, agro-ecological, and regenerative forms of agriculture. Now, alternative agriculture is a mainstay of local agriculture in the Finger Lakes. Yanna and her husband are part of this wave of agriculture, having bought their farm in the early 2000s, and choosing to use organic farming methods. Like many of the farmers I interviewed, they are white, middle-class, owned their land, and following the USDA definition of small-scale agriculture (earning below $250,000 annually and farming below 231 acres) are classified as small-scale growers. While Yanna's and her husband's experience is one extreme case of the threats of increasingly erratic weather to small-scale farmers, their general worry about their farm's health and economic persistence and their commitment to agricultural adaptation was a constant in my fieldwork.

The rising threat of climate change has become resoundingly clear in NYS and has become a chief concern for the state and farmers alike. From 1970 onwards, the average annual temperature across the state has experienced a rise of approximately 0.6°F per decade. Furthermore, during the winter season, the rate of warming has surpassed 1.1°F per decade (DEC, 2021). Simultaneously, there is a trending increase in annual precipitation mainly concentrated in the winter and springtime, with a rapid uptick in extreme weather events. Storms like the one that devastated Yanna's fall crop, described to me by many farmers as a "one in a hundred-year event," have occurred three times between 2017–2020, a stark indicator of a shifting baseline of normalcy. The generalized warming has led to a shift in seasons and seasonality, with winter snow cover becoming less dependable, and spring beginning at least a full week earlier than decades prior. Farmers described to me how the seasonal nature and timing of farming are changing. For example, tomato crops are struggling with excessive rain, perennial crops like apple trees are harmed because they are budding too early before the pollinators are out, and early summer crops are destroyed by late-season frosts. And that is just the observed present. For the future, New York's climate is expected to be comparable to that of Arkansas by 2080, opening questions about how agriculture in the region may shift to accommodate such a change.

Adaptations: Adjustment, Everyday, and Transformative Adaptation

While agricultural adaptation efforts all seek to respond to the impacts of climate change, they are not a monolith. Instead, adaptation varies across different actors based on their relationship to land, economic incentives, and ultimate motives for adapting. Agricultural adaptation embodies different economic goals, values, and politics for different agricultural actors. In the following three sections, I describe three different forms of agricultural adaptation I saw during my fieldwork time which I call *adjustment adaptation, everyday adaptations*, and *transformative adaptation*. These forms each hold promise for successfully dealing with climate change, although each form—along with the "adaptation" paradigm as a whole—faces clear limitations. These empirical sections demonstrate that adaptation, as a concept and as a practice, is not settled. Instead, as shown, the ways agricultural adaptation is conceptualized and implemented is diverse and each articulates different priorities and approaches when it comes to responding to climate threats.

Adjustment Adaptation

In 2013, Cornell University, which oversees the statewide cooperative extension network, initiated a Climate Smart Farming (CSF) program, founded as a voluntary initiative to help farmers in NY increase farm resilience to extreme weather events. The CSF program consists of extension specialists with expertise in climatic risk management who host information events, webinars, and train local extension agents. CSF works in tandem with the NYS Soil Health Initiative (NYSH), to provide soil health programming to farmers for adaptation and resiliency purposes. To gain insight into the perspectives of extension agents, I attended various agricultural outreach events, including field days, conferences, and workshops, to understand their agricultural imaginaries and understanding of agricultural adaptation.

For extension agents, a central focus of adaptation outreach was improving economic productivity. Such a productivist orientation to adaptation originates from the foundation of soil science and extension outreach to increase soil fertility and yields. Across field days and workshops, a focus on soil health management, including the use of cover crops, reduction of tillage, increasing the inputs of soil organic matter (such as manure and compost), row cover and tarping, and crop rotation, was posited as ways farmers could adapt to increasing extreme weather events while simultaneously improving agricultural productivity and profits. In this framing, climate adaptation was presented as a business prospect in which farmers can "minimize the risks and capitalize on opportunities" (CSF, 2020).

Specifically, CSF extension agents emphasized soil health as the primary mechanism for adaptation, highlighting how healthy soil can improve crop yields, reduce fertilizer requirements, increase carbon sequestration, minimize erosion, leaching, and runoff, and enhance pasture quality. Central to these educational outreach programs, soil health was communicated to enhance soil's productive character for profits. In more explicit words, as one program handout describes, "climate preparedness makes good business sense" (CSF, 2020). Appealing heartily to the business identity of farmers, implementing the best practices for soil health was often explained as a means of maximizing their bottom line, as depicted through the one slogan scrawled across one extension vehicle "Soil Health = Continued Wealth for New York Farms." In the CSF workbook (2019), climate adaptation was broadly outlined as "the adjustment in practices to moderate harm or exploit beneficial opportunities of current and future climatic conditions." Adaptation was seen as strategies that could readily be added to farmers' existing systems, specifically strategies that could provide economic benefits.

In this imaginary of agricultural adaptation, the locus of responsibility and action is scaled onto the farmer who is called on to change their practices. This form of adaptation aligns most closely with what Bassett and Fogelman (2013, p. 50) have termed *adjustment adaptation*, which ultimately "draw attention to responses to climate change rather than the social causes of vulnerability." In what some critics have termed a band-aid approach to adaptation, adjustment approaches understand society in a state of equilibrium and understand the purpose of adaptation is to return to this state, reinforcing the status quo (Ribot, 2011).

This recasting of climate change, a political issue, in technical terms to be "addressed" via soil health practices conceals productivist agriculture's link to climate change. While this depoliticization may not be overtly intentional, by centering productivity and profit, this approach maintains a capitalist agricultural status quo that has contributed to climate change in the first place. It is important to note, in this case of soil health, as the educators described, the benevolence of appealing to a farmer's business identity to help farmers improve their practices is not coming from a space of environmental malintent, but rather a genuine place of goodwill. Yet, through such a presentation, they uphold a historic relation to soil that perpetuates soil's inherent link to climate change. For some critics, upholding this productivist orientation via adaptation maintains a system that exacerbates the very vulnerability soil health is attempting to redress. Thus, while an adaptation approach such as CSF may be beneficial for the state and the farm's bottom line, it may perpetuate the current agrarian status quo by depoliticizing agriculture's links to climate change. In this next section, I move to small-scale alternative farmer's own notions and conceptions of adaptation as an everyday practice, which center on the farmer's stewardship and care for soil.

Everyday Adaptations

While productivity and profitability were important to the farmers I interviewed, they also emphasized that adaptation was an iterative, daily practice informed by their practical knowledge about their individual circumstances along with their deep care for the land they work. Many farmers indicated there was not one set of "best practices" to adaptation. Rather, adaptation would be different from farmer to farmer, based on their relation to their soil and farm ecosystem. Adaptation to them meant understanding their connectivity and responsibility to their land, an everyday practice where they were attentive to the needs of their soil. Importantly, this approach centered on their identity as stewards of the land, rather than their business identity.

As many farmers pointed out to me, adapting was an intrinsic part of farming, a constant negotiation between the farmer, the soil, and the weather. Adaptation for them was not something new, not a discrete direction—it was a way of life, many described as a "reflex" necessary to work the land (see Taylor, Chapter 25, this volume). As the following farmer described, adaptation was part of the farmer's lifestyle and care ethic grounded in the embodied practice of working the land every day.

Farmers are the stewards of the environment. We're everyday out in the field, getting our hands in the soil and staying grounded to the earth. And I think with that comes the responsibility of taking care of not just the land but also the air and the water. And it's not like I just get done working for the day and then stop, you know, caring about the environment … it's more of a lifestyle.

This lifestyle is an example of what Clay (2023, p. 256) terms *everyday adaptations*, defined as "land use and livelihood adjustments that are woven into the fabric of daily life." In one farmer-led workshop at the Northeast Summer Conference, farmers cautioned that the framework of soil health as used by CSF puts forth a set of practices that support soil generically. Instead, the workshop described how a farmer needs to sensitize themselves and respond to the specific needs of their individual soil. As one farmer explained, this is a fundamental difference in the approach to soil that he and other alternative agricultural farmers take as opposed to climate smart soil approaches. He explained that understanding agriculture as a set of best practices puts the complexity of agriculture into a scientific vacuum and "strives to break it all down, compartmentalize everything." Instead, he describes farming as a co-creative process, an approach which he described as a mutual dance, "if you go dancing, you're not like 'you dance this way and then you do that'." You have to flow with things. Use your intuition. "Oh, I think I might need to put more of this down or maybe more chicken manure." According to this farmer, through this synchronicity, the farm is adaptive to change and ultimately more resilient in the face of climate shocks.

This resulting imaginary of climate adaptation contrasts sharply with the business-centric imaginary of adaptation disseminated by CSF, as the following farmer describes. Through this iterative and responsive practice, the farmer becomes responsible for the land, learning that what might be economically good for them may not be good for the land.

> as opposed to that attitude of domination and monetization which I think is like at the heart of our entire economy and way of life and even our brains probably, but certainly at the core of most of agriculture. You know, to me, a good farmer is a farmer that is... just actively engaged in observation by constantly observing what is going on... and doing their best to respond.

As farmers demonstrated their soil practices to me, it became clear that these practices were not static, following a prescribed set of "best practices," but rather, a dynamic constantly amended through what they learned from their lived experience with soil. Some regenerative farmers took this one step further, centering their role as stewards as part of a broader political shift in their farming ethic to transform agricultural production. I start the next section with a vignette from one such farm.

Transformative Adaptation

Reaching to grab a peach off a particularly heavy branch, I recoiled as my hand touched a tacky and unknown substance. A neon green tree frog sat watching me, toe pads gelled to the peach. I am working at a regenerative farm to get a sense of the day-to-day practicalities of farm care, and today my task is to thin out the excess peaches on each branch to allow for larger peaches to grow. I had never seen a frog like this in the region, despite having grown up only 30 minutes away from the farm. Unnerved, I reported this finding to Jim, the farmer of this regenerative perennial farm. He beamed, explaining that it had taken years for this creature to find a home in his trees, an indication of the overall health of the farm's ecosystem. Arriving on this land in 2008, the farm was bare and degraded soil, left damaged from years of continuously growing soybeans and corn. Now, as Jim describes,

> There's like ten times more species here than there were when we started, there was almost nothing here. Just crop species and very few earthworms and maybe that one pest species for that crop like a moth. Now, tree frogs are a significant thing here and there's turtles and higher order birds and hawks … and unfortunately there's loads of voles which is a problem, but there's foxes and blue herons, and the soil life is a lot livelier!

His focus, from the beginning, was on creating a biodiverse farm from the "ground up." Jim implemented regenerative practices that increased soil life and soil organic matter, centering soil at the heart of his system. These included initially covering the soil with diverse grasses to "protect the soil from the top down," intercropping, incorporating perennial crops such as apple trees, making sure to keep some plant roots in the ground year-round, and even the integration of livestock into his tree row alleys for their manure. Soil informed his farm management decisions, and, as he explained was, "the best tool for designing landscapes for uncertain climate times." The treefrog was an indicator to Jim that the farmland itself was "healing" from regenerating the soil, the increase in soil organic matter meaning it was more resilient to drought or flooding, ensuring as he explains "a buffer" for the eventual "bad times" to come.

A growing number of farmers, including Jim, are embracing regenerative agriculture—an umbrella term encompassing various agricultural practices such as no-till, cover cropping, holistic grazing, composting, silvopasture, perennial systems, agroforestry, and agroecology. These practices are explicitly designed to address and mitigate climate change by restoring soil organic matter and enhancing soil microbial biodiversity. Regenerative agriculture not only adapts to climate extremes but also tackles agriculture's contribution to greenhouse gas emissions by facilitating soil carbon sequestration for climate mitigation.

For these regenerative farmers, rather than centering a business or stewardship identity, their imaginary of adaptation centers on a political identity. The regenerative farmers emphasized that regenerative agriculture is a verb, not a noun, a political practice that rethinks agriculture. For these farmers, regenerative agriculture represents more than just a different type of farming—it is a political movement seeking to transform the current agricultural paradigm. It embodies transformative adaptation, which recognizes the causal structure of vulnerability in different socio-economic and environmental contexts as the basis for adaptation planning (Bassett and Fogelman, 2013). Regenerative agriculture directly confronts the link between agriculture and climate change, making it a transformative rather than merely an adjustment approach. They cautioned that regenerative agriculture was increasingly becoming co-opted by the agricultural status quo, and that true regenerative practices put the health of the farm, and the soil, above profit.

The farmers I interviewed explained that adaptation to rising extreme weather was not simply a change in their soil practices as a type of reactive measure, like adjustment adaptation, nor a change in how they worked the land as a daily practice, such as everyday adaptation, but rather a proactive shift in their relations to the farm as a system, starting from the soil. Identifying as "soil farmers," or as one joked, "soilmates," their approach was soil-centric. Farmers noted that certain soil practices rooted in boosting productivity and quick return not only exacerbate a farm's vulnerability to changes in the climate but also are a contributing force. A healthy farm ecosystem arises from healthy enlivened soil, something that cannot be achieved through shortcuts, but, rather, by thinking long term.

These farmers adopted a long-term perspective, forsaking shortcuts in favor of prioritizing the well-being of their soil over immediate economic gain. Importantly, this approach has the co-benefit of enabling lively, regenerated soil to sequester carbon effectively, transforming agriculture into a net sink of emissions rather than a significant source. Consequently, these farmers viewed agricultural adaptation as a political avenue to reshape the relationship between agriculture and the climate, moving beyond simple adjustments or everyday responses to climate change.

Conclusion

From extension agents and soil scientists to farmers, agricultural adaptation has become central to agriculture in the Finger Lakes. To account for diverse approaches to adaptation in this case, I introduced three typologies of adaptation: adjustment adaptation, everyday adaptations, and transformative adaptation. Certain forms of adaptation center agricultural productivity through economic logic. *Adjustment adaptation* may help farmers make minor adjustments to their farming operations, but it also maintains an agricultural status quo. Such an approach may help farmers in the short run, but ultimately it fails to address the root cause of agricultural vulnerability to extreme weather events and a shifting climate. Indeed, it may put a more politicized approach to agricultural adaptation further out of reach. In contrast, *everyday adaptations* take the everyday act and practice of stewardship seriously, sometimes at the expense of a farm's bottom line. While this care-centric approach helps farmers such as Yanna "cope with" extreme weather events and respond to the needs of their land, it still does not directly address the root relation between agricultural production and climate change. Last, *transformative adaptation* uses the opportunity of adaptation to rethink agricultural systems, following Jim's words, shifting the agricultural system from the "ground up." However, such an approach that takes a long-view perspective to adaptation may be limited in application by its time scale. This chapter shows that agricultural adaptation is not just one thing in the Finger Lakes, but, rather, is a diverse socio-culturally constituted practice. Agricultural adaptation means and looks different across various actors. In this way, agricultural adaptation holds different political and social significance, which, in turn, inform distinctive trajectories of agrarian change and agrarian futures amidst climate change.

Further Reading

Eriksen, S.H., Nightingale, A.J., and Eakin, H. (2015) "Reframing adaptation: The political nature of climate change adaptation." *Global Environmental Change, 35*, pp. 523–533.

Guthman, J. (2019) *Wilted: Pathogens, Chemicals, and the Fragile Future of the Strawberry Industry* (Vol. 6). Berkeley: University of California Press.

Paprocki, K. (2021) *Threatening Dystopias: The Global Politics of Climate Change Adaptation in Bangladesh*. Ithaca, NY: Cornell University Press.

Taylor, M. (2014) *The Political Ecology of Climate Change Adaptation: Livelihoods, Agrarian Change and the Conflicts of Development*. New York: Routledge.

Watts, M.J. (2015) "Now and then: The origins of political ecology and the rebirth of adaptation as a form of thought." *The Routledge Handbook of Political Ecology*, pp. 19–50. New York: Routledge.

References

Bassett, T.J. and Fogelman, C. (2013) "Déjà vu or something new? The adaptation concept in the climate change literature." *Geoforum*, (48), pp. 42–53.

Bonsteel, J., Fippin E., and Carter, W. (1905) "Soil survey of Tompkins County, New York." Field Operations of the Bureau of Soils. NRCS.

Clay, N. (2023) "Uneven resilience and everyday adaptation: making Rwanda's green revolution 'climate smart'." *The Journal of Peasant Studies*, 50(1), pp. 240–261.

Climate Smart Farming (2020) "Climate preparedness makes good business sense." [Online]. Available at: http://climatesmartfarming.org/changing-climate/. (Assessed: June 15, 2019).

DEC (2021) "Observed and Projected Climate Change in New York State: An overview." [Online]. Available at: www.dec.ny.gov/docs/administration_pdf/ccnys2021.pdf. (Accessed: June 20, 2023).

Di Napoli, T. (2019) "A Profile of Agriculture in New York State." Office of Budget and Policy Analysis NYS.

Ribot, J. (2011) "Vulnerability before adaptation: Toward transformative climate action." *Global Environmental Change*, (21), pp. 1160–1162.

Stockmann, U., Adams, M. A., Crawford, J. W., Field, D. J., Henakaarchchi, N., Jenkins, M.,... and Zimmermann, M. (2013) "The knowns, known unknowns and unknowns of sequestration of soil organic carbon." *Agriculture, Ecosystems & Environment*, (164), pp. 80–99.

27 Climate Adaptation, Methodology, and the Case Study

Sarah E. Vaughn

Extreme weather reminds us that we now live with vulnerabilities that are unlike those of the past. Alongside scientific assessments, news outlets such as *The New York Times* as well as documentaries including *Before the Flood* (2016) have served as vital platforms for exposing the destruction brought about by climate change. Levies, once an assumed form of coastal protection, are now more prone to collapse from intense storms in places as culturally distinct as Louisiana and Bangladesh. Severe droughts have depleted water reservoirs in Australia. Freak snow storms reveal that power lines need to be (further) weather-proofed in Texas. Rising ocean temperatures may impact the boundaries of marine conservation areas in Belize and the Philippines.

Such events have galvanized a new sense of urgency in the global debate about climate change. The debate has shifted from whether climate change is occurring to concern over what can be done to address climatic impacts and risks, and how quickly these actions should be taken. A central actor in these debates is the Intergovernmental Panel on Climate Change (IPCC). A governing body of the United Nations, the IPCC provides policymakers with scientific assessments about climate change. An important policy recommendation has focused on **climate adaptation**, which involves taking actions that make people and places less vulnerable to the consequences of climate change.

Most of us would expect that the resources governments, charities, and humanitarian organizations provide after a disaster are enough to help rebuild and restore our sense of confidence. Many also assume that disasters can be avoided if we simply "prepare" for extreme weather. But this belief is shortsighted. The concentration of greenhouse gases in the atmosphere due to fossil fuel emissions has yet to stabilize or decrease to levels deemed safe. The environmental damage climate change is now causing will continue to have long-term effects. This is why climate adaptation is central to survival. Climate adaptation provides a roadmap for addressing the challenges of living with a planet in peril.

Defining Climate Adaptation

Political leaders from across the Global North and Global South have advanced climate adaptation projects through the United Nations Framework Convention on Climate Change. The 2010 Copenhagen Convention and the 2015 Paris Convention were significant in that participants vowed to limit global carbon emissions to 1.5°C and launch funding programs for climate adaptation. The Global Climate Change Alliance (GCCA) is one such example. It advises Global North countries about how to provide technical assistance to Global South countries that do not have the money or technical capacity to adapt on their own. To date, funding for climate adaptation projects has focused on the enhancement of infrastructure systems, **ecosystem services**, the

DOI: 10.4324/9781003409748-35

varied benefits natural resources and habitats provide humans and related economic and social institutions, information networks, housing, sanitation, biodiversity conservation, and health care.

Because climate adaptation projects intervene on a wide range of problems, they often require collaboration. Of particular importance in climate adaptation projects is how groups of
experts use scientific models to communicate with one another. However, this is tremendously
difficult to achieve while also considering local communities' needs, existing (in)inequalities,
and access to resources. Not all scientific models adhere to the same norms or conventions for
representing information about (un)certainty and causality. In addition, some scientific models
may be more influential than others for reasons including budgetary and computing constraints,
project leadership, and local perceptions about the urgency of a given environmental threat.

Questions about how to prioritize climate adaptation projects are urgent ones, as well as how
to create pathways for public education about climate change. Many of these questions are best
illustrated through a methodological approach known as the **case study**. Both the sciences and
humanities organize case studies around a shared set of facts about a given event.[1] Following the
insights of historians of science, there's no timeless interpretation of facts because of innovations
in scientific experts' workplace practices, ethos, and technologies. In climate adaptation projects,
these innovations are made most apparent through the ways experts collaborate across disciplinary divides, with the goal of collecting *both* scientific and humanistic facts about vulnerability.
In the remainder of this chapter, I offer a case study of mangrove restoration in Guyana and how
transdisciplinary dialogue unfolds as engineers attempt to protect people and coastal life.

Climate Adaptation and Shorelines under Threat

The majority of my anthropological research has focused on climate adaptation in Guyana, a
low-lying country located along the Atlantic northeast coast of South America. Climate change in
Guyana is expected to materialize as a 4.2-degree increase in average temperature, a 10-millimeter
decrease in rainfall, and a 40-centimeter sea-level rise by the end of the century (Solomon et al.,
2007). Specifically, I have been interested in the work of geotechnical and hydraulic engineers
adapting sea defenses. For centuries they have struggled to maintain seawalls and mangrove forests and groynes (a low wall or barrier built to slow wave action) alongside patches of mangrove
forest. Engineering efforts to address the fate of mangroves is not, however, exclusive to Guyana.

Across the planet, mangrove forests are under siege. Satellite data reveal that between
2001 and 2012, 192,000 hectares of mangrove forest were lost to disasters and rampant human encroachment.[2] Agriculture, fisheries, and urbanization have traditionally posed a risk to
mangroves. Recent estimates indicate that mangroves now occupy only 15 million hectares of
shorelines worldwide. Under the simultaneous threat of climate change and development, mangroves are a vital but vulnerable resource to continued human coastal settlement.

In 2010 the Guyanese government announced the launch of the GCCA-sponsored Guyana
Mangrove Restoration Project (GMRP). As a response to the threat of sea-level rise, the GMRP
looks to mangroves for storm protection. The GCCA committed $4.165 million euros for the
GMRP's initial three years (2010–2012). Expressing goodwill through science, the GCCA also
complements United Nations climate treaties by supporting its Millennium Development Goals.

Much of Guyana's coastal region is inhabitable only thanks to the continual maintenance of
an intricate network of dams, canals, sluices, seawalls, and groynes that keep the sea and rivers
at bay. Engineers refer to a timetable based on the tide, season, and farmers' irrigation needs, to
release water from this network. The seawalls also have sluice gates that allow for the drainage
of water from heavy rains and waves that crest the seawall.

The importance of this infrastructure designed to keep land dry becomes clear when we
look at where people in Guyana live. Guyana has a land area of roughly 83,000 square miles

and a population of around 777,000. Between 85–90 percent of the population lives along its low-lying 268-mile-long Atlantic coast. Portions of the coast sit from 19.7 inches to 39.4 inches below sea level, with 25 percent protected by seawalls, 60 percent by mangroves, and 15 percent by sandbanks or mudbanks. Despite these arrangements, the persistent deposits of river silt and erosion of land make flooding from the sea a constant threat. Ambitious state planning in shore zone management has provided a rationale for sea defense policies regarding everything from housing to roads, commerce, energy, and wildlife protection.

Climate Adaptation and Modeling

For engineers more familiar with seawalls and groynes than mangroves as sea defense, the GMRP has entailed a broad-based research program into mangrove ecosystems. The GMRP focuses on exploring how and why processes of coastal erosion impinge on forests. These research questions have put Guyana's engineers in unorthodox partnerships with other experts, including French geoscientists analyzing erosion and local beekeepers developing apiaries. This is not to say that their combined efforts always prove effective—there are obstacles to be found on the shoreline like waves, mud, and debris from decades of inconsistent waste management and shoreline development. Thus, flooding, erosion, and climate change are processes that have combined in unlikely ways to contribute to experts' encounters with mangrove forests. Looking to rework metaphors about pristine nature, the GMRP produces knowledge about deforestation with the hopes of establishing new systems and interventions of care. Climate adaptation requires expertise that can reconcile humans' and mangroves' mutual vulnerability.

During the initial phases of their research, GMRP participants produced a report describing Guyana as a difficult place for mangrove research. In the face of a call for improved communication between engineering and agricultural agencies, the report highlighted the complex dynamics of erosion that gives the country its distinctive "muddy waters." This cycle is supported by the Orinoco and Amazon rivers, which during the rainy season fill with copious amounts of sediment, some of which can transform into mile-long mudbanks.

Every 30 years, these mudbanks travel down river and migrate southeast by northwest along the coast, dampening waves and providing habitats for mangroves and other marine life. The areas between mudbanks are prone to severe erosion which can cause mangroves to uproot. Over time, the rivers' silt counteracts this dynamic by contributing to the build-up of sandy beaches, and the cycle eventually repeats itself.

Relying on a third-generation quantitative wave model called SWAN, geoscientists have provided engineers with detailed analyses of how mudbank-caused erosion affects the re-growth of mangroves. Their collaborations show how influential the global circuits of localized engineering knowledge are to climate adaptation. With the aid of satellites at the international space station, launched from the Space Centre in neighboring French Guiana, the GMRP began building an archive of mudbank images that spans the regional mudbank migration route from the shores of northern Brazil to southern Venezuela.

The GMRP's modeling efforts underscore that both engineers and geoscientists utilize a case study approach to analyze mangrove regeneration. Importantly, this case study approach draws on the insights of wave behavior and geomorphology. They used models to form a shared knowledge and understanding of erosion at specific locations of mangrove forests along the Guyanese shoreline. Specifically, they contend that when mudbanks migrate to the outfalls of the coast they become habitats for mangroves, creating fertile pockets for their seedlings to take root. At the same time, the SWAN model indicates that many existing mangrove forests in the path of these migrations may become uprooted by mudbanks.

The model shows that some forms of erosion are *not* predicted by the 30-year mudbank migration cycle. In turn, there is much scientific uncertainty about mangrove regeneration as erosion cycles become affected by sea-level rise or changes in wave patterns. To compensate for the models, the GMRP turned to mangrove-planting activities and the knowledge of botanists and local beekeepers to enhance their understanding about the life expectancy of mangroves.

Negotiating Scientific and Humanistic Facts in Climate Adaptation

As engineers and geoscientists continued to model, mangrove-planting activities commenced. Environmental consultants with backgrounds in botany and affiliated with Guyana's Ministry of Agriculture were responsible for implementing planting activities. All but one of the consultants were Guyanese, with prior experience working on state-sponsored coastal environmental projects and, therefore, some experiential knowledge of mudbanks. Their efforts initially focused on planting a species of tree called "black mangrove" (*Avicennia germinans*) because it was the most prevalent species found on Guyana's Atlantic coast. They sourced seedlings from the ministry's nursery, a decision they hoped would ensure sturdier roots than if seedlings were collected from wild forests.

With planting underway, the environmental consultants were eager to collaborate with another group of local experts, beekeepers. Their apiaries had the potential to increase mangrove populations. Instead of interpreting the life span of mangroves as mainly contingent on mudbanks, the beekeepers prompted the GMRP to rethink mangroves' life cycles through pollination as well. For instance, black mangroves have a particular scent, nectar type, and color of flower that contribute to what Guyanese beekeepers call "tar honey" pollination.

One local beekeeper explained to me in an interview the trial and error he had handling Africanized bees (*Apis mellifera*) in black mangroves. When Africanized bees were introduced, most of the indigenous population of stingless bees went extinct. Unexpectedly, the new species pushed beekeepers out of business because they did not know how to "smoke out" the bees' aggressive behavior. By the late 1980s, Edward and his colleagues had adapted to the new species, but it was already too late. Much of the standing mangrove forest where he had worked was gone due to people burning trees for fuel and housing development. Edward explained:

> By the time the mangroves were destroyed I really began to learn the bees' flight, how to read them. I could tell you if the rain was going to fall, how long, when the trees are blossoming … If I visited my hive, I saw them working. They sing and move to make it plain to me that mangroves are healthy despite whatever cycle of erosion.

The beekeepers' concern about their participation in the GMRP, thus, revolved around whether they would have enough resources to acclimate bees to the coast for the long term. In turn, they relied on a case study approach of the life cycle of a given bee colony, to identify patterns of mangrove regeneration. Unlike the engineers and geoscientists, their case study approach drew on not only observations of "natural" erosion but also tactile knowledge and "folk knowledge, of the *changes* in bees' movement in response to urban development.

The beekeepers' efforts are best summarized as combining the scientific and humanistic facts about vulnerability to climate adaptation. Specifically, they claimed that it would take months to adjust their bees to a new landscape. The transition period, they argued, was an intuitive matter as much as a technical one, a point that a well-seasoned beekeeper knows all too well. They also requested private land plots to accommodate their bees' four- to five-mile flight patterns. Yet environmental consultants turned a deaf ear to the beekeepers, encouraging them to provide more scientific evidence for their requests. Citing the engineers' and geoscientists' work with the SWAN model, the environmental consultants warned that even if land officials granted plots, they were not confident about where land would eventually erode. Ultimately, they asked bee-keepers to adjust their methods to smaller areas in reserves along the coast.

The expert focus on pollination created awareness within the GMRP about subtle variations in mangrove regeneration, disrupting certain assumptions on which the SWAN model relied. At the same time, pollination is indifferent to the expert modes of organization, case study approaches that the GMRP envisions as crucial to mangroves' protection. Bees and mangroves do have a co-evolutionary relationship with mudbanks. But bees have their own everyday lives of communication and occupation that have often simply ignored mangroves.

But even as apiary development and planting became integrated into the GMRP's climate adaptation agenda, new knowledge disputes have since arisen amongst Guyanese engineers. Their credibility is now attached as much to the SWAN model for the purposes of knowing mangrove ecologies as to their willingness to collaborate with non-engineering experts to respond to erosion. Some engineers speak about the turn to systematic mangrove planting as "common sense" and long overdue. This professional tension reminds us that climate adaptation can take on multiple meanings and may appear self-evident only after repeated failures of other technical interventions or time wasted. In turn, some engineers have expressed concern about whether mangroves are a reliable form of sea defense given Guyana's complex erosion cycles.

In interviews with me, retired engineers and senior staff at Guyana's Ministry of Public Infrastructure offered critiques of the GMRP through autobiography. They detailed stories about when they were children encountering "courida bush" (black mangrove; *Avicennia germinans*) one day near a groyne, trapping sediment, and a year later the courida bush was gone. Others recounted the ministry's past failed attempts to plant mangroves to manage deforestation caused by urbanization. By the inauguration of the GMRP in 2010, retired engineers formalized such "mangrove autobiographies" in editorials to local newspapers. Far from proving a model of nature as infrastructure, engineers' indoctrination into mangrove restoration further revealed the uneven advancement of climate adaptation projects in engineering professional circles. In many respects, the sharing of these autobiographies demonstrates that engineers also adhered to a hybrid humanities-sciences case study approach to climate adaptation that echoed the concerns of beekeepers.

Moreover, the GMRP included public awareness campaigns for the further promotion of mangroves as sea defense. A senior engineer at the ministry in the early years of the GMRP gave interviews on local television news. She delivered ample information about the ministry's past initiatives to engage public support for mangrove protection, including anti-squatting and mangrove education programming.

One way in which the minister's concerns have been addressed is through the GMRP-supported ecotourism venture. The tour, in operation since 2014, was originally planned around a "mangrove heritage trail" that included stops at some of the GMRP's experimental mangrove-planting sites. On the trail, tourists can visit mangroves, learn about their ecological-cultural value to Guyana, and during some seasons, participate in planting seedlings. A few tour guides have also been trained as rangers authorized to monitor forests to stop people from cutting trees and dumping garbage.

Guides also check on faulty sea defenses. At times, guides have found derelict groynes near mangrove replanting sites. For the guides, these derelict groynes now serve as tour attractions, a technical artifact that broadens the appeal of the mangrove heritage trail for educating tourists in the coastal engineering sciences. Whereas for engineers, derelict groynes further underscore that tour guides, along with other GMRP participants, are allies that help keep them informed to advance new measures when needed for climate adaptation.

Conclusion

Mangrove restoration in Guyana is a case study of climate adaptation. But it also reminds us that climate adaptation has consequences for how we value the sciences and humanities in solving problems. This is a point often overlooked in popular and scholarly accounts of the ways climate adaptation projects are produced through (inter)national diplomacy and policy-making arenas. Transdisciplinary dialogue matters to climate adaptation gaining traction and viability across the planet. Not only have Global South countries such as Guyana experienced record-breaking droughts, flooding, storms, and temperatures but so too have Global North countries. Climate adaptation knows no boundaries. Identifying the competing demands, modes of expert communication, and conventions of modeling, would help those in the academy and beyond, promote more equitable climate adaptation.

The world's uneven responses to the COVID-19 pandemic, the Ukraine war, and rising tides of populism, have only reinforced the need for climate adaptation. These recent events have spurred contestations over suitable models of expertise and modes of civic participation for a more livable planet. Envisioning survival may not only depend on critical assessments about the "drivers" of climate change—capitalism, fossil fuel consumption, and rogue states. Climate adaptation, instead, reminds us that our future must make space for shared knowledge and accountability.

Notes

1. For arguments about the value of case study approaches in the sciences see Lorrain Daston (1998), engineering design see Bruno Latour (1996), and in the humanities see Lauren Berlant (2007).
2. For a discussion on the use of satellite imagery and data for mapping the planet's mangrove coverage see Strong and Minnemeyer (2015).

Further Reading

The references below represent work written by scholars in the disciplines of anthropology and history. They provide examples of how "case studies" are theorized and deployed to analyze climate change and adaptation across geographies, cultures, and time periods.

Callison, C. (2014). *How Climate Change Comes to Matter: The Communal Life of Facts.* Durham: Duke University Press.

Coen, D. (2018). *Climate in Motion: Science, Empire, and the Problem of Scale.* Chicago: University of Chicago Press.

Vaughn, S. E. (2022). *Engineering Vulnerability: In Pursuit of Climate Adaptation.* Durham: Duke University Press.

References

Berlant, L. (2007). "On the case." *Critical Inquiry*, 33(4), 663–672. Available at: www.journals.uchicago.edu/doi/abs/10.1086/521564

Daston, L. (1998). *Wonders and the Orders of Nature*, 1150–1750. New York: Zone Books.

Latour, B. (1996). *Aramis, or The Love of Technology*. Cambridge: Harvard University Press.

Solomon, S., Qin, D., Manning, M., Marquis, M., Averyt, K., Tignor, M. M. B., Miller Jr., H. L., and Chen, Z. (Eds.) (2007). *Climate Change 2007: The Physical Science Basis; Contribution of Working Group I to the Fourth Assessment Report of the Intergovernmental Panel on Climate Change*. Cambridge: Cambridge University Press.

Strong, A. and Minnemeyer, S. (2015). "Satellite data reveals state of the world's mangrove forests." *All Insights: World Resources Institute Blog*. Available at: www.wri.org/blog/2015/02/satellitedata-reveals-stateworld%E2%80%99smangrove-forests

Part IX

Art, Infrastructure, and Climate

Introduction

Mark Vardy

Walter Benjamin, an enigmatic and engaging figure whose life ended in tragic circumstances in 1940, left a wealth of inspiring essays about topics including art, philosophy, and society. In a short text about the philosophy of art, which was unpublished in his lifetime, Benjamin (1996) says that you can direct questions to philosophy that you cannot ask of art. He said this because art, for Benjamin, evokes a metaphysical realm that exists beyond language. For this reason, we can feel moved by art in ways that escape words, while philosophy offers a vocabulary to discuss both the material and metaphysical aspects of art. Benjamin (1996) makes this point by asking readers to imagine that they meet a "handsome and attractive" person who "seems to be harbouring a secret" (p. 219). He says that if you want to know more about this compelling individual, who symbolizes art, it would be "tactless" to approach them directly. However, as he goes on to say, you could find out if they have a sibling, who is philosophy, and direct your inquiries to them. In many ways, the chapters in Part IX do exactly this.

The fact of climate change requires us to rethink how we conceptualize ourselves as humans because we are now acting with what postcolonial scholar Dipesh Chakrabarty (2012) calls "geophysical agency." That is, for the first time in history, humans are impacting the Earth on a planetary scale. And we know, cognitively, that our daily habits and practices are wrapped up with the emission of greenhouse gasses that will continue acting on the Earth long after they are released. But the vast temporal and spatial distances between our greenhouse gas-emitting actions and their effects mean that geophysical agency is beyond our capacity to actually experience with visceral immediacy. In an effort to bridge this divide between cognitive knowledge about climate change, on the one hand, and the embodied experience of what it means to be a living breathing human being who is endowed with geophysical agency, on the other hand, the authors in Part IX turn to interconnections that link the material world with its artistic representations.

In Chapter 28, Dominic Boyer begins with a discussion of the futures that are depicted in scientific scenarios of greenhouse gas emissions. While it is crucial that humans mitigate emissions to avoid the worst impacts of climate change, there is already a certain amount of warming – and consequent sea level rise – that the world is committed to even if emissions were to cease immediately. Drawing from his experience living and working in Houston, Texas, which has been flooded by at least four so-called 500-year storms in the past decade, Boyer shows how cities can be reimagined as living with water. Employing the playful metaphor of the amphibious as that which travels between and blurs the modernist binaries of nature-culture and wet-dry, Boyer discusses the ways he has worked with artists to reimagine Houston as an urban space that has come to terms with its essential wetness.

Whereas Boyer uses the amphibious to slide between the binary of wet-dry, Désirée Förster uses atmospheres to rework the old dichotomy of subject-object. Atmospheres belong to both

DOI: 10.4324/9781003409748-36

subject and object, mediating and connecting embodied experiences with the material world. In Chapter 29, Förster uses the example of two different installations – both of which feature algae – to discuss responses to climate change. On the one hand, responses to climate change can recycle old ways of orienting to the natural world by dominating nature and forcing it to serve human needs through concepts such as ecosystem services. On the other hand, they can open up space for new ways of connecting with the complexity of the world. To assist with understanding the latter, Förster's chapter develops a theory of "processual aesthetics," which provides a vocabulary to make connections between the physicality of climate change, art, and experience. Just as Walter Benjamin argued that art provides a way of connecting with a world that is beyond that of language, so too does Förster; processual aesthetics engages with "pre-reflective, pre-conceptual interrelatedness."

In Chapter 30, writing from her perspective of both a practicing artist and a scholar, Karolina Sobecka applies the ancient philosophical concept of "averted vision" in new ways to consider how artistic practices can help us think through what an ethical response to climate change might look like. But this is not an easy task. In order to grapple with climate change, artists must recognize and respond to the histories of colonialism and capitalism that are not only related to the production of climate change but also to the production of both science and art. Sobecka uses clouds as material to think with as she considers these issues, illustrating how different historical concepts of uncertainty have been formed in relation to clouds. Given that the planetary heating associated with climate change is intensifying conditions of uncertainty, it is important to consider how we can respond to surprises in inventive and creative ways that celebrate playfulness, not domination.

Taken together, the chapters in Part IX remind us of the abundance of human creativity. In an era when climatic events are increasingly likely to take us by surprise, the wellspring of creativity provides insight into how we – individually and collectively – can build good relations with human and more-than-human others.

References

Benjamin, W. (1996) "The theory of criticism," in Bullock, M. and Jennings, M.W. (eds.) *Selected Writings, Volume 1: 1913–1926*. Cambridge, MA: Harvard University Press, pp. 217–219.

Chakrabarty, D. (2012) "Postcolonial studies and the challenge of climate change," *New Literary History*, *43*(1), pp. 1–18.

28 Amphibious Cities

Dominic Boyer

Even if the world gets serious about climate action in the next couple of decades, the next few centuries will very likely involve steadily increasing rates of sea level rise. In this chapter, I discuss Houston, Texas, as an example of an amphibious city to show how we can respond to sea level rise. How much water is coming exactly? It's a complicated question to answer because a lot depends on how quickly humans can stop adding to the concentration of greenhouse gases in the Earth's atmosphere. So, instead of certainties, climate scientists tend to offer possible scenarios and then rate those scenarios in terms of their likelihood of occurring. One of the most important sets of climate scenarios comes courtesy of the United Nations Intergovernmental Panel on Climate Change (IPCC), a collection of many of the most accomplished climate experts in the world. The IPCC scenarios are called Representative Concentration Pathways (RCPs) and they take into account many factors including energy use, population growth, economic activity, and technological development. The RCPs blend all these factors together into a measure of **radiative forcing**, which tracks roughly how much solar energy is being trapped by greenhouse gases in the earth's atmosphere and not reflected back into space.

That energy imbalance is calculated in watts per square meter so when you see RCP 4.5 in the climate literature it means roughly 4.5 watts per square meter are being added to the energy's energy system by solar radiation. RCP 4.5 is described by the IPCC as a relatively moderate scenario—some scientists go as far as to say it's the most likely scenario—in which annual global greenhouse gas emissions peak around 2040 and then begin to decline. RCP 8.5 is the highest emissions scenario that IPCC offers and it's sometimes described as the "business as usual" scenario, in which the world doesn't strongly commit to energy transition away from burning fossil fuels and in which energy use and economic activity are allowed to grow at current trajectories.

What does this all have to do with sea level rise? Quite a lot actually. Something even climate scientists don't discuss very often is that even RCP 2.6, the most optimistic IPCC scenario in which the world acts swiftly and decisively to lower its carbon footprint, assumes atmospheric CO_2 equivalent concentrations around 430–480 parts per million (ppm) in the year 2100. That's higher than today's concentrations but not by very much.

The trouble is that according to paleoclimatologists—the scientists who study the history of the Earth's climate systems—the last time the Earth experienced an extended period of time above 400ppm CO_2 was during the Pliocene Epoch, three million years ago, when the global average sea level was 20 to 25 meters (65 to 82 feet) higher than it is today. It takes a while for glacial masses and ice sheets to catch up with atmospheric conditions so we're not going to receive all of that melt within our lifetimes. The current high-end projections for sea level rise in 2100 are only 2.2 meters (7.2 feet). But even that level of rise will be catastrophic for many of America's coastal cities as they currently exist. Much of coastal South Florida, from Naples to Miami to Ft. Lauderdale will be underwater with seven feet of sea level rise. Ditto for

DOI: 10.4324/9781003409748-37

Galveston, TX, New Orleans, LA, Charleston, SC, and Atlantic City, NJ. The next time you are on a web browser, take a look at the National Oceanic and Atmospheric Administration's Sea Level Rise Viewer, which is available at: https://coast.noaa.gov/slr/. This interactive tool allows users to zoom into coastal areas and see what areas will be flooded with varying amounts of sea level rise. For example, it can show you what the Boston area looks like a 7 feet of sea level rise, and the picture isn't pretty. And if you look at sea level rise projections and their impact on cities globally, especially in Asia, 7 feet of sea level rise means potentially displacing literally hundreds of millions of people. The C40 Cities Climate Leadership Group projects that some 800 million people living in 570 coastal cities will be at high risk from sea level rise and coastal flooding by 2050 and that is less than 30 years from now.

So, what can be done beyond the obvious goals of ending fossil fuel use as quickly as possible and downscaling energy and resource use to sustainable levels? What climate science is telling us is that the world needs to start preparing, and quickly, for the reality of what might be called **amphibious cities**. Environmental anthropologist Caspar Bruun Jensen (2017) writes,

> After a few centuries where terrestrialization was in the ascendant, the amphibious is gaining new life. In many parts of the world, water now seems to be flowing back into land, submerging coastal areas on a semi-permanent basis or creating recurrent floods, making the insufficiency of terrestrial responses increasingly apparent.
>
> (p. 225)

What Jensen means is that coastal cities are increasingly going to have to learn to live with the routine presence of water. They will need to learn how to adapt to tidal incursions, more intense storms and rainfall, increasing salinization of freshwater resources, and more frequent floods.

That term "flood" is crucial; long before the worst impacts of sea level rise are felt, floods will become a more constant companion of coastal urban life around the world. But what exactly is a "flood"? On the one hand, the answer seems quite simple. A flood means that water is somewhere it's not supposed to be. But landscape architects Anuradha Mathur and Dilip da Cunha (2001) have challenged the seeming obviousness of this idea. They argue that the concept of flooding is a symptom of colonial cartography. That is to say, it is difficult to disentangle the idea of flooding from the historical, often colonial work of controlling wetness, of confining it to certain abstractly determined river landscapes, thus rendering all other space as "dry" and fit for human ownership and occupation. Rivers, according to them, are always at least partly human creations as humans try to maintain a fixed boundary between dry land and watercourse. Rather than thinking about dry land and watercourse as stable, opposite entities, Mathur and Da Cunha challenge us to think about planetary life as an **ocean of wetness** in which dry land is always an exception and never the rule (see Howe, Chapter 36, this volume).

That kind of perspective—a perspective which is often foregrounded by Science and Technology Studies (STS) scholars who trouble the way that distinctions are drawn between nature and culture—could be valuable at a time when coastal urban amphibiousness is on the rise. Unlike the more comfortably terrestrial cities of the Holocene period, the amphibious cities of the Anthropocene will have a blurrier boundary between water and dry land. But what if we tried to reconsider that blurriness not as a bug but as a feature? Amphibians do not need wetness to survive, they thrive in wetness. What if we could similarly consider wetness not as a condition of urban life to be battled but as a condition that could and should be embraced? What new modes of urban life could emerge from that attitude adjustment?

Much of my own imaginative work on urban wetness has focused on my adoptive home of Houston, a megacity residing in ancestral wetlands, a muddy place since its inception and a site of constant flooding. Built over coastal prairie, woodlands, and swamplands, Houston's search for dry land has been a constant yet precarious enterprise since the beginning. As Houston architect Larry Albert explains (1997), efforts to "divide swampland into solid ground and watercourse" have been the central infrastructural struggle of the city's history; "to live, we separate something dry and something wet from the undifferentiated muck" (p. 144).

In Houston, the ocean of wetness has intensified in recent years. After three so-called "500-year storms" visited the city within the space of 24 months between 2015 and 2017, the stakes of this struggle heightened. Hurricane Harvey alone resulted in $125 billion worth of damage and trillions of cubic feet of flood waste. At one point during the storm, 18 inches of water covered 70% of the surface area of Harris County, home to more than 4.5 million people. Floodwaters damaged 204,000 homes—75% of them outside the official floodplain. In the storm's aftermath, Harris County voters approved an unprecedented $2.5 billion dollar flood bond to pay for 181 stormwater infrastructure projects to help reduce flood risks. The projects ranged from home buyouts to widened, channelized watercourses, new bridges, expanded upstream detention systems, and so on. This sounds impressive and it may even reduce flood risks temporarily for some residents. However, the largest stormwater infrastructure project Houston has seen since the 1930s—Project Brays—cost $550 million and took over 20 years to complete. And yet Harvey inundated the neighborhoods it was meant to keep dry all the same.

Houston will likely not become a truly coastal city in this century. But if Pliocene-like atmospheric conditions are allowed to endure for a period of centuries, Houston will eventually become a coastal city at 20–25 meters of sea level rise. Meanwhile, more urgent threats are looming from storm-related watery precarity. Houston is home to a deepwater port and the largest petrochemical industrial assemblage in the United States. On any given day, Houston bunkers 9 billion gallons of petrochemicals in aging, poorly regulated containers, most of them just a few feet above sea level and situated near the Houston Ship Channel, which flows all the way down to the Gulf of Mexico. Disaster modelers have determined that if a Category 4 or Category 5 hurricane were to make landfall at Galveston Bay and push a storm surge back up the Ship Channel, it is very likely that Houston would experience a massively catastrophic petrochemical spill, probably the worst such disaster in U.S. history. That could happen this year, it could happen next year. Without major changes to Houston's economy and infrastructure, it may be that Houston doesn't live to see its fully amphibious future.

Still, while fighting for change today, we try to remain optimistic about what the future might bring, what it might mean to live happily with our muck. Working together with my partner Cymene Howe (see Chapter 36, this volume) and local artist Ilse Harrison we tried to imagine what a fully amphibious Houston of the future might look like. We challenged ourselves to look past today's Houston and its current relationship to water and wondered how an amphibious city might evolve and what its physical and cultural forms might look like. What kinds of amphibious sensibilities will emerge as hopeless projects of "defense" against water yield to acceptance of wetness as coastal fluidity grows? Figure 28.3 offers a glimpse of a possible amphibious Houston.

This amphibious Houston has long left its ecocidal investment in fossil fuels behind it. Those infrastructures—pipelines, tankers, storage containers—are little more than curious reefs on the fringe of a city that now experiences coastal tides and dense mangrove forests at its periphery. Houston's characteristic 20th-century sprawl is also now a thing of the past. There is no more terrestrial automobility since dry land is scarce. Cul-de-sac suburbanism makes no sense in a place without cars and lawns. This Houston has adapted its urban form by becoming denser and more vertical to lower its resource use and demands upon local ecosystems. A complicated system of massive stilts and flotation apparatuses allows homes and businesses in the core of the

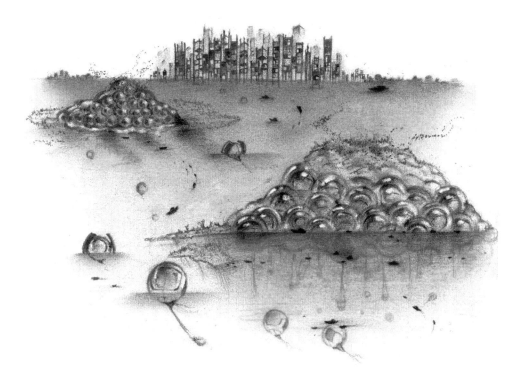

Figure 28.1 Speculative design for an amphibious Houston.
Source: Ilse Harrison.

city to either rise above the water or to ebb and flow with it. Most transportation around town is aquatic, with sailboats and solar-powered motorboats being the most popular. At the periphery, many structures, especially those invested in Houston's new dominant industry—sustainable aquaculture—are modular. Spherical structures that look like fish eggs venture out to harvest from aquafarms in the larger, warmer Gulf of Mexico. When supercyclones threaten, the modules can band together for protection and float in packs. In this world, a storm surge can redistribute the floating elements of Houston temporarily but it can never fully inundate or destroy them. The dominant theme of amphibian Houston is to move with water rather than against it. The ocean of wetness always has the final word.

In another collaboration with the Egyptian artist (and current Houston resident) Ganzeer we explored the intermediary form of Houston as a stilted Aquapolis. This vision of another possible Houston has accepted the presence of water even though it has gone to great lengths to recreate the norms of dry urbanism by raising the whole city into the sky. We view this possible future as in some ways more likely than the amphibious embrace discussed above. But even Aquapolis will no longer be able to think about floods in the same way. Water is not an invasive force in Aquapolis, rather an ambient condition.

These kinds of speculative imaginative exercises are important. If you live in a coastal or near-coastal city yourself you might want to try this out for yourself. What could an amphibious New York look like? Or Guangzhou? Or Manila? You can become a designer of amphibious cities, creating memories of the future, breadcrumbs for us all to follow forward.

Figure 28.2 Aquapolis Houston.

Source: Ganzeer.

The other dimension of amphibious urbanism that is important to cultivate now is what Charlotte Lemanski calls **infrastructural citizenship**. Infrastructure is about enablement; something is an infrastructure only to the extent that it enables something else to happen. Current infrastructure tends to enable the reproduction of the status quo but new infrastructure could enable new trajectories. Infrastructural citizenship is civic attention to and engagement with infrastructure in ways that allow us to lay the groundwork (and waterwork) for more sustainable futures. Nikhil Anand has called the kind of infrastructural citizenship that concerns itself with water, **hydraulic citizenship**. Amphibious cities will need active and imaginative hydraulic citizenship to survive and thrive.

To return to Houston's flooding, I was recently speaking to a well-known landscape architect, Keiji Asakura, who offered me a fundamental way of rethinking the problem. Once upon a time, the legendary Harris County public infrastructure czar, Art Storey, told Keiji that if every building in Houston had an adjacent rain catchment or rain garden it would put the local flood control district out of business. A rain garden is a very humble infrastructure that consists of digging a hole or trench in the ground a few feet deep. Into the dugout, you place logs, branches, sticks, leaves, mulch, pretty much anything at hand. And then you fill back in the soil and plant it over, ideally with local coastal prairie vegetation whose root systems can run meters deep and are excellent at sponging up water. As a rain garden ages, the logs and leaves decompose creating new, excellent soil that can be harvested in a periodic process of rain garden renewal for projects like raised bed urban farms. Meanwhile, the rain garden prevents rainwater from becoming runoff by holding it until it can absorb into the soil. This addresses a large part of Houston's current problem; the city is covered by too much impermeable concrete while the underlying soil has a lot of dense clay in it, which needs more time to absorb wetness. Impermeable surfaces are a major cause of flooding in urban areas across the world.

So together with Keiji and local leaders in the Kashmere/Trinity/Houston Gardens neighborhoods of Northeast Houston, I have been working to create rain gardens as an alternative to conventional stormwater infrastructure. The amazing thing about rain gardens is their relatively low bar to entry, which is especially important in Northeast Houston, an underprivileged area of the city that has suffered greatly from Houston's long history of environmental racism. Conventional stormwater infrastructure is immensely expensive to make, uses a lot of high-energy materials like concrete, takes forever to build, and, as such, is a domain that political and financial elites control. It's also not a domain that offers effective solutions given the rapid pace of climate change. Within only 20 years, Houston's 500-year floodplain became a 100-year floodplain rendering many conventional stormwater interventions obsolete. Green infrastructure like rain gardens is a different thing entirely, taking very little technology, time, and expense to accomplish. All the tools that are needed to make a rain garden are no more than medieval technology: shovels and wheelbarrows. Depending on the size of the project, a rain garden can take as little as a few hours or as much as a few days to create. The main cost is finding people willing to dig and fill and plant.

So, here's a revolutionary infrastructure idea. What if Houston were to declare a rain garden week (or month or even a year) and ask its citizens to do nothing other than dig and fill and plant the green areas around their buildings? At the end of that period of time, Houston's flooding problem would largely be under control, all without channelizing bayous and installing giant storm sewers and digging massive detention ponds, and, most importantly, without waiting for decades for a concrete and steel engineering solution that will never come. A Houston jeweled by a million little wet and wild sanctuaries would not only be a much better place to live now but it would also be a place that might still steal a future from the jaws of its toxic history. Imagine the sense of possibility that would emerge from having made civic power the real solution.

Revolutionary infrastructure projects are experiments in creating new relations and enabling alternative future trajectories to those that have brought us to the brink of ecological emergency. Projects of revolutionary infrastructure are diverse, locally attuned, and typically invisible to conventional infrastructural politics. The radical rain garden plan outlined above has no traction yet in mainstream Houston politics, at least not for the moment. Yet, because it is hard to make something out of nothing, revolutionary infrastructure often captures and redistributes the materials and energies within existing infrastructural ecologies to do its work. The modern shovel was designed as a tool for the mastery of the soil. But in the case of a rain garden, you can feel how those same shovels are now meshing differently into ecological relations to try to create more balanced, respectful, and sustainable alliances between human and nonhuman forces.

The path ahead toward amphibious cities will have its challenges but it will also offer many opportunities for positive change. The Anthropocene is going to demand new kinds of civilization across the world. And it would be in our collective human and nonhuman interest to develop a civilization that seeks to live with, rather than to wage war against, natural forces. In places like Houston, the wetness is coming whether we like it or not. But rather than strike a defensive posture this is a time for creativity and organization. Happily, amphibiousness offers a more expansive and undetermined vision of coastal urban futures than dry cities ever had. Plus amphibious infrastructure will not make you wait, endlessly, for its payoff. You can do it right now. You should do it right now. Rather than try to tell you what amphibiousness could mean in your environment, I would encourage you to try to figure it out together with your allies. And then tell me. I want to hear about your rain gardens!

References

Albert, L. (1997) *Houston Wet*. Unpublished Masters of Architecture, Rice University.

Anand, N. (2017). *Hydraulic City*. Duke University Press.

Jensen, C. B. (2017). 'Amphibious worlds: environments, infrastructures, ontologies.' *Engaging Science, Technology, and Society*, 3, pp. 224–234.

Lemanski, C. (2019). *Infrastructural Citizenship*. Routledge Press.

Mathur, A. and da Cunha, D. (2001). *Mississippi Floods: Designing a Shifting Landscape*. Yale University Press.

29 Aesthetic Encounters with the More-Than-Human

Désirée Förster

Introduction

Scholars in Science and Technology Studies (STS) argue that climate change challenges the very way in which strict lines are drawn between the subjective experience of humans on the one hand, and the objectivity of scientific knowledge on the other. Put another way, climate change disturbs the habitual ways in which humans make distinctions between their inner world of emotions and thoughts, on the "inside," and the world of nature, on the "outside." Building on these insights, I suggest that climate change requires us to formulate new ways of understanding ourselves as humans in relation to each other and to the world. To this end, this chapter proposes a turn toward aesthetic experience that makes it possible to understand the human being from a new perspective: not as an element opposed to nature, but as one element among many, linked to nature in multiple ways. More specifically, I argue that what I call "processual aesthetics" can overcome the separation between nature and culture, or the more-than-human and human, by foregrounding the continuous metabolic exchanges between bodies and their environments. In doing so, I explore how specific environments that intentionally stage specific encounters with more-than-human entities, for example in art installations or urban shopping malls, can heighten our sensitivity to ecological interdependencies in the external world and, at the same time, the internal responsiveness of our bodies. This responsiveness provides new ways of perceiving and understanding how the internal and external worlds are interrelated in an era of climate change.

In what follows, I first discuss the example of an air quality intervention called *PhotoSynthetica* that was installed in an urban shopping mall. I then discuss an art installation called *The Molecular Ordering of Computational Plants*. I use these two examples to illustrate the different conceptual frameworks of phenomenology, ecosystems services and onto-sympathy, before returning to the concept of processual aesthetics. The very first thing that I want to do, however, is begin this chapter by situating the atmosphere as an artistic medium.

Atmospheres as a Medium of Aesthetic Environments

The term "atmosphere" goes back to the Greek *atmos* (vapor, steam) and *sphaira* (sphere), and refers to the layer of gas that surrounds the Earth. Most people are familiar with how the Earth's atmosphere is the subject of scientific inquiries, for example in the field of meteorology, which treats atmospheres as physical and chemical systems. However, atmospheres are also studied in philosophy and, in the last 60 years, have been increasingly explored as a theme in both art and architecture. In these fields, atmosphere can be understood as the spatial quality of emotions and moods. Using this approach, recent studies have linked affect with the flows that envelop humans and other beings in space. For the philosopher Gernot Böhme (1993), atmospheres mark

DOI: 10.4324/9781003409748-38

the in-betweenness of subjects and objects. That is, he argues that atmospheres belong neither to subjects nor the physical environment alone, but instead blur the sharp distinctions between the two.

In this chapter, I explore atmospheres as media, as that which mediate biochemical processes that are both inside bodies and outside in the environment that surrounds our bodies. Understanding atmospheres as media foregrounds awareness of metabolic bonds between humans and the messy and complex more-than-human world that often remain unnoticed. This approach is grounded in an STS perspective that makes it clear how – rather than maintain the old dichotomies of subject-object, inner-outer, human-nonhuman – we can begin thinking about how the purposeful blending of living and non-living entities changes our environments and how this impacts our subjectivity.

Example 1: PhotoSynthetica

PhotoSynthetica (2018) is the title of a system designed by the UK-based design team Ecologic Studio to reduce carbon concentration in the atmosphere within indoor spaces like shopping malls. It is one example of so-called *bio architecture* that uses natural materials and their properties as part of urban design. The system looks like an additional green layer attached to an indoor space and combines several components: an urban carbon tool used to measure the carbon concentration in the air and the performance of the overall system, an algae species that metabolizes carbon into oxygen, and a digital management system composed of sensors that measure the microclimate to ensure the best conditions for algae growth in relation to human behavior. The site-specific air is pushed into the system and enters the algae culture, where it is designed to be filtered and then reintroduced into the atmosphere. For the algae to perform as a filtering system, it must receive enough sun exposure. Additionally, the structure provides nutrition composed of bioplastic raw material, biofuel, fertilizers and super-food.

Example 2: The Molecular Ordering of Computational Plants

The piece, *The Molecular Ordering of Computational Plants*, by artist Andreas Greiner and composer Tyler Friedman brings together algae, sound, and science fiction to create what Greiner calls a "living sculpture." What makes the sculpture "living" is that it highlights processes instead of displaying a static object. When exhibited at a gallery called basis in Frankfurt in 2018, the installation featured an aquarium filled with algae of the species *pyrocystis fusiformis*, a species that glows with bioluminescence when it is disturbed. In this case, the algae were disturbed – and its bioluminescence triggered – by soundwaves. More specifically, the algae illuminated rhythmically in response to sound coming from subwoofers mounted underneath the aquarium. The sounds were part of a recorded story describing a future in which humans enter a symbiotic relationship with algae. The installation itself gave the visitors the possibility of an experience approximating the symbiotic bond between algae and visitors.

Over the passage of time, the visitors became part of the artwork itself, part of a collective super-consciousness that the audio recording described. This happened in two ways. On one hand, the sounds produced by the visitors became part of the sound composition that lit up the algae. On other hand, they became resonance bodies for the rhythmical illuminations of the algae and audible vibrations. They entered, so to say, into the first phase toward a symbiotic relationship with algae as described in the audio recording. This first phase consists of the sharing of a rhythm. It is important to emphasize that this relationship, this becoming part of a larger organism, here symbolized by the exhibition, does not happen through a cognitive choice or a decision on the part of the visitor. Rather, the relationship is established through the biological constitution of the bodies present and how they respond to audio-visual signals in the environment.

Three Ways of Configuring Relationality: Ecosystem Services, Phenomenology and Onto-sympathy

Ecosystem services is an ecological term that summarizes commodities and services provided by ecosystems for human benefit. This includes certain ecosystems that provide products for humans, such as wildlife, or resources and raw materials such as timber, but also regulatory services such as the purification of air and water through, for example, photosynthetic processes in plants and algae. In cities, urban design strategies often install ecosystem services in order to support the wellbeing and security of human inhabitants, such as vertical gardens that supposedly clean the air of pollutants. Sometimes known as "habitat enhancement techniques" or "resiliency infrastructures," these strategies tend to regard the human habitat as vulnerable in the face of external environmental influences in a particular way. That is, this distinction reiterates the absolute divide between inner subjectivity of humans and outer objectivity of nature that I want to challenge. I think this is an important challenge to launch because this way of framing the issues suggests that climate change is inevitable and yet manageable. In other words, while the concept of ecosystem services can have positive effects on biodiversity, it remains selective and reductionist in the way it highlights particular characteristics of living organisms that are deemed beneficial within a larger, human-centered system.

PhotoSynthetica can be seen as a typical case of an ecosystem service that functionalizes certain properties of ecological systems to facilitate human life styles. The metabolism of a living organism is embedded into a context of anthropocentric consumerism: the algae are bred, monitored, and controlled in order to ensure the best atmospheric condition for human needs. Such a reduction of living organisms to forces of labor, without the potential for spontaneous, difference-generating processes, is typical of many approaches to environments. This is evident in ecosystem services, which renders living organisms *smart* and *adaptable* within a predefined, closed feedback loop. Such a reduction of life to labor force extends the biopolitics of neoliberal systems to nonhuman animals and vegetal life, further normalizing the appropriation and exploitation of life itself. What is needed to counter such a reduction of biological processes to a utilizable essence and economic terminology, I argue, is an aesthetics that highlights potential relationships with environments and others we are sharing environments with, without reducing them to manageable conditions. If biological processes are shown to generate differences, in the form of surprising patterns of behavior for example, an alternative to the essentializing appropriation of life in ecosystem services comes to light. That is, if biological processes are shown to develop in ways that go beyond or even counter what made them utilizable in the first place, aesthetic renderings of biological processes can point to the limits of the appropriation of life.

PhotoSynthetica, as an example of ecosystem services, instrumentalizes nonhuman entities through a reduction of their photosynthetic activity, which is made productive in service of human wellbeing. In contrast to the instrumentalization of life seen in *PhotoSynthetica*, the art installation *The Molecular Ordering of Computational Plants* shows how bodies are always both the subject of sensing and object of being sensed. To explicate this in more detail, I will briefly turn to the phenomenology of Maurice Merleau-Ponty. Phenomenology is a philosophical movement that is concerned with human experience. As such, it develops ways to explore sense perception and consciousness. Merleau-Ponty, who was born at the beginning of the 20th century, is one of the most influential philosophers associated with this movement. One of his major contributions to phenomenology was his concept of **intersubjectivity**. Oftentimes, subjectivity is regarded as if it were an individual property of one human isolated from others. But Merleau-Ponty argued that subjectivity actually emerges from the ongoing engagement between the sensitive and sensual body, its surroundings and others with whom we share these surroundings.

Merleau-Ponty's approach to perception as embodied and mediation as an embodied practice provides an alternative to the instrumentalization of nature seen in ecosystem services. As I have remarked critically, algae in *PhotoSynthetica* are essentialized and identified with their capacity to metabolize carbon into oxygen. An embodied phenomenology sets out differently. Regarding bodies as always already embedded in the world, they are understood to expand sensually into the world, whereby they enter into relationships with other bodies. In the aesthetic situation of *PhotoSynthetica*, it is the shared metabolic capacity of algae and humans that compose a relation through which both bodies (ex)change. In algae, it is through photosynthesis that carbon dioxide is metabolized into oxygen. Humans need oxygen for their own metabolism. Metabolism is thus a basic activity of energy production shared by all living organisms and as such can be regarded as that which connects all living organisms. Bodies in this non-deterministic view are not defined as separate from other bodies but are instead recognized in their openness to affect and to be affected.

Drawing from Merleau-Ponty and his understanding of intersubjectivity, attuning to a shared rhythm suggests an intimate relationship with others. The sharing of a rhythm stands in as sharing a "manner of handling the world" (Merleau-Ponty, 2012, p. 370). This observation lies at the heart of intersubjectivity, by which a subject extends one's own intentions into space and time, thereby creating a rhythm that can be recognized and taken up by others: "[I]t is precisely my body that perceives the other's body and finds there something of a miraculous extension of its own intentions, a familiar manner of handling the world" (Merleau-Ponty, 2012, p. 370). Intersubjectivity then relies on modes of perception that enable the perceiving subject to recognize a familiarity in the way others are handling the world.

Merleau-Ponty was mainly concerned with human perception and at the center of intersubjectivity is a human subject. In the following, I propose that we can extend intersubjectivity toward nonhuman others and that aesthetic experiences such as those offered in Greiner's and Friedman's piece gesture toward such a more-than human intersubjectivity. To help make this argument, I draw from Jane Bennett's concept of **onto-sympathy**, which foregrounds a more extensive understanding of our account for otherness. Onto-sympathy renders the world "as full of propositions waiting to be registered by interested bodies" (Bennett, 2017, p. 91). In order to register those propositions in the world, we have to "become alert to nonhuman-human affinities" (Bennett, 2017, p. 103). Her way of doing so is in parsing the operational process of sympathy into phases: the *gravitation* of bodies toward other bodies, the *cooperating* of bodies whereby those bodies are transformed, and *annunciation*, the recognition of others.

If we translate the human experience of Greiner's and Friedman's art installation into Bennett's concept of onto-sympathy, the first phase of gravitation could be visitors leaning toward the algae and becoming aware of their specific light intensity and illumination patterns. The second phase of cooperating bodies could be the transformation that takes place as the sounds of the installation stimulates the skin receptors of visitors whose bodies viscerally detect the rhythmical changes in force and pressure. At the same time as the sound is resonating in visitors' bodies in this way, the illumination of the algae is captured by their eyes and transmitted to the nervous system. The visitors literally take the shared rhythm under their skin. The third and final phase, annunciation, could be the becoming-familiar with the nonhuman other, the recognition of a mechanism of life as it responds to the environment. An aesthetic encounter that joins the experience of intersubjectivity and onto-sympathy does not start from the human subject or the nonhuman other as an individual, nor their surroundings as a base for manipulation. Instead, it opens up toward the processuality that enables and limits the relations between organisms and their surroundings – processes from which the human subject and the nonhuman other as individuals as well as their surroundings are continuously composed.

Processual Aesthetics

In the first example above, algae and their photosynthetic activity are represented as *ecosystem services*, which leads to a reduction of the nonhuman other to benefit human consumption. In contrast, the second example shows how the negotiation processes between algae and their environment are represented in a way that preserves their absolute difference, but still suggests a possibility of connection. This raises the question: how can we experience the relation to our surroundings as processual and at the same time meaningful, not from a perspective that is solely instrumental but as the expression of meaning that lies in the nature of our relational being itself? I argue that processual aesthetics allows us to describe the different effects of biochemical and atmospheric processes on our subjective experience. This concerns, on the one hand, processes inside human bodies and their effects on sensation, perception, and action. These bodily processes include, for example, our dependence on oxygen availability in the surrounding atmosphere. On the other hand, it concerns the impact that changes in our environment have on how we feel and act. Aesthetic practices that foreground these interdependencies do so by enabling an experience of these effects, such as a change in air quality or humidity on one's own body as well as the immediate environment. In particular, they foreground how we relate to our environments and others before we render them an object that we could further utilize. In this way, our pre-reflective, pre-conceptual ways of being-in and -with the world, start to matter.

The Molecular Ordering of Computational Plants displays a processual aesthetics that shows how humans can relate with the more-than-human in ways that allow the otherness of nonhuman others, with all their differences, to remain intact. In including other sense modalities besides the usually more dominant, visual sense, different phases of experience are highlighted, including the intermingling of vibrations that come from sound waves from different sources. The aesthetic situation enables an encounter in which the nonhuman other is not simply subordinated to a functional logic. Instead, the sharing of a rhythm offers ways of attuning to otherness, of becoming attentive to a resonance formed within the encounter. This resonance cannot easily be objectified, and its allusive, processual character remains intact, enabling an aesthetic experience of a pre-reflective, pre-conceptual interrelatedness.

Conclusion

In exploring two aesthetic environments through the lens of process, I have shown how the aesthetic rendering of the nonhuman other can enable different experiences of being-in and being-with environments. The objectifying representation of the nonhuman other as part of a functional apparatus (ecosystem services) was contrasted with an encounter with the nonhuman other that emphasizes the potential for change, for affecting, and being affected. I have proposed in this chapter an aesthetic approach that investigates how different modes of perception, cognitive and biochemical processes impact the relationship between humans and nonhuman animals, which I call processual aesthetics.

A processual aesthetics foregrounds the ways bodies affect and are affected by their environments in different, mostly unnoticed ways. Developing a sensitivity for these often-disregarded processes, for the different forces that lie behind shifts in intensity, changes in form, or inscriptions, could contribute to developing a more comprehensive understanding of the complexity of the present and to examining one's own options for action accordingly. Processual aesthetics, in shifting our attention away from things in their being-so toward their becoming, creates an openness toward alterity – as found in our own bodies as they extend toward their environment, and the bodies of others. While design projects that employ ecosystem services emanate from a more-or-less given human subjectivity that comes with an array of needs, projects like *The Molecular Ordering of Computational Plants* might provide perspectives that move beyond our immediate self-interest, and provide a context to articulate human subjectivities and non-ordinary forms of representations of others.

A processual aesthetics, as I have described it here, gathers together elements in the environment such as air, light, but also sensing technologies, which can mediate and express different temporalities of actions in space. Ultimately, developing a processual aesthetics with atmospheric media, I propose, can highlight that meaning-making is not only reserved for conscious thought but happens on a pre-reflective layer as well. In addition to making us aware of the interdependencies between human bodies and the environment, processual aesthetics can thus show how our pre-reflective embodied knowledge is already full of meaning-making. To sense this processuality as the ongoing relation between inside and outside, between pre-reflective, subjective knowledge and rational thinking, makes it possible to understand the human being from a new perspective: not as an element opposed to nature, but as one element among many, linked to nature in multiple ways. While a processual aesthetic cannot help simplify complex processes such as climate change, it can lead to a shift in perspective and give individuals the opportunity to position themselves in relation to the experience of change and complexity: in their own bodies as well as in their environments.

Further Reading

Brannigan, J., Ryfield, F., Crowe, T. and Cabana, D et al. (2019) "'The languo of flows': Ecosystem services, cultural value, and the nuclear legacy in the Irish Sea," *Environmental Humanities*, 11(2), pp. 280–301.

Coccia, E. (2019) *The Life of Plants: A Metaphysics of Mixture*. Medford, MA: Polity.

Förster, D. (2021) *Aesthetic Experience of Metabolic Processes*. Lüneburg: meson press.

Horn, E. (2018) "Air as medium," *Grey Room*, 73, pp. 6–25.

References

Bennett, J. (2017) "Vegetal Life and Onto-Sympathy," in Keller, C. and Rubenstein, M. (eds.) *Entangled Worlds*, New York: Fordham University Press, pp. 89–110.

Böhme, G. (1993) "Atmosphere as the fundamental concept of a new aesthetics," *Thesis Eleven* 36(1), pp. 113–126.

Merleau-Ponty, M. (2012) *Phenomenology of Perception* (trans. D.A. Landes). Abingdon, Oxon: Routledge.

Photosynthetica. (2023) Photosynthetica. [Installation] Available at: www.photosynthetica.co.uk/ (Accessed January 12, 2023).

The Molecular Ordering of Computational Plants. (2023) The Molecular Ordering of Computational Plants [Installation] Available at: www.andreasgreiner.com/works/molecular-ordering-of-computational-plants/ (Accessed January 12, 2023).

30 Averted Vision

Karolina Sobecka

Introduction: Art and Other Techniques of Knowing Otherwise

"Have you ever had the experience of looking at the stars at night, seeing a faint point of light out of the corner of your eye, and then having it disappear as soon as you look at it directly?" asked the presenter. A philosopher and historian of science, he was talking to art students and faculty in an art academy. It was a lecture on the economy of attention. A final discussion centered on a technique for observing dim objects in the night sky. Called "averted vision," it was already used by Aristotle in 325 BCE to make astronomical observations. The technique is based on the phenomenon of faint objects being perceivable only through peripheral vision. The phenomenon is explained by how the human eye works. The fovea, the center of our retina, is primarily made up of cells known as cones, which detect bright light and colors. The cells surrounding the center are largely rods, which have a higher sensitivity to light but detect only luminosity. Very faint light can be detected by our rods, but as soon as we move our focus, and our fovea full of cones, to look at it directly, we stop being able to detect it. "There are certain kinds of things that you only see when you look elsewhere," concluded the philosopher.

Being part of that particular audience, I realized how well the averted vision technique resonated with this group, serving as a very good metaphor for understanding how art functions as a knowledge practice. Being "indirect" might be artists' most fundamental mode of operation. Getting at something obliquely is our method of working. Art, along with the various philosophical traditions of "unknowing" – or knowing by rejecting reason and tuning into experience, presence, emergence – has always offered an alternative to a rational, scientific, and direct way of knowing. This is perhaps what makes art of interest today when scholars urgently call for ways of knowing *otherwise* in the planetary crisis. With a recognition that this crisis has roots in the modern mindset and its prerogatives of scientific development, technological advancement, and progress, interest in art as a knowledge practice has grown. Motivated by a need to reconsider the relevance of existing methods in both natural and social sciences, scholars from across STS, human geography, feminist technoscience, media theory, and other approaches have proposed developing methods that borrow from or employ the practice-based methods of art. Moreover, "artistic research," actively and creatively engaging with an object of investigation, is starting to be considered as something that can expose the limits of existing disciplines, revealing the "messiness" of any actual process of research, as well as its "performativity," understood as the fact that the very act of representing something contributes to bringing it into being. Artistic research implies that art and research can be one practice rather than two distinct ones, a practice that enables a critical reflection on the value, status, and significance of knowledge today.

There are many ways through which art can be relevant to climate change: from providing a reflection on social imaginaries of nature and their world-making power, to offering experiences

DOI: 10.4324/9781003409748-39

of material embeddedness in environments, to generating visibilities of knowledge infrastructures and destabilizing them, to creating counter-narratives and speculative scenarios. But in this chapter, I limit my discussion to what art has to offer as a way of knowing that functions much like the averted vision technique. The established scientific methods, just like the cones in our retina, are capable of detecting only some things, while obscuring others. Art sensitizes us to the fact that there is more to be known beyond what can be captured instrumentally, quantified, described, or recorded, with the current scientific tools we have.

Just as science is entangled with histories of Enlightenment science, colonialism and capitalism, so too is art. In order to engage critically with the climate crisis – understood not simply as a physical phenomenon, but as a complex socially engendered set of processes rooted in those histories – art must engage with contemporary legacies of those logics. Artists have to consider what happens to art's agency, and its potential for innovation, subversion, connection, or critique if it is enacted from within a socio-economic system structured by those histories. If it carries a transformative and transgressive potential, creativity is attractive to both those for whom novelty represents a countering of the existing arrangements, and to those for whom it is simply more grounds for expansion, welcomed within a system that feeds on diversity. Can artistic interventions bring about new ways of being and doing, while escaping capture by contemporary forms of neoliberal capitalism, which can turn new ways of being into territories enclosed in markets?

The "averted vision" technique suggests an approach to meet these challenges. For the contemporary philosopher Jacques Rancière, when art acts obliquely, it can challenge the existing order. It has the capacity to alter the "distribution of the sensible" by altering, as averted vision does, what becomes perceivable. When art is too direct, when it consists of overtly political messaging, Rancière suggests, it loses its capacity to reorder, because it simply participates in the established discourses and institutions. When art works obliquely, it doesn't contest the world revealed and constructed by dominant ways of knowing. Rather, it develops an ensemble of tools that might enable us to reveal, or construct, another world on the margins.

Clouds and Indirect Knowledge

I came to these issues through a series of art projects that I developed on the topic of clouds. The projects included developing devices for cloud making or cloud collecting and organizing events around sending them into the atmosphere, creating a crowdsourced collage of the sky over an entire city, tasting clouds, and recreating historical clouds based on their chemical composition. Clouds are interesting objects to think *about*, as well as to think *with*, because they have been connecting knowledge, representation, control, and governance for thousands of years. They have been another tool that is akin to averted vision, that "offers access to truth by way of a simultaneous veiling and unveiling," as scholars Peters and Piechocki (2021) recently argued. Since ancient times, clouds were seen as a challenge to knowledge. Lacking a stable form, they were understood as provisional and uncertain by ancient philosophers and poets. As such, they were a challenge to enumeration and to representation-based governance.

Over 2,000 years ago, the Greek playwright Aristophanes said that clouds can turn "into anything they want," and therefore, they could only be the object of a speculative, poetic language. Cloud-like imagery has been associated with the various traditions of "unknowing," from the *via negativa* which insists that God, understood as a "dazzling darkness," is indescribable and has to be referred to only in terms of what he is not, to the Cloud of Unknowing, the fourteenth-century mystic text which suggests that the way to know God is not through knowledge and intellect but through contemplative immersion, stripped of all thought. In these ways and more, clouds have been reminding us since antiquity that there is always more going on than what humans can comprehend using reason. A cloudy – speculative, poetic, or artistic – approach can convey something of those additional "goings on," without losing its multivalency and ambiguity. Additionally, it does so in a way that avoids the trap of itself becoming the erasing or obscuring force.

For Rancière, the focus on obliqueness of art came from examining nineteenth-century struggles and utopias that ended up contributing to or turning into the totalitarianisms that they were opposing. Art too, he suggests, can be too often caught up in patterns that circulate the same practices it purports to critique, ending up as "media images that denounce the media, spectacular installations that denounce the spectacle" (Rancière, 2007). But subtle gestures in the peripheral vision, a using-otherwise of senses, a multiplicity of small ruptures, small shifts, builds a practice that can "refuse the blackmail of radical subversion" (ibid). Such small-scale, oblique art interventions are also important for a feminist STS scholar Donna Haraway. She describes them as part of a "modest project" of "a little re-tooling," in contrast to the construction of a grand utopia (Haraway, 2004).

But there is yet another reason why the oblique workings of artistic methods are embraced today. They are seen as one of the potential responses to the demands for epistemologies that are adequate to a world that consists of incalculable entanglements and is therefore "ontologically opposed to definitive capture" (Neimanis, Åsberg, and Hayes, 2015). In other words, art might be one type of response to the various calls for knowledge "without mastery," without directionality, or without "knowing-in-advance." Among such approaches are media ecology thinkers that follow Michele Serres' "errant materialism" to articulate notions of turbulent, circumstantial way of thinking, human geographers speculating on new methods that "cause thinking to take place in ways that are not always given in advance" (McCormack, 2015), or eco feminist writers calling for knowing that rejects "master rationality" and respects limits (Plumwood, 2009). Here atmospheric thinking or thinking-with-clouds can again be helpful, to reflect on the legacy of the logics that humans have developed for managing uncertainty, on limits of knowledge and control.

Clouds and Uncertainty

Clouds were of special interest to Descartes, a French philosopher and mathematician who is widely considered to be one of the fathers of the Enlightenment or the Age of Reason. According to Descartes, if meteorology would enable humans to understand the physics of clouds, then everything else that was mysterious, marvelous, or unknowable could also be explained by science. But a lesson we learnt from clouds in the last half century – through the chaos and complexity sciences – is the opposite: it is not just the clouds but the whole world that is provisional and uncertain. Far from being something that operates mechanically, a theatre of scenographic techniques, as Descartes suggested, nature is inherently mysterious and unknowable.

This shift in thinking was set in motion in 1963, when mathematician and meteorologist Edward Lorenz, having developed an early numerical simulation of weather, realized that in the atmospheric system, a very small difference in initial conditions can cascade into large changes, rendering it fundamentally unpredictable. Lorenz' insight challenged Descartes' mechanistic clockwork universe and had a profound effect on all domains concerned with prediction and anticipation of the future, fundamentally changing notions of action, agency, and control. The revolutionary thought was this: since uncertainty cannot be prevented, it should be factored into all the futures and managed. This new way of approaching the future led to developments of computationally supported mathematics of probability, as well as to the invention of techniques of "scenario planning" by Hermann Kahn at the Rand Corporation think tank. These techniques went beyond mathematics of chance, and relied on intuition, and flexibility and adaptability of a creative mind. "What these strategists sought to develop," writes political scientist Melinda Cooper, "was a mathematically rigorous method that nevertheless built on the non-quantitative and emotive dimension of our relation to the future" (2010). One of the insights tied to recognizing the indeterminacy of the world was that envisioning particular futures contributes to bringing them into being. That is, the future became pliable; it could be imagined to be a "blank slate," open to creative intervention.

This insight was the basis of two competing conclusions. On one hand, uncertainty started to appear as the technique of entrepreneurial creativity, promoted by proponents of the new market-based modernity. It led to the assumption that one is free to manufacture the world of one's choosing, regardless of planetary material constraints. Against this, the various feminist post-humanist approaches have been developing understandings of uncertainty that relinquishes control of the future and embraces not-knowing-in-advance, which relies on immersive and situated sensing of the world here and now. It is critical to distinguish between these two types of responses to uncertainty. One leverages the incalculable for the "creative destruction" of neoliberal economics, while the other can "welcome the unknown, the surprise, the other" without trying to control it (Bergoffen 2003). This distinction is all the more important because atmospheric heat will be a massive primary source of uncertainty in the coming decades. In other words, understanding and reframing the understandings of uncertainty we have inherited from the last century is crucial to human ability to respond to the climate crisis.

Conclusion

In principle, all practices of research, including creative activity, require opening up to not knowing where one is going. "The goal of the research process is to produce results that by definition cannot be produced in a goal-directed way. The unknown is something that cannot be approached straightforwardly precisely because one does not know what is to be approached," writes philosopher of science Hans-Jörg Reinberger (1994). But how to intentionally build in the "room for surprise" into one's experimental system? How can we open ourselves to what we do not yet know? One mode of being that recognizes and responds to the inherent indeterminacy of the world, that tests and responds to what unfolds, is an attitude of play.

The novelist E. L. Doctorow once said that writing a novel is like driving a car at night: you can see only as far as your headlights, but you can make the whole trip that way. I, for one, completely identify with such a process of art-making (or research). My projects which used atmospheric balloons were perhaps an especially good model of a process of setting something in motion, following it, and engaging with what unfolds. I am not the only one who adopts this way of working: a number of scholars have been developing research methods based on engagements with atmospheric processes and materialities under the umbrella term of "atmospheric theory." Examples range from Sasha Engelmann and Derek McCormack's essay "Sensing atmospheres" (2018) to Ben Anderson and James Ash's extensions of affect theory to think about the diffuse and collective nature of affective life (2015). In addition, I encourage you to look at the art and research projects listed in the "Further Viewing" section below. In many of the interventions listed there, the most interesting things that happen emerge not by design but through negotiation of the thousand small contingencies that shape the project's materialization. Through an attempt at doing one thing – an often playful, absurd proposition – they accomplish something quite different that only in hindsight can be even identified. Art is at its best when it is a complex tangle of relations out of which "things come out" (Marres, 2016) or can be "worked out" (Butler, 2021), a process of constant improvisation.

Whether it is setting off a balloon, creating a ballet of trash trucks, or attempting to grow animal muscle in a petri dish, one of the common threads between the models of art-as-research referred to in the Further Viewing section is an attitude of play. Play might be one way to stay astray, in constant improvisation, to avoid the trap of directionality. Play might also be a way to do research not only *about* nature but *like* nature. "Play captures a lot of what goes on in the world," writes Donna Haraway.

> There is a kind of raw opportunism in biology and chemistry, where things work stochastically to form emergent systematicities. It's not a matter of direct functionality. We need to develop practices for thinking about those forms of activity that are not caught by functionality, those which propose the possible-but-not-yet, or that which is not-yet but still open.
>
> (Haraway, 2019)

Today, as climates are on the move, as ecologies are disrupted and reshaped, as co-existence is transformed by planetary heat, surprise needs to be embraced without recycling destructive modes of being. By being open to not-knowing, we can allow ourselves to engage with the emergent materialities of the world that are, to repeat Haraway, "not caught by functionality."

Perhaps the most prominent impact of play, as seen in the artistic projects discussed in this chapter, lies in the ability to disrupt and transform sites, institutions, and practices involved in knowledge practices and their governance – whether that is in a biotechnology lab, museum, or municipal agency. Some of the oblique effects of such projects have to do with the formation of new publics that do not map onto existing social groupings. *The art intervention then becomes a platform and a stage on which a public, knowledge, and their relations emerge in different forms.* The projects become mediators that politicize sites and institutions of technoscience – and alternative ways of knowing – and the practices of making them public. Dimitris Papadopoulos (2018) describes what is further necessary for such new publics to take hold. He says they can be built by engagement with materialities and nonhuman forces, and are ways of connecting and building a world that can be then integrated with organization and work of existing social movements to achieve "thick justice" or justice that is "done" with matter rather than mere discourse (Papadopoulos, 2018). This aspect, of communal collectivity, is where I can circle back to the lecture on attention that I started this chapter with. Next to the "averted vision" technique that artists might immediately connect with, there was another attention practice that the philosopher discussed. It is maybe not as attention-catching as averted vision but just as importantly a mode of operation that art, socially-engaged art that generates the aspects of communal collectivity discussed above. It was a practice of *paying attention together*, a practice of creating a shared internality. Practices of averted vision, when performed collectively, becomes yet another way of knowing. It is perhaps more radical for how simple yet transformative it can be.

Further Viewing

Open-weather, led by Sophie Dyer and Sasha Engelmann,

> is a feminist experiment in imaging and imagining the Earth and its weather systems using DIY community tools. … Open-weather encompasses a series of how-to guides, critical frameworks and public workshops on the reception of satellite images using free or inexpensive amateur radio technologies.
>
> (Centre for Research Architecture, 2023)

Mierle Laderman Ukeles has been an artist-in-residence at the New York Department of Sanitation since 1977. Her performances, such as *Touch Sanitation*, in which she shook hands with each of the 8,500 sanitation workers in New York City, or *Social Mirror*, in which she covered a garbage truck with a mirrored surface, brought the city's waste management system into public visibility in a way it never before has been, and that invited the NYC citizens to consider the questions related to waste and material circulation (Brooklyn Museum, 2023; Ronald Feldman Gallery, 2023).

The Natural History Museum, a project initiated the group Not An Alternative, describes itself as a "traveling, pop-up museum that highlights the socio-political forces that shape nature [but are excluded from traditional natural history museums]" (Natural History Museum, 2023). The Natural History Museum (NHM) focuses on the institution of the museum, arguing that it has a unique position of shaping and disseminating cultural narratives. In many of its projects, NHM says it publicly "interprets environmental history according to new coordinates – connecting local threats and movements to protect the environment, public health, and local cultures to the history of museums, the legacy of colonialism, and to ongoing concerns about cultural and environmental heritage" (Natural History Museum, 2023).

Oron Catts and Ionat Zurr's *Disembodied Cuisine* (2003) was an early bioart investigation of a lab-grown meat. It required particular institutional arrangements that provided access for the artists to the necessary labs, equipment, technologies, and expertise. Catts and Zurr developed their project through a lab at SymbioticA (2000), an art school and artistic research centre that they were instrumental in establishing at the University of Western Australia. It advises Masters and PhD candidates who focus on biological arts as well as offering research residencies for artists, designers, and other researchers. At the time of writing this chapter, SymbioticA was facing an uncertain future generated not by the inherent uncertainty of the world, but by potential funding cuts (SymbioticA, 2023).

References

Bergoffen, D. (2003) "Failed friendship, forgotten genealogies: Simone de Beauvoir and Luce Irigaray," *Bulletin de La Société Américaine de Philosophie de Langue Française* 13(1), pp. 16–31.

Bernstein, P. L. (1998) *Against the Gods: The Remarkable Story of Risk*. New York: Wiley.

Brooklyn Museum (2023) "Mierle Laderman Ukeles." Available at: www.brooklynmuseum.org/eascfa/about/feminist_art_base/mierle-laderman-ukeles (Accessed June 27, 2023).

Butler, J. (2021) *The Force of Nonviolence: An Ethico-Political Bind*. New York: Verso Books.

Centre for Research Architecture. (2023) "Open-weather: DIY satellite ground station workshop." Available at: https://research-architecture.org/Open-Weather (Accessed July 7, 2023).

Cooper, M. (2010) "Turbulent worlds," *Theory, Culture & Society* 27(2–3), pp. 167–90.

Dizikes, P. (2011) "When the butterfly effect took flight," *MIT Technology Review*.

Hacking, I. (1981) "From the emergence of probability to the erosion of determinism," in Hintikka, J., Gruender, D., and Agazzi, E. (eds.) *Probabilistic Thinking, Thermodynamics and the Interaction of the History and Philosophy of Science: Proceedings of the 1978 Pisa Conference on the History and Philosophy of Science Volume II*. Synthese Library. Dordrecht: Springer Netherlands, pp. 105–23.

Haraway, D. (2004) *The Haraway Reader*. New York: Psychology Press.

Haraway, D. (2019) "Feminist cyborg scholar Donna Haraway: The disorder of our era isn't necessary" [Interviewed by Moira Weigel], *The Guardian*. Available at: www.theguardian.com/world/2019/jun/20/donna-haraway-interview-cyborg-manifesto-post-truth (Accessed July 7, 2023).

Marres, N. (2016) *Material Participation: Technology, the Environment and Everyday Publics*. New York: Springer.

McCormack, D. P. (2015) "Devices for doing atmospheric things," in Vannini, P. (ed.) *Non-Representational Methodologies*. London: Routledge, pp. 89–111.

Natural History Museum (2023) Natural History Museum: About. Available at: https://thenaturalhistorymuseum.org/about/ (Accessed March 26, 2023).

Neimanis, A., Åsberg, C., and Hayes, S. (2015) "Post-Humanist Imaginaries," in Bäckstrand, K. and Lövbrand, E. (eds.) *Research Handbook on Climate Governance*, Cheltenham, UK: Edward Elgar Publishing, pp. 480–90.

Papadopoulos, D. (2018) *Experimental Practice: Technoscience, Alterontologies, and More-Than-Social Movements. Experimental Practice*. Durham, NC: Duke University Press.

Pellizzoni, L. (2011) "Governing through disorder: Neoliberal environmental governance and social theory," *Global Environmental Change, Symposium on Social Theory and the Environment in the New World (dis)Order*, 21(3), pp. 795–803.

Peters, J. N. and Piechocki, K. (2021) "Early modern clouds and the poetics of meteorology: An introduction," *Romance Quarterly*, 68(2), pp. 65–78.

Plumwood, V. (1993) *Feminism and the Mastery of Nature*. London: Routledge.

Rancière, J. (2007) "Art of the possible: An interview with Jacques Rancière" [Interviewed by Kelsey Fulvia and John Carnevale], *Artforum*, Available at: www.artforum.com/print/200703/fulvia-carnevale-and-john-kelsey-12843 (Accessed July 7, 2023).

Reinberger, H. J. (1994) "Experimental systems: Historiality, narration, and deconstruction," *Science in Context*, 7(1), pp. 65–81.

Ronald Feldman Gallery. (2023) "Mierle Laderman Ukeles." Available at: https://feldmangallery.com/artist-home/mierle-laderman-ukeles (Accessed June 27, 2023).

SymbioticA (2023) "SymbioticA: About." Available at: www.symbiotica.uwa.edu.au/home/about (Accessed March 26, 2023).

Part X

Climate Engineering

Introduction

Zeke Baker

This Part addresses Science and Technology Studies (STS) approaches to **geoengineering**, or climate engineering, generally defined by research, development, and deployment of large-scale techniques in deliberately modifying the planetary atmosphere. Before engaging this Part further, take about 10 minutes to complete the following simple tasks, which are less about gaining an understanding of geoengineering techniques and programs, and more about gaining some vivid reference points for how geoengineering has been taken up in media and public discourse. First, get started by Googling "geoengineering." After scrolling through a few entries, perform an "image" search. Upon doing so, what do you see? What patterns can you pick out in the images? Jot some notes down. Next, conduct a "news" search (again, still using Google or another search engine). What words or phrases seem to be repeated in news headlines? If you open up a few recent news articles on the topic, try to discern how geoengineering is being reported on and discussed.

Although other patterns may have emerged (and are worth exploring), one that probably stands out is framing that depicts geoengineering as controversial. In short, geoengineering is a *debate*. The topic exhibits a whole lot of "should" and "shouldn't" language. It is something like the death penalty or the right to abortion that one can apparently be "for" or "against." Like these social issues, the lines in the sand may on the surface appear clear, yet the controversies are indeed complicated because they involve crisscrossing issues of science, ethics, politics, economics, and culture.

One mode of intellectual reasoning, when faced with a debate of planetary proportion, is to engage it head-on: to use reason, evidence, and values to inform positions of public decision-makers, publics, researchers, and other stakeholders. A range of scholars have done just this. David Keith has publicly defended his team's geoengineering work. Atmospheric scientist and public scholar Katherine Hayhoe has in turn debated Keith and encouraged her audience to take a more skeptical position. In 2014, geographer Mike Hulme, in his book *Can Science Fix Climate Change?* answered "no" to the question of engineering climate, reasoning that such efforts would problematically simplify climate issues to matters of a manipulated global thermostat, entail major unknown risks, and are impossible to effectively govern. And historian of science J.R. Fleming has pointed to the longstanding hubris of technical "control" agendas (which frequently came up short or failed miserably) to argue that contemporary geoengineering research is a technocratic *déjà vu*. Large professional organizations, including for example the American Meteorological Society, have issued formal statements on geoengineering research or programs, some more supportive than others. And STS scholars have joined efforts at creating science policy and research norms for conducting geoengineering research, for example, the drafting of the "Oxford Principles for Geoengineering" (see Rayner et al., 2013).

STS authors in this Part do something different. Each steps out of the prevailing terms of a "for or against" debate, which is still raging in professional and media outlets. Instead of directly informing debate-like positions, they provide new ways of explaining or considering the

DOI: 10.4324/9781003409748-40

logics, assumptions, and challenges that geoengineering presents. Schubert does so historically; Schäfer, by comparing geoengineering to other climate science, technology, and policy efforts; Low, by turning to the question of how geoengineering research is assessed; and Hansson, by problematizing how geoengineering technologies have become presumed parts of climate policy solutions.

Let us pull out some key issues and questions that each author raises and draws upon their empirical research to help answer. Notice how each author's key questions neither begin nor end with whether or not geoengineering is a "good" or "bad" idea.

First, Schubert asks: where did geoengineering come from? How are recent ideas and research programs linked especially to relationships between scientific and governmental institutions? In particular, Schubert argues that "visions of control" have long been a central discourse linking climate knowledge to public decision-making regarding climate. Although climate engineering may have novel features, Schubert thus shows how the modern context is a deeply historical one, centrally tied up with climate (including modification) research developed over the latter half of the twentieth century.

Second, Schäfer asks: what is so unique about climate engineering? Are not what he calls "planetary geopolitics" present in other areas of climate science and policy? By comparing geoengineering to other areas of science and policy, Schäfer provokes us to consider if science about climate is not inherently also a "science of climate intervention."

Third, Low brings attention not to geoengineering research and development, but rather to *assessment*: what counts as "good" geoengineering work? How do different groups, for example geoengineering researchers and local communities, assess climate engineering programs with different interests, languages, and concerns in mind? How might analysis of assessment practices help us understand how highly consequential authoritative knowledge gets established? Through analysis of assessment, might STS work help inform a more socially-responsible (if not democratically accountable) approach to geoengineering research and climate policy?

Finally, Hansson explores a basic paradox: how is it that yet-to-be-developed, large-scale carbon dioxide removal (CDR, a form of geoengineering) technologies have made their way into climate science reports intended to inform climate policies? Rather than exploring this paradox as a mistake, Hansson brings a detailed view of the **co-production** of climate science/technology and climate policies, including in this case policies that presume geoengineering will need to be used to cut emissions down to levels that conform with existing policy targets. Hansson takes readers behind the scenes of the 2018 Special Report of the IPCC, which aimed to outline scientifically sound ways that society could keep global warming to 1.5°C. In the communications that Hansson studies, report writers and commentators effectively set boundaries around what concerns or topics (including the paradox about CDR feasibility) were taken up in the final report.

Through engaging each of the chapters in this Part, readers can expect to learn more about past and current geoengineering science/technologies and consider controversies within the field of geoengineering research and policy. Readers can also expect to come out of the Part with new questions that may help to inform and deepen their engagement with climate intervention – a topic that will undoubtedly remain a major scientific issue and salient public debate for years to come.

References

Hulme, M. (2014) *Can Science Fix Climate Change? A Case Against Climate Engineering*. New York: Polity Press.

Rayner, S., Heyward, C., Kruger, T. et al. (2013) "The Oxford Principles." *Climatic Change*, 121, pp. 499–512. https://doi.org/10.1007/s10584-012-0675-2.

31 The Politics of Climate Engineering Research

Julia Schubert

What Is at Stake?

In December 2019, a small budgetary decision made headlines in the United States. "For the first time," online news sources announced, the government had "authorized funding to research *geoengineering*" (Temple, 2019). The articles introduce us to the spectacular visions that linger behind this label. They talk of "controversial methods to cool the Earth" (Fialka, 2020), of "schemes to engineer the climate" (Pontecorvo, 2020), and of different approaches that would "reflect heat away from the planet" (Temple, 2019). The point of this, we learn, is to prepare for a Plan B, in case "the U.S. and other nations fail to reduce global greenhouse gas emissions" (Fialka, 2020).

For the past decade, the buzz over geoengineering has been growing louder. This confronts policy makers with a challenge. As attention to these measures increases, calls for governing geoengineering have followed. But what is it, precisely, that is to be governed here?

Geoengineering, also referred to as climate engineering or climate intervention, is commonly defined as the "deliberate large-scale intervention in the Earth's climate system, in order to moderate global warming" (Royal Society, 2009). This definition suggests that what is at stake here, is quite a lot. It raises the question of how to address an issue of global societal significance, of how to make our collective future on this planet, and how to live on it. While some hope for geoengineering to provide an additional tool for avoiding the worst impacts of a warming world, others fear that geoengineering further distracts from desperately needed emission cuts while also epitomizing human hubris and its potentially dangerous consequences.

On the other hand, climate engineering does not exist yet. There are currently no technically established means to "cool the Earth" by reflecting sunlight back to space or by drawing huge amounts of carbon from the air. Climate engineering is an **emerging technology**, a technology in the making so to speak. It therefore concerns a lot of basic research questions at the moment. This is how David Fahey, the scientist who received the controversial research funds, likes to see it: "It's not so much to study geoengineering," David Fahey suggests, "it's to understand the stratosphere as it exists today" (Pontecorvo, 2020). He speaks of observing and measuring things that you would be surprised scientists are not measuring already, like what kinds of materials are in the stratosphere and how they interact with one another. Fahey heads the Chemical Sciences Division at the National Oceanic and Atmospheric Administration (NOAA). "Geoengineering is this tangled ball of issues," he suggests, and "one of the things I'm interested in doing is let's separate the science out" (Fialka, 2020).

The challenge with the current debate over climate engineering is that it is necessarily a debate in which basic research is charged with grand visions of the future. As I am writing this chapter, there are at least two major petitions for and against research on solar climate

DOI: 10.4324/9781003409748-41

engineering circulating. One side suggests we need to know much more to inform meaningful decision-making on these measures. The other side calls for a non-use agreement and the need to regulate and even halt research, worrying that more research will provide a slippery slope into deployment. Both sides have reasonable arguments. Yet, their positions seem hardly reconcilable. So, the question is, how to make sense of this conflict?

What is at odds here are different expectations of how science informs policy-making. At the core of this conflict lie fundamentally different visions regarding the role that science plays, or should be playing, in informing decision-making and shaping society at large. Insights from STS thus seem crucial here. To overcome the current impasse in the debate over climate engineering research, we need to disentangle and scrutinize these different roles of science in society and reconsider how they may inform robust and fruitful governance around the discussed measures.

A Question of Science in Society

"Solar geoengineering is scary—that's why we should research it," says climate scientist Kate Ricke in a recent commentary published in the journal, *Nature* (Ricke, 2023). The need for more research on solar geoengineering primarily rests on the notion that we need to know much more to make an informed decision on this speculative technology. Drawing on Brian Wynne, we might say that scientific research serves to address questions of **propositional truth** in this context (Wynne, 2003). Research is mobilized, in other words, to settle matters of fact. So, essentially, to determine if solar geoengineering would work: What happens if we inject aerosols into the stratosphere? How would we even go about doing this? Would the particles behave as our models suggest? And how would we know? What kinds of instruments would we need to monitor deployment and assess its effects? What would the potential risks be? By providing answers to these kinds of questions, research is expected to establish a much-needed evidence base for politics.

This also means that research and decision-making on solar geoengineering are expected to be independent from one another: Arguing for the need for research *does not* imply arguing for deployment. In fact, it is often suggested that research might contribute to the opposite conclusion. "We need to develop … a firm [scientific] understanding of what it means and what the risks are so that we can decide if it's something we want to use or not," Representative McNerney argues in defence of a national research program on solar geoengineering in the US (McNerney qtd in Pontecorvo, 2020).

The expectation that scientific research provides important insights to then inform complex technical decision-making is, of course, perfectly reasonable. It makes sense to gather knowledge *before* making such decisions. And even if research is not able to settle all of the uncertainties, it will certainly be paramount to settling a lot of issues. What this perspective fails to acknowledge, however, is that "it" does not exist without the research that we supposedly need more of. Climate engineering, in other words, is not a given thing in the world. It is a vision that we have and continue to create precisely based on the kinds of research programs that we are developing.

This brings to the fore, then, a much more powerful role that science plays in informing policy making and shaping society at large. That is, science helps us to make sense of the world. It informs how we look at and respond to issues such as climate change. Quite importantly, science is able to fulfill this powerful role not by delivering a uniform body of "truth" or facts, but by providing a wealth of different and oftentimes conflicting disciplinary, methodological, and theoretical perspectives. These different perspectives enable us to make sense of what certain issues "mean" by providing multiple angles to various sets of people. This, in fact, made Dan

Sarewitz argue that science tends to "make environmental controversies worse"; that is, it tends to stoke rather than settle conflict and uncertainty (Sarewitz, 2004).

We might describe this latter and powerful role of science in society as informing a kind of **collective sense-making process**. At the moment when an issue of technical decision-making is formulated, this collective sense-making process is already concluded. In our case of climate engineering, for example, once we are able to ask what kinds of equipment we would need to successfully deliver reflective particles to the stratosphere or to properly monitor their behavior, quite a lot has already been decided. This in itself, of course, may not necessarily be a problem. Arriving at a collective understanding of what is at stake here seems like a great achievement. It becomes a problem, however, when the very prospect that we have arrived at is commonly deemed "scary." This is precisely what has happened in the case of so-called climate engineering proposals.

The relevant question, then, is not so much how we might best deliver the particles to the stratosphere. Reducing uncertainty on these matters would indeed be great, but it fails to answer a bigger question that is at stake here, which is: how did we get here? Why are we discussing a prospect that is advanced as "a bad idea" in the first place? How and why have we come to make sense of the problem of climate change as an issue of sun shield design and atmospheric carbon accounting despite the fact that even climate scientists describe this prospect as "scary"?

Whatever the answers to these questions might be, they will surely be important to advance a more constructive engagement with so-called climate engineering proposals. Understanding the dynamics that have brought us here will be essential to moving forward.

In the following, I will paint a bigger picture of the current climate engineering debate in three sequential episodes (for a detailed account, see Schubert, 2022). This account suggests that proposals to modify or engineer the climate hardly emerged from the fringes of climate policy. Despite being framed as a last resort against urgent climate change, such proposals provided a central motive for political interest and support of the climate science field for the past 60 years. Climate engineering, in other words, is deeply rooted in the way that we have chosen to mobilize science in order to make sense of the climate change issue.

Early Visions of Control

The first episode that I want to turn back to in order to make sense of the current debate over climate engineering brings us to the 1960s and the beginnings of modern climate science. This episode is relevant because it shows that political hopes for "climate modification" importantly defined atmospheric research at the time. It points us to the joint origins of today's problem and response, illustrating that it is not only science that informs politics, but also politics that informs science.

To understand why this political interest in climate modification was so important to the nascent field of climate science, we have to consider where this field came from. Particularly in the United States, the formation of the climate science field was closely linked to interests of the state (Baker, 2017; Edwards, 2010; Harper, 2017; Hart and Victor, 1993). The military need for better weather forecasts turned meteorology and oceanography into a national priority during the first half of the twentieth century. The Second World War essentially fostered the professionalization of meteorology in the United States.

The end of the war therefore confronted scientists with a challenge, namely how to sustain this vital support by the state. The emerging field of "Weather and Climate Modification" promised help in this context (Harper, 2017). Although many meteorologists of the time were rather skeptical of the prospect, this vision of modifying and even controlling weather and climate catered to national strategic concerns of the time. And therefore, it promised continued state support. So, in order to mobilize political support and funding for their research, scientists couched their questions into the prospect of Weather and Climate Modification. The result was a rather chaotic research program.

When the National Academies of Sciences were tasked with reviewing the field in 1966, their summary was grand, to say the least. Weather and Climate Modification, they suggested, is

> concerned with any artificially produced changes in the composition, behavior, or dynamics of the atmosphere. Such changes may or may not be predictable, their production may be deliberate or inadvertent, they may be transient or permanent, and they may be manifested on any scale from the microclimate of plants to the macrodynamics of the worldwide atmospheric circulation.
>
> (NAS 1966, p. 1; see also NSF, 1965, p. 7)

In essence, the field was concerned with human ("artificial") modifications of all scales, forms, and outlooks. What is particularly noteworthy from today's perspective is the distinction between *deliberate* and *inadvertent modification*. This distinction is relevant because it suggests that the notion of climate modification provides a joint origin of what we discuss today as climate change (inadvertent, at least initially) and for what we discuss today as climate engineering (deliberate).

Yet, rather than relating the two sides of this distinction as crisis and remedy, they appear here as two sides of the same coin, as both challenge and opportunity. The reason for this was the particular connection of scientific and political interests that I hinted at earlier. Different lines of scientific inquiry, like the carbon cycle and atmospheric modeling communities, had long been interested in a better understanding of human impacts on climate and the impact of carbon dioxide on climate dynamics. Such scientific interests, however, garnered minimal political attention. So, in order to gather political support for this research, scientists linked their interests to the promise of concrete application, namely deliberate modification of weather and climate. As a result, the notion of Weather and Climate Modification was one of the central motives by which meteorology managed to connect basic research questions to the interests of the state.

From Modification to Change: A Problem Breaks into Politics

The second episode that is worth revisiting in light of current controversies over climate engineering is the early politicization of global warming during the 1970s and 1980s. While scientists had been concerned with the greenhouse effect for some decades already, it was only during this period that their discoveries reached widespread recognition. This meant that human impacts on the climate no longer made sense as both challenge and opportunity, that is, as *deliberate* and *inadvertent modification*. Instead, they were increasingly seen as an ecological problem of global societal significance, namely as *global climate change*. This shift from modification to change marks what Zeke Baker calls "a fracture" of established alliances between climate science and the state (Baker, 2017). It meant that climate science no longer appeared as a possible tool of control at the hands of the state. Instead, and somewhat ironically, it now pointed to a problem that questioned precisely such hopes for control and techno-scientific optimism.

Despite this "fracture," climate science remained politically relevant. When the problem of climate change arrived on the political agenda between the 1970s and 1990s, it was primarily the geophysical sciences that successfully defined the issue in the political realm. Climate change was established here as "the CO_2 problem" or "the greenhouse problem," as an issue of atmospheric chemistry and thermodynamics. The relevant question was now: "What *should* the atmospheric carbon dioxide content be over the next century or two to achieve an optimum global climate?" (NAS, 1977).

This perspective on the issue of climate change implied that visions to deliberately modify the climate now appeared as a potential response to "the greenhouse problem" (Keith, 2000). Experts were rather pessimistic about the prospects of these response measures. But posing the climate change problem as "the greenhouse problem" stoked speculation over means, for example, to "control CO_2 emissions" by sequestering carbon dioxide at smokestacks or by planting trees, as well as means to "offset greenhouse warming" by "reducing the amount of solar radiation that penetrates the troposphere" (Seidel and Keyes, 1983, pp. 6–13). This particular problem definition of greenhouse warming thus united two sets of otherwise fundamentally different techno-scientific concepts under one umbrella: namely methods to remove carbon dioxide from the atmosphere and methods to modify incoming solar radiation. Here, we thus find the roots of a distinction that continues to define the debate over climate engineering to this day under the labels of Carbon Dioxide Removal (CDR) and Solar Radiation Management (SRM).

This second historical episode suggests how early versions of what we discuss today as climate engineering have roots in some of the earliest high-level policy frameworks on the issue of climate change, which emerged over the period between the 1970s and the 1990s. While climate modification proposals were not seen as particularly viable approaches to the issue of climate change, they were implicated in the very formulation of the problem as it emerged.

Peak Shaving and Overshoot: Climate Intervention as Last Resort

For the final episode, let us fast forward to the first decades of the new millennium. This is when climate engineering emerged as a "bad idea whose time has come" (Kintisch, 2010). Climate engineering gained renewed political traction during the early 2000s not because of surprising scientific or technological advancements. Climate scientists had not suddenly changed their mind about the prospect of these measures. In fact, a prominent climate scientist released a list of "20 reasons why geoengineering may be a bad idea" in 2008 (Robock, 2008).

If not because of a breakthrough in climate engineering research, then why the sudden policy interest in the topic? Rather than positive visions of techno-scientific innovation, it was daunting visions of the climate crisis that pushed public attention and political interest toward climate engineering over the past two decades. Concerns over climate tipping points and a perceived lack of effective climate policy tools at hand made experts suggest a rather dire prospect: "I know this is all unpleasant," Hugh Hunt, engineering professor at the University of Cambridge argued in a conversation with the New Yorker.

> Nobody wants it, but nobody wants to put high doses of poisonous chemicals into their body, either. That is what chemotherapy is, though.... This is how I prefer to look at the possibility of engineering the climate. It isn't a cure for anything. But it could very well turn out to be the least bad option we are going to have."
>
> (Specter, 2012)

Especially the Paris Agreement of 2016 advanced these kinds of claims and pushed climate engineering into the policy spotlight by highlighting a dilemma. The agreement states the goal of holding "the increase in the global average temperature to well below 2°C above pre-industrial levels." A Special Report on Global Warming of 1.5°C, published by the IPCC in 2018, served to scientifically support this target, warning of the severe impacts to be expected from warming even beyond 1.5°C above pre-industrial levels. The report, however, also suggests that this goal is not compatible with the national commitments to emission reductions that the Parties have submitted under the Paris Agreement.

Now, in order to produce emission reduction pathways that would be compatible with the 2°C target, despite the lack of existing emission reduction commitments, experts introduced into their models so-called **negative emissions technologies** (NETS), defined by their capacity to draw carbon emissions out of the atmosphere. This helped to give rise to the concepts of "net-emissions" and "net-zero emissions," which have since become central frames for long-term climate governance. Furthermore, the role of NETS in emission reduction pathways pushed Carbon Dioxide Removal (CDR) to the heart of climate policy, even though it had received rather little policy attention up to that point (Beck and Mahony, 2017). Integrated Assessment Models, as they are called, incorporated CDR as a means to complement emissions reductions by other means (like reducing greenhouse gas emissions). Incorporating CDR thus provided a way for climate scientists to envision climate policy targets as capable of being met.

Notably, the reason why this seemed feasible was not a sudden scientific breakthrough. Rather, the economic assumptions built into the models, and particularly the goal of long-term cost-optimization and choice of discount rate, made the introduction of CDR feasible *despite* the speculative nature of deploying these technologies at scale (Lund et al., 2023; van Beek et al., 2022; Sapinski, Buck, and Malm, 2020).

Thus far, Solar Radiation Management (SRM) approaches remain much more controversial than CDR and are still largely ignored in mainstream climate policy contexts. This, in fact, is why many have called for a categorical separation between CDR and SRM approaches. They propose to only define SRM as climate engineering and to treat CDR as a regular form of climate mitigation.

Yet, some scientists have suggested that it might become necessary to deploy SRM measures in order to stay within the temperature targets postulated by the Paris Agreement (Irvine et al., 2019). Specifically, these scientists argue that SRM might become necessary to "buy time" for societies to develop the emissions reductions and carbon removal capacities that would be necessary to stay within these temperature targets (Long, 2017; MacMartin et al., 2018). In so-called "overshoot" or "peak-shaving" scenarios, SRM is deployed to allow for a temporary overshoot of the global "carbon budget". Under these scenarios, SRM compensates for the potentially dangerous levels of warming that results from such an "overshoot", by figuratively "shaving the peak off" the resulting warming curve. Scholars in the field of science studies have drawn critical attention to the fact that SRM is mobilized in such scenarios to "escalate climate debt" (Asayama and Hulme, 2019; Asayama, Hulme, and Markusson, 2021). They suggest that SRM serves here to merely mask, rather than erase, dangerous warming levels all the while assuming the successful built-up of massive (and thus far speculative) CDR capacities.

Outlook: What Kind of Science for What Kind of Politics?

It is arguably true that we would need to know quite a lot more to determine if a "global thermostat" would work and what it would look like. But is this really the relevant question that should guide decision-making on climate engineering research? Is this a meaningful and convincing objective for future research programs on the matter? I am not so sure.

The need for research on climate engineering is often invoked in a quest to answer what Brian Wynne called questions of propositional truth. Research is needed, it is argued, to settle matters of fact. It is needed, in other words, to properly *inform technical decision-making* regarding a speculative and "scary" technology. With this chapter, I hope to complicate this picture. Drawing on the longer history of climate engineering, I want to suggest mobilizing a different and much more powerful role of science in society when addressing the question of research in this context of climate engineering. Namely the power to *inform a collective sense-making process*.

The longer history of climate engineering, as briefly outlined above, suggests that hopes to modify or even control the Earth's climate have provided a central motive for political support of the climate science field for the past 60 years or so. In a sense, this controversial response measure was born before we even knew we had a problem.

This longer history, then, is relevant to the current controversy around climate engineering research in at least two respects. First, it highlights political agency in crafting the very prospect of climate engineering that we are confronted with today. We cannot properly make sense of this history of climate engineering by focusing on internal scientific developments alone. This is not simply a story of scientific "progress" or "discovery." Rather, it is a story of how science sought to respond to ever-changing societal concerns and the respective political agendas. Proposals to modify or engineer the climate evolved according to the defining issues of their time. As historian Jim Fleming aptly put it, "each generation … has had its own leading issues for investing in technologies of control" (Fleming 2010, p. 265).

Secondly and consequently, this historical perspective calls for a change of course. If climate engineering is indeed a scary prospect, then the problem is not so much that we don't have all the answers yet, but rather that we have been asking the wrong kinds of questions. Understanding the role of science in society as one that serves to inform collective sense-making rather than technical decision-making would imply a different approach to governing climate engineering research. It would imply striving towards a more meaningful vision of climate engineering than "peak shaving" and "overshoot". It would imply focusing on gaining a better understanding of how climate engineering can (and cannot) respond to the various concrete problems that climate change implies for concrete sets of people. It would imply radically diversifying and pluralizing research perspectives on climate engineering. Arriving at a constructive societal engagement with these proposals is not a question of more or less science. It is a question of what kind of science for what kind of politics (see also Schubert, 2021; Hulme, 2012).

References

Asayama, S. and Hulme, M. (2019) "Engineering climate debt: Temperature overshoot and peak-shaving as risky subprime mortgage lending," *Climate Policy 19*(8), pp. 937–946.

Asayama, S., Hulme, M., and Markusson, N. (2021) "Balancing a budget or running a deficit? The offset regime of carbon removal and solar geoengineering under a carbon budget," *Climatic Change, 167*, p. 25.

Baker, Z. (2017) "Climate state: Science-state struggles and the formation of climate science in the US from the 1930s to 1960s," *Social Studies of Science, 47*(6), pp. 861–887.

Beck, S. and Mahony, M. (2017) "The IPCC and the politics of anticipation," *Nature Climate Change*, 7(5), pp. 311–313.

Edwards, P. N. (2010) *A Vast Machine: Computer Models, Climate Data, and the Politics of Global Warming*. Cambridge, MA: MIT Press.

Fialka, J. (2020) "NOAA gets go-ahead to study controversial climate plan B," *Scientific American*. www.scientificamerican.com/article/noaa-gets-go-ahead-to-study-controversial-climate-plan-b/

Harper, K. C. (2017) *Make it Rain: State Control of the Atmosphere in Twentieth-Century America*. Chicago: University of Chicago Press.

Hart, D. M. and Victor, D. G. (1993) "Scientific elites and the making of US policy for climate change research, 1957–74," *Social Studies of Science, 23*(4), pp. 643–680.

Hulme, M. (2012) "What sorts of knowledge for what sort of politics? Science, climate change and the challenge of democracy," Science, Society and Sustainability (3S) Research Group. Norwich: University of East Anglia. https://mikehulme.org/wp-content/uploads/2011/09/12_05-Copenhagen-script_web.pdf

Irvine, P., Emanuel, K., He, J., Horowitz, L. W., Vecchi, G., and Keith, D. (2019) "Halving warming with idealized solar geoengineering moderates key climate hazards," *Nature Climate Change*, 9(4), pp. 295–299.

Keith, D. W. (2000) "Geoengineering the climate: history and prospect," *Annual Review of Energy and the Environment*, 25(1), pp. 245–284.

Kintisch, E. (2010) *Hack the Planet: Science's Best Hope-or Worst Nightmare-for Averting Climate Catastrophe*. London: Wiley.

Lund, J. F., Markusson, N., Carton, W., and Buck, H. J. (2023) "Net zero and the unexplored politics of residual emissions," *Energy Research & Social Science*, 98.

MacMartin, D. G., Ricke, K. L., and Keith, D. W. (2018) "Solar geoengineering as part of an overall strategy for meeting the 1.5°C Paris target," *Philosophical Transactions of the Royal Society A*, 376, 20160454.

Miller, C. and Edwards, P. (Eds.). (2001). *Changing the Atmosphere: Expert Knowledge and Environmental Governance*. Cambridge, MA: MIT Press.

NAS, National Academies of Science (1966) *Weather and Climate Modification: Problems and Prospects*. Washington, DC: National Academy of Sciences-National Research Council.

NAS, National Academies of Science (1977) *Energy and Climate: Studies in Geophysics*. Washington, DC: National Academies Press.

NSF, National Science Foundation (1965) *Weather and Climate Modification*. Washington, DC: National Science Foundation.

Pontecorvo, E. (2020) "The climate policy milestone that was buried in the 2020 budget." *GRIST*. https://grist.org/climate/the-climate-policy-milestone-that-was-buried-in-the-2020-budget/

Ricke, K. (2023) "Solar geoengineering is scary—that's why we should research it," *Nature*, 614(7948), p. 391.

Robock, A. (2008) "20 reasons why geoengineering may be a bad idea," *Bulletin of the Atomic Scientists*, 64(2), pp. 14–18.

Royal Society (2009) *Geoengineering the Climate: Science, Governance and Uncertainty*. London: The Royal Society.

Sapinski, J., Buck, H. J., and Malm, A. (2020) *Has it Come to This?: The Promises and Perils of Geoengineering on the Brink*. New Brunswick, NJ: Rutgers University Press.

Sarewitz, D. (2004) "How science makes environmental controversies worse," *Environmental Science & Policy*, 7(5), pp. 385–403.

Schubert, J. (2021) *Engineering the Climate: Science, Politics, and Visions of Control*. Manchester, UK: Mattering Press.

Schubert, J. (2022) "Science-state alliances and climate engineering: A 'longue-durée' picture," *WIREs Climate Change*, 13(6), p. e801. https://doi.org/10.1002/wcc.801

Seidel, S. and Keyes, D. L. (1983) *Can We Delay a Greenhouse Warming? The Effectiveness and Feasibility of Options to Slow a Build-up of Carbon Dioxide in the Atmosphere*. Washington, DC: EPA.

Specter, M. (2012) "The climate fixers," *New Yorker*. www.newyorker.com/magazine/2012/05/14/the-climate-fixers

Temple, J. (2019) "The US government has approved funds for geoengineering research," *MIT Technology Review*. www.technologyreview.com/2019/12/20/131449/the-us-government-will-begin-to-fund-geoengineering-research/

van Beek, L., Oomen, J., Hajer, M., Pelzer, P., and van Vuuren, D. (2022) "Navigating the political: An analysis of political calibration of integrated assessment modelling in light of the 1.5°C goal," *Environmental Science & Policy*, 133, pp. 193–202.

Wynne, B. (2003) "Seasick on the third wave? Subverting the hegemony of propositionalism: Response to Collins & Evans (2002)," *Social Studies of Science*, 33(3), pp. 401–417.

32 The Intervention of Climate Science

Stefan Schäfer

Introduction

In 2020, the United States Congress ordered the US National Oceanic and Atmospheric Administration (NOAA) to set up a new research initiative. Designed to improve knowledge of "Earth's radiation budget," the initiative funds projects that inquire into the balance between, on the one hand, energy that reaches Earth from the sun, and, on the other hand, energy that leaves Earth, in the form of heat emitted from the planet's surface or when sunlight is reflected back into space. "Earth's radiation budget," the initiative's website informs, "determines the climate of the Earth and makes our planet hospitable for life." Funded research projects use the standard repertoire of climate science methods, such as computer modeling and observation from aircraft and satellites. By 2022, a total of 23 million USD had been committed to the initiative.

In the journal *Science*, an atmospheric scientist at NOAA is quoted saying that the initiative constitutes "very basic science" (Voosen, 2023, p. 628). Indeed, most if not all the projects seem like they might just as well be funded under any climate science funding program. Earth's radiation budget is not at all a new field of study. And yet, the ERB initiative, as NOAA calls it, is different. It is different in that it presents itself as directed expressly at the study of "climate intervention," a term NOAA uses to describe strategies for actively changing the composition of the atmosphere. Such changes may increase the fraction of incoming sunlight that is reflected away from Earth and back into space, for example by making clouds brighter and increasing their longevity, or by adding reflective particles to the atmosphere. By studying the effects of particles and clouds on Earth's radiation budget, scientists are thus also studying how an out-of-balance radiation budget might be intervened in for it to once again become balanced. What is, on the one hand, "very basic science" studying processes that have kept climate scientists busy for decades, is, at the same time, also a set of projects developing strategies for "climate intervention." What is different about the ERB initiative thus is not necessarily a qualitative difference in the kind of research it funds. Rather, it stands apart from other research initiatives in announcing the study of climate intervention as its intention. We may thus take the ERB initiative as a prompt for asking: what, exactly, *is* a science of climate?

What the ERB initiative displays is the intimate relationship between climate science and climate intervention – in Ian Hacking's terms, between **representing** and **intervening** in the science of climate. This intimacy is often overlooked. It is more common to think of climate science as primarily concerned with representation. In this view, climate models, satellite observations, and other methods, such as the counting of tree rings or the analysis of ice cores, allow scientists to represent with increasing accuracy the global climate and the mechanisms that govern its behavior, independent of any intervention into that object. Intervention is thought of as standing apart. Intervention is what humans may decide to do, or not to do, with the theoretical

DOI: 10.4324/9781003409748-42

knowledge that science provides. But, I argue, such a strict separation between representation and intervention does not hold up. Questioning it has important consequences for how we might think about climate science and about the forms of collective life that we build with it.

"The harm comes from," writes Hacking in his classic text, *Representing and Intervening*, "a single-minded obsession with representation and thinking and theory, at the expense of intervention and action and experiment" (Hacking, 1983, p. 131). Hacking's "harm" is an idealist philosophy in which humans cannot know what really exists independent of human thought. In this chapter, I use the example of climate science to show how lines of inquiry from STS and neighboring fields – lines of inquiry that, like Hacking, emphasize the importance of "intervention and action and experiment" – can help us rethink the politics of science. My concern in doing so, however, is different from Hacking's. Unlike Hacking, I am not primarily interested in the ontological question of what is real, or the epistemological question of how we can know. Instead, I am interested in the political question of what is *good*.

In what follows, I will elaborate three ways in which an interventionist redescription of climate science can show us that climate science is both more and less than we conventionally take it to be. In its most basic formulation, my argument is that climate science is a form of "intervention and action and experiment" at the scale of the entire planet, or in even less words: that climate science writ large is a science of geoengineering. Thus, as we commit to a form of life built out together with a science of planetary climate, we also commit to a world in which intervening in the planetary climate is not a contingent, but a necessary feature.

Experiment

Contemporary knowledge of the planetary environment, and with it of the planetary climate, was importantly developed from interventions into that environment. These interventions often were highly destructive: acts of large-scale destruction and contamination could be studied as experiments whose traces became evidence of planetary interconnectedness and systematicity. First and foremost, in the mid-twentieth century, it was the tracing of the spread of radioactive material released in nuclear explosions ("tests") through the atmosphere and ocean and through plants, animals, and humans, which formed the basis for a scientific knowledge of the "planetary environment" as we understand it today – as all-encompassing, total, systematic, everything connected to everything else. Global warming itself can be seen as another such act of destruction, in which the massive-scale extraction and burning of hydrocarbons provided new opportunities for studying ocean, atmosphere, and land as an integrated, volatile, and total planetary environment.

Scientists were quick to use the language of "experiment" in connection with these destructive interventions into what, during the 1980s, came to increasingly be called the "Earth system." Oceanographer Roger Revelle and chemist Hans Suess concluded a now-famous 1957 paper on the accumulation of carbon dioxide in the atmosphere from industrial activity, by stating that "Human beings are now carrying out a large scale geophysical experiment of a kind that could not have happened in the past nor be reproduced in the future" (Revelle and Suess, 1957, p. 19). Similarly, the physicist Stephen Schneider gave his popular science book the title "Laboratory Earth," in which he wrote that "Much of what we do to the environment is an experiment with Planet Earth, whether we intend it to be or not" (Schneider, 1996, p. 8). The experimental opportunities that come with changes to the planetary environment are also on display when scientists study volcanic eruptions as "natural experiments" that can inform about the complex chemistry and physics of the atmosphere. The scientists funded under NOAA's ERB initiative, too, are open to what the already-mentioned article in *Science* calls "missions of opportunity": science projects in which volcanic eruptions or massive wildfires would be used to understand how the climate system works – and how, consequently, it can be manipulated.

As scientists learn to study nuclear explosions, hydrocarbon burning, volcanic eruptions, and wildfires as planetary-scale "experiments," and as they record and interpret these "experiments" in all their similarities and differences with ever greater technical sophistication, they develop new possibilities for planetary intervention. These possibilities are contained in the equations of climate science. From undirected "experiments" in planetary intervention, a *scientific* form of planetary intervention – a science of geoengineering – can be developed: climate science.

Geoengineering

The interventionist and often destructive origins of contemporary planetary environmental science are well-known and documented. However, most existing scholarship assumes that this origin leaves no trace on the scientific knowledge thus generated. Once established, science does not carry with it baggage from its provenance. This should by now seem surprising. My argument is that, instead, originating in planetary-scale intervention, today's climate science is a science of intervention at the scale of the entire planet.

Today, when people speak of "climate intervention," they often mean, like NOAA in the ERB initiative, what others (including in this volume) have referred to as "climate engineering" or "geoengineering." This is not a sharply defined category and what one will subsume under it is, as with any act of categorization, a decision that is at once scientific and political (Schäfer and Low 2018; see Schubert, Chapter 31, this volume). For a while, a prominent categorization held that "geoengineering the climate," as the UK Royal Society called it in an influential 2009 report, has two subcategories: carbon dioxide removal and solar radiation management. Carbon dioxide removal could range from planting trees or combining bioenergy production with carbon capture and storage to the filtering of ambient air or the manipulation of oceanic biochemistry. Solar radiation management could include placing mirrors in space, brightening clouds, or introducing reflective particles into the stratosphere. It is solar radiation management that NOAA means when it says "climate intervention."

"Climate engineering" or "geoengineering" is usually considered apart from other approaches to addressing climate change, such as renewable energy technologies or legal and financial mechanisms designed to curb carbon dioxide emissions. There are certainly differences between all of them. But if we examine them from the standpoint of climate science, we also see a crucial similarity: each approach is justified as a strategy for managing a changing planetary climate and they all rely on climate science for their claims of effectiveness, benefits, and risks. We might thus also say that climate science describes an object in need of management – the changing planetary climate – and at the same time supplies strategies for effecting such management. By extension, we might reasonably conclude that approaches to managing global climate – from renewables to carbon markets to sunlight-scattering particles, and whatever may in the future come to be included in the list – constitute variants of geoengineering, and that such geoengineering today is a dominant concern in collective social life. The differences between these variants of geoengineering are internal differences; choosing one over the other does not mean leaving the world of geoengineering.

The most familiar shape that this takes in the present is in the attempt to manage the composition of the planetary atmosphere by reducing the number of planet-heating molecules, carbon dioxide prime among them, that gets released from furnaces, smokestacks, vehicles, and other sources. Ambitions to massively build out infrastructures of renewable energy, to establish new financial markets for carbon trading, or to implement policy instruments such as carbon taxes, form part of this attempt. They are justified by the same logic that is also used to justify carbon dioxide removal or solar radiation management. In this sense, critics of what NOAA calls climate intervention, who instead insist on the massive-scale production of wind or solar energy or on the expansion of natural gas production as a "bridge technology" or on the transition to a "hydrogen economy" – these critics, too, participate in the form of social life that makes geoengineering one of its important organizing logics.

But schemes for intervening in the planetary climate date back to the very origins of climate science, long predating contemporary worry about anthropogenic climate change. Already when the scientific idea of a planet-spanning order of heat distribution first took shape in the late 18th century, it emerged together with schemes for how that order might be manipulated (Schäfer and Mauelshagen, 2021). As Johann Gottfried Herder wrote in his "Reflections on the philosophy of the history of mankind": "Now it is no question that, just as the climate is the embodiment of powers and influences that both plant and animal contribute to and that serves all life in mutual connection, humans are made masters of the Earth also in that they may change it through art" (Herder, 2002 [1784–1791], p. 244, my translation). Another thinker of the day, the French natural philosopher Comte de Buffon, noted widespread puzzlement over the fact that Paris and Quebec experienced such different weather despite occupying roughly the same latitude on the globe. His explanation was that

> Paris and Quebec are about at the same latitude and at the same elevation on the globe; Paris would thus be as cold as Quebec, if France and all the countries that neighbor it were as bereft of people, as covered with forest, as bathed in waters as are the lands that neighbor Canada. To cleanse, to reclaim and people a country, is to provide it with warmth for many thousands of years.
>
> (Buffon, 2018 [1781], p. 126)

To Buffon, a hospitable climate – hospitable to a European sense of comfort – would come with the "cleansing," "reclaiming," and "peopling" of a country so characteristic of settler colonial expansion.

Scientific representations of climate were, from the very start, thus always also blueprints for how climate – and with it, people – might be manipulated. How, precisely, such manipulation might be effected has differed across history and geography. At the same time, and as is evident in the quotes from both Herder and Buffon, knowing and manipulating the global climate were deeply implicated in the colonial expansion of European powers into the Americas (see Mahony, Chapter 2, this volume). As Prussian geographer Alexander von Humboldt noted in a discussion of the historical advancement of climate science: "The progress of 'Climatology' has been remarkably favored by the extension of European civilization to two opposite coasts, by its transmission from our western shores to a continent which is bounded on the east by the Atlantic Ocean" (Humboldt 1858, 318). A science of climate at the scale of an integrated planetary whole was possible only with geographical expansion at planetary scale. Consequently, as more than one power came to lay claims to global representation, *whose* science of climate would produce the binding representations of planetary environment became a matter of politics – a **planetary geopolitics** in which dominance meant the ability to devise prevailing strategies for geoengineering the planetary environment.

Geopolitics

Climate science is thus always also a geopolitics. Consider computer modeling. In the late 1980s, a controversy broke out when Western scientists insisted that in the assessments of the newly founded Intergovernmental Panel on Climate Change (IPCC), predictions of future climate should be made exclusively from computer models. This was opposed by Soviet scientists who specialized in the paleo-analog method. In the paleo-analog method, future climates were known in analogy to past climates, based on the record that ancient atmospheres have registered in soil, ice, trees, and corals. A crucial difference between computer modeling and the paleo-analog method was that Western computer modelers saw a potentially catastrophic future in their models, while Soviet paleoclimatologists concluded from the paleo-analog method that, while the climate was certainly changing under human influence, what that change would look like was highly uncertain and possibly quite positive. A meeting was called to resolve the controversy. The invitation letter read: "This is not just an academic issue; the forecasts from the two techniques are different and may have different policy implications, and so it is important that the issue is resolved amongst those who are best equipped to deal with it" (Skodvin, 2000, p. 140; compare also Doose, 2022).

At the meeting – held on November 20–21, 1989, less than two weeks after the Berlin wall had fallen – computer modelers unequivocally declared the planetary past too idiosyncratic to serve as a tool for predicting the future. This would henceforth be the sole prerogative of computer models. The modeler's position and their victory over the specialists of the paleo-analog method was confirmed in the IPCC's first assessment report, and even in its publication history. Bert Bolin, Swedish meteorologist and first chairman of the IPCC, referred to the controversy by noting that "scientific as well as practical difficulties arose" from the Soviet Union's leadership in the IPCC's Working Group 2: "This was finally resolved when one of the co-chairmen of the working group, M. Tegart, Australia, together with some of his Australian colleagues took on the task of compiling and editing the final report" (Bolin, 2007, p. 65). As a result, the leader of the Soviet IPCC delegation and chair of Working Group 2, Yuri Izrael, was not named as an editor on the cover of the Working Group 2 report. The Working Group 1 report, edited by computer modeler and initiator of the November meeting John Houghton, declared upon its publication in 1990 that "In conclusion, the paleo-analogue approach is unable to give reliable estimates of the equilibrium climatic effect of increases in greenhouse gases." In the IPCC's predictions of future climate, the report continued, paleoclimatology's main use would be to "provide useful data against which to test the performance of climate models" (Houghton, Jenkins, and Ephraums, 1990, p. 159).

At stake in this controversy was the very question of who would define the problem of climate change and devise solutions to it. With the end of the Cold War, one particular science rose to global dominance: Earth System Science, a cybernetic science of planetary systematicity developed in the US during the 1980s and based on computer modeling and remote sensing with satellites. It is in Earth System Science that the possibility of Earth system management appears in particularly stark relief. In 1999, the physicist John Schellnhuber wrote of a "geo-cybernetic task" that confronts Earth system scientists, a task that "can be summed up in three fundamental questions. First, what kind of world do we have? Second, what kind of world do we want? Third, what must we do to get there?" Schellnhuber proceeds to offer a "menu from which humanity can select its master principle, or suitable combinations thereof, for Earth-system control," a manual to be further elaborated and put into operation by fellow Earth system scientists (Schellnhuber, 1999). Earth System Science's cybernetic concept of a volatile and highly interconnected Earth system subject to "feedback" while offering "leverage points" for intervention is largely uncontested today. It is within this conceptualization of the planetary environment that geoengineering, in its many contemporary variants, today seems most plausible and necessary.

Questions about the geopolitics of geoengineering are thus always also questions about the geopolitics of climate science – whose climate science provides the basis for making and attacking claims about geoengineering's necessity and its risks and benefits and their distribution? But at a more basic level, loosening the grip that geoengineering has on contemporary politics would require a rethinking of what it would look like to worry about climate change without the technoscientific interventionism of climate science controlling all possible responses. While some commentators argue about whether solar radiation management can or cannot be governed at all, it might thus be just as pertinent to ask: Can climate science be governed at all?

Conclusion

In this chapter, using the example of climate science, I have shown some possibilities that arise within an STS that denies a strict separation between representing and intervening and how, from an interventionist redescription of climate science, we can develop new political questions about the forms of social life and social order with which a science of planetary climate is **co-produced** (Jasanoff, 2004).

To study this form of life in which a science of planetary climate plays such an important organizing role, we must position ourselves outside of the presuppositions that underwrite it, and thus outside of the presuppositions of climate science. We must approach climate science as one form that human creativity takes, a form of creativity that participates in shaping the kind of collective social life that humans build out today, while itself being shaped by prevailing forms of social life and social order. An investigation of this kind would not ask about the necessity or riskiness of different strategies for intervening in the planetary climate. Nor would it hinge upon uncertainty or a formally calculated ratio of costs to benefits at the global scale. It would instead ask why and how those questions arise in the first place, and on whose terms. It would interrogate who has the power to raise such questions, and who has the power to answer them. It would critically assess what questions and answers are possible, and which are impossible. Finally, it would consider what might be gained and what might be lost as a science of planetary climate comes to define the possibilities and impossibilities by which humans organize their collective life. What are the hierarchies of harm that a form of life organized around a science of planetary climate establishes for itself? What tradeoffs does it accept? What is expendable to it, and what must be preserved at all costs? In short: is it *good*?

Further Reading

Hacking, I. (1983) *Representing and intervening: Introductory topics in the philosophy of natural science.* Cambridge: Cambridge University Press.

Jasanoff, S. (ed.) (2004) *States of knowledge: The co-production of science and social order*. London: Routledge.

Masco, J. (2021) *The future of fallout, and other episodes in radioactive world-making.* Durham: Duke University Press.

Stilgoe, J. (2015) *Experiment earth: Responsible innovation in geoengineering.* London: Routledge.

References

Comte de Buffon, G.-L. L. (2018 [1781]) *The epochs of nature.* Chicago: University of Chicago Press.

Doose, K. (2022) "Modelling the future: Climate change research in Russia during the late Cold War and beyond, 1970s–2000," *Climatic Change* 17(6), pp. 1–19.

Hacking, I. (1983) *Representing and intervening: Introductory topics in the philosophy of natural science.* Cambridge: Cambridge University Press.

Herder, Johann Gottfried (2002[1784–1791]) *Ideen zur Philosophie der Geschichte der Menschheit.* München: Carl Hanser Verlag.

Humboldt, A. (1858) *Cosmos: A sketch of the physical description of the universe, Vol. 1.* New York: Harper & Brothers.

Houghton J., G. Jenkins, and J. Ephraums (1990) *Climate change: The IPCC scientific assessment*. Cambridge, UK: Cambridge University Press.

Jasanoff, S. (2004) "The idiom of co-production." In S. Jasanoff (ed.) *States of knowledge: The co-production of science and social order*. London: Routledge, 1–12.

Revelle, R. and H. Suess (1957) "Carbon dioxide exchange between atmosphere and ocean and the question of an increase of atmospheric CO_2 during the past decades." *Tellus* 9(1), 18–27.

Schäfer, S. and S. Low (2018) "The discursive politics of expertise: What matters for geoengineering research and governance?" In Trentmann, F., A.-B. Sum and M. Rivera (eds.) *Work in progress: Environment and economy in the hands of experts*, Munich: Oekom, pp. 291–312.

Schäfer, S. and F. Mauelshagen (2021) "Die technologische Kolonisierung des Klimas." *Dritte Natur* 3(1), pp. 39–55.

Schellnhuber, H. J. (1999) "'Earth system' analysis and the second Copernican revolution." *Nature* 402, C19–C23.

Schneider, S. (1996) *Laboratory earth: The planetary gamble we can't afford to lose*. London: Orion Publishing Group.

Skodvin T. (2000) *Structure and agent in the scientific diplomacy of climate change: An empirical case study of science-policy interaction in the intergovernmental panel on climate change*. New York: Kluwer Academic Publishers.

Voosen, P. (2023) "United States tiptoes into solar geoengineering research," *Science* 379(6633), pp. 628–29.

33 Making the 1.5°C Aspirational Climate Target Tangible with Carbon Dioxide Removal and Boundary Work

Anders Hansson

Introduction

How are visions of future climate mitigation measures created, and how do these visions matter? These are foundational questions often raised and addressed by social scientists, and especially STS scholars concerned with climate change and the interactions between science and politics. Some of these scholars direct their interest toward how climate mitigation methods are established as feasible or contested tools to manage climate change. Climate engineering methods have in recent years become more and more discussed and assessed in political and scientific arenas. Some methods, primarily solar radiation management, are still highly controversial and are widely held as speculative or too risky to qualify for implementation. Compared to solar radiation management, however, methods for carbon dioxide removal (CDR) have gained traction over the last few years and are now often assessed alongside more conventional methods of mitigating global warming, for example by reducing carbon emissions from fossil fuels (IPCC, 2018). Many countries have presented visions and plans for reaching national climate targets and they often rely heavily on CDR to compensate for hard-to-abate emissions, or in the worst case for insufficient climate policies. However, despite the heavy reliance on CDR to project future mitigation targets, most CDR methods are only modestly implemented today. In fact, only a few demonstration and pilot plants exist for the CDR method that has received the most political and scientific attention, namely bioenergy with carbon capture and storage (BECCS). In short, BECCS is a method for capturing carbon dioxide from the flue gases at a point source when combusting biogenic fuel and preventing the gases from reaching the atmosphere. The method can only be applied at large point sources, which means that it is primarily pulp and paper mills, ethanol plants, district heating plants, and coal power plants converted to biomass that can be targeted. When the carbon dioxide is captured and compressed, which is a very costly process, the gas is transported via pipelines or ships to permanent geological storage.

If the entire process is properly managed and supplied with sustainable biomass, BECCS theoretically results in a net reduction of carbon dioxide in the atmosphere. According to the more recent IPCC reports, grand-scale use, on the gigaton scale, of CDRs, and primarily BECCS, is now a necessity for the drastic emission reductions needed for reaching the 1.5°C aspirational climate target and for balancing the probable overshoot of emissions in the shorter term. It is important to note that the IPCC and many other scientific reports have been criticized for relying too heavily on CDRs, and that the main motive for that is not that CDRs at that scale actually are deemed to be feasible, but instead simply that otherwise the mitigation puzzle cannot be resolved. The IPCC's mitigation scenarios and pathways have envisioned that, on average, 5–20Gt of carbon dioxide must be managed annually by the mid-century in order to reach those

DOI: 10.4324/9781003409748-43

targets. For reference, this figure corresponds to an infrastructure system in the order of magnitude of the current oil industry (Cointe and Guillemot, 2023).

In this chapter, I will argue that scientific climate mitigation scenarios and pathways help create visions of the future, and that scenarios of a certain centrality or political and scientific legitimacy end up influencing the kinds of future developments deemed to be desirable or achievable. The argument will be supported by an overview of relevant research and is illustrated by an empirical study I conducted a couple of years ago with colleagues at Linköping University (Hansson et al., 2021). I will illustrate and discuss the construction of visions in the context of the IPCC reports. Engaging STS concepts, I reveal aspects of how relevant science is established. A point of departure, which I share with many STS scholars, is that scientific knowledge is negotiated and influenced by social processes and often co-constructed with politics (van Beek et al., 2022; Schenuit, 2023). I want to stress that such a point of departure does not mean that the ambition is to discredit the scientific validity of the IPCC. On the contrary, it is unavoidable that science is situated in a cultural and political context. The boundaries between what is and is not considered scientific knowledge are blurry. It must be negotiated, especially when the scientific object is complex, in our case future development of the global climate in interaction with global society. Those interactions are inherently characterized by deep uncertainties and unpredictable social dynamics. Nevertheless, scientific communities have been tasked with the herculean mission to compile and analyze current understandings of climate change. Furthermore, they are called upon to convey structured and transparent scenarios or pathways on how various futures may unfold. The main tools for that are integrated assessment models (IAMs) (IPCC, 2018). Therefore, my ambition is that STS-informed analyses can help society to open up and also improve scientific efforts to understand the future.

In what follows, the background section will briefly present the work process of the IPCC, a selection of STS concepts, and an explanation of IAMs, which underpin the work done by the IPCC when assessing future mitigation efforts and their impacts. The background section will be followed by the presentation of the empirical case and a few concluding remarks.

Background

First, it is important to characterize the IPCC reports and get a basic understanding of how they are compiled and written. A challenge is that they must balance policy relevance with scientific validity. Those ambitions are often contradictory. This tension makes it appropriate to analyze that process of negotiation as **boundary work**. The simplest way to define boundary work is the rhetorical work of distinguishing one thing, intentionally or unintentionally, from another (Gieryn, 1983). In the context of scientific practices or processes, it is about distinguishing relevant knowledge and science by drawing boundaries between science and politics, the negative from the positive, the subjective from the objective, and the valid from the invalid – in other words, filtering out what is established as relevant and legitimate. In a paper on how the feasibility of BECCS is assessed in IAMs (Low and Schäfer 2019, 2), the concept is defined as the process in which "stakeholders advance their separate areas of authority by defining the objects and terms of debate in ways responsive to their own expertise and agendas – this is boundary work."

The procedural work structure of the IPCC, that is, the scientists' work instructions, is clear and transparent and follows a workflow that facilitates boundary work at every turn. Already, the work instructions broadly point out what types of texts to include or exclude (for example guidelines for how to delimit scientific literature), appropriate terminology to use, and the general scope and aim of a report. However, the instructions are often not case-specific or very detailed and must be interpreted by the authors and reviewers. This leaves room for individual actors to argue for certain boundaries and contest others (Hansson et al., 2021). The process by which an IPCC report is reviewed usually consists of three steps: (1) review of first order drafts, (2) government and expert review of the second order drafts, and (3) government reviews of the summaries for policymakers (SPMs), overview chapters and the final synthesis report. The reviewers are either self-nominated and finally selected by the IPCC or directly invited. Each chapter has designated review editors who must coordinate with lead authors in order to manage responses to reviewers' comments in the two review rounds. The draft of the SR1.5C report, which we studied, had 224 authors and review editors from 44 countries that were selected from 541 nominations. In sum, 796 individuals and 65 governments acted as reviewers. In total, reviewers delivered more than 40 000 comments (Cointe and Guillemot, 2023; Hansson et al., 2021; IPCC, 2018).

Before I address the results of the study in greater detail, it is important to understand a basic feature of IPCC science, namely the IAM. According to the IPCC glossary, an IAM "combines results and models from the physical, biological, economic and social sciences and the interactions among these components in a consistent framework to evaluate the status and the consequences of environmental change and the policy responses to it." The IAMs combine and process inputs from many disciplines and therefore bring together researchers who otherwise rarely meet and communicate science. The interdisciplinary systems perspective and the ambition to manage and systematize huge amounts of data are the main reasons why IAMs are privileged tools. These models are for that reason used to answer the grander questions raised by the IPCC on how climate systems and societies interact over time, and to explore feasible mitigation options.

The IAMs and their associated climate mitigation scenarios were key features already in the IPCC's second assessment report, published in 1995, and their importance has increased in later years. One reason for the centrality of IAMs in these contexts is the IAM communities' explicit policy-advising ambitions. STS scholars have emphasized the ambiguity of providing policy-relevant results while maintaining scientific integrity. There is an obvious challenge of striving toward pure and objective science while also integrating assumptions on how society works, for example, how markets are structured and behave, how quickly innovation diffuses in society, what the costs are for certain methods over time, and how people and actors behave and make decisions (Beck and Mahony, 2018; Hansson et al., 2021).

Yet another argument in line with the previous claim is that modeling practices cannot easily be separated from the political spheres. After the Paris Agreement, with its more stringent temperature targets, a further critique has been forwarded, namely that IA modelers are tempted to deliver politically palatable results, often at the expense of striving for scientific realism. It is argued that the rationale is that modelers want to remain included as experts and not risk losing research funding (Geden, 2015).

The **co-production** process of science and politics matters because modeling assumptions and modeled outcomes may set the frames for policy deliberations about what seems feasible or worth striving for (Low and Schäfer, 2019). A major challenge is when conflict arises between striving to construct realistic mitigation pathways or scenarios and staying policy-relevant. The boundaries between these two modes are often blurred. For example, to even theoretically reach the more stringent aspirational 1.5°C target – a target set by politicians, not scientists – modelers had to include large amounts of net negative emissions of carbon dioxide via CDR methods. The inclusion of CDR was not because these methods had become more accessible in recent years, but rather because they became mathematically necessary to counterbalance the expected overshoot of carbon dioxide emissions (Cointe and Guillemot, 2023; Hansson et al., 2021). Arguably, the models risk depicting the 1.5°C temperature target as more feasible than would be the case if the modeling activity were guided exclusively by traditional scientific ideals. With this serious problem in mind, my colleagues and I set out to explore the boundary work performed in the drafting of the SR1.5C report (IPCC, 2018), finalized in October 2018 as part of its sixth assessment cycle, with a focus on BECCS, the role of IAMs, and the 1.5°C temperature target. I present aspects of that analysis below.

Analysis

To trace the process of boundary work in the construction SR1.5C draft we did not apply predefined analytical categories for the empirics. We both scrutinized how critical comments were dealt with in general and in the arguments for not including or assessing a certain claim, text, or argument. This allowed us to systematically expose the dialogue and interactions between the authors and reviewers, because every specific reviewer's comment is clearly addressed and paired with an author's response and made publicly available (IPCC, 2019a, b, c). Based on our reading of the authors' rebuttals and partial rebuttals of critical comments, we constructed four boundary work modes, although we did not manage to make them fully mutually exclusive. A fifth mode, negligence of a comment, could have been constructed as well. However, we decided to only mention negligence when that occurred (Hansson et al., 2021):

(1) *Remitting or referring to a limited scope or capacity.* This means that the already selected literature is explained to be in line with a predefined scope, or that certain issues are too complex to be dealt with, or remitted due to space restrictions.
(2) *Claiming to be beyond the mandate: subjective and policy prescriptive.* This means that an issue is not within the mandate, or that it would be too subjective or policy prescriptive to deal with a certain issue. It can also mean that a forwarded critique is not reflected in assessments already covered in a certain selection of literature. For that reason, it should be disregarded.
(3) *Restricting and defining what legitimate science is.* This means that a certain field of literature defines what is scientifically relevant or that alternative literature does not meet scientific criteria.
(4) *Relativizing uncertainties.* This means that a specific problem, challenge, or obstacle for a certain method is acknowledged and accepted, but the problems are considered to be even worse for alternative options. For that reason, the method with relatively fewer problems is preferred.

The two following sections present snapshots and illustrations of the empirics in Hansson et al. (2021), yet they help to illustrate how boundary work ultimately shaped what is (and is not) in the resulting report. Let us first summarize the reviewers' comments, followed by the authors' responses.

The Reviewers' Comments

An often-recurring reviewer critique was that the second order draft of SR1.5C had a strong bias in favor of BECCS and that major ecological, technical, and social concerns were downplayed. This resulted in, according to some reviewers, a misleadingly favorable picture of BECCs' real potential that was not reflected in science. In some cases, reviewers even forwarded that the draft conflicted with the ideal of providing comprehensive and unbiased assessments. Many reviewers forwarded that the upper limits of BECCS' potential were only feasible because massive

Table 33.1 Boundary work modes with associated coded rhetoric.

Boundary work mode	Examples of rhetoric and coding
1) Remitting or referring to a limited scope or capacity.	Selection of literature argued to be not in line with the scope of the chapter; impacts on a certain SDG are outside the scope of the report or chapter; the issue is too complex for the authors; space restrictions.
2) Claiming to be beyond the mandate: subjective and policy prescriptive.	Land-use issues are within the mandate of another forthcoming IPCC report; excluding BECCS from the analysis is policy prescriptive/subjective; the mandate is to reflect assessments in the scientific literature only.
3) Restricting and defining what is legitimate science.	A certain field of literature (i.e. the pathway literature) defines what is relevant to include, cannot conduct an analysis beyond what is already conducted in the relevant literature; the reviewer's suggested literature does not meet scientific criteria.
4) Relativizing uncertainties.	A specific problem/challenge/obstacle is also valid, or worse, compared to another alternative; a global energy transition will involve a large land footprint regardless of whether BECCS is implemented or not; most CDRs are untested.

Source: Adapted with permission from Hansson et al. (2021).

challenges related to land-use effects, economic costs, and other systemic risks were neglected. Other critical comments addressed that the IPCC analysis relied too heavily on IAM literature, and for that reason only reflected the methods that particular scientific fields had picked up. As a result, the IPCC consequently marginalized other methods, specifically natural CDR methods (e.g. soil carbon sequestration, afforestation/reforestation, restoration of wetlands).

Reviewer critique often targeted lack of transparency and comprehensibility in the IAMs, unrealistic or flawed key assumptions, and how the natural world was modeled, including for land-use effects and land-use trade-offs, hydrological and water aspects, and nutrient loss from soils, biomass productivity rates, and crop yields. Also, how societal aspects like technological learning and the effects of the influence of economy of scale were simulated in the models were often contested by the reviewers. Other lines of critique emphasized the draft's internal inconsistencies, for example, that the draft's concluding remarks on BECCS large-scale feasibility contradicted other findings in the very same draft. On the contrary, it was claimed by some reviewers that the findings in fact pointed toward a very limited potential and a slow implementation of BECCS. Other major concerns revolved around assumptions underpinning the IAM models, in particular, that as much as 25–46% of the available land globally could be dedicated to BECCS systems. Reviewers questioned such high numbers, specifically because the draft did not elaborate on the consequences for forest degradation, social tensions, loss of biodiversity, land-use conflicts, and land degradation. In the next section, I summarize how the authors responded to the critique.

The Authors' Responses

Most author responses acknowledged the validity of the reviewer comments, and in most cases they complied, at least to some extent. Many of the main concerns, such as BECCS being unproven at scale, uncertain feasibility, the downplaying of caveats, and the need for additional clarifications were accepted as relevant. The authors also admitted the mistake of presenting BECCS and the natural CDR method afforestation as equally feasible, and declared an ambition to reformulate sections that may be misleading. To sum up, the authors' responses often entailed the addition of text and clarifications rather than subtractions. But some of the more profound critical comments were deferred through what we identified as boundary work.

The four boundary work modes were prevalent in the author's responses to critiques. Regarding how land-use issues and conflicts were dealt with, authors claimed that the issues were too complex for them to manage. Authors also cited space restrictions (boundary work mode 1: remitting or referring to a limited scope or capacity). A similar response was that the issues were considered to be very important by the authors and warranted some minor adjustments. Yet, authors tended to reject the possibility of major revisions. Instead, they indicated that such issues may be dealt with in the then-forthcoming IPCC Land-use Report to be published in 2019. The authors' rebuttal of critique along these lines is a combination of mode 1 (limited scope/capacity) and referring to a restricted mandate mode 2 (subjective and policy prescriptive). The request to decrease the scale of BECCS because it may not be feasible was rebutted by a combination of claims, namely that it would be difficult to assess and too subjective, since it was not reflected in what was considered relevant literature for such an analysis.

A recurring author response was to argue that sustainability concerns will be addressed in other chapters, or by a specific type of literature, in this case, the pathway literature (i.e. the IAM literature). The pathway literature had already defined which climate mitigation options to assess, which consequently meant that the referred literature defined what lay beyond the scope of the report or not. This is an example of boundary work mode 3, which restricts and defines what is legitimate science. The authors emphasized that they must make solid and objective assessments and were therefore not allowed to favor certain methods or prescribe policies. Regarding the request to include more of the constraints for large-scale BECCS, the authors replied that incorporating these constraints would be policy prescriptive because they were not supported by the literature they were restricted to review.

Mode 3 boundary work is also clear in authors' justifications regarding why natural CDRs were only sparsely included in the models. One author mentioned that it was a pity that the reviewed literature, explicitly the integrated pathways literature, did not assess these natural methods, and for that reason was disqualified from inclusion in the report. The authors admitted that it was a flaw that the only qualified CDR options were direct air capture, BECCS, and afforestation.

Finally, we have a few examples of mode 4 boundary work (relativizing uncertainties). Many reviewers argued that natural CDRs imposed lesser ecological risks and were more proven than BECCS. Those claims were often acknowledged, but many times the authors' responses were followed by a comment that stressed that, also, the natural CDRs demanded equal or even larger land areas and had other constraints. Thus, the authors acknowledged the constraints of BECCS, but responded to that critique by emphasizing that all forms of bioenergy use come with similar and fundamental land-use challenges. The authors furthermore emphasized that any global energy transition, no matter with or without BECCS, would entail large land footprints. Along these lines the authors often responded that it may be true that BECCS was untested at scale, but that fact also applied to many of the other CDR methods.

Discussion

The authors generally accepted and complied with the critical comments and acknowledged their relevance. However, the recurring comments questioning how viable or feasible BECCS is on a gigaton scale were often deflected through the boundary work modes instead of being incorporated in the report. The review process is definitely open for critical scrutiny, but the pathway/IAM literature and interpretations of the IPCC's scope are often used to justify the filtering out or deflection of certain fundamental critiques. If the caveats and limitations of the underpinning science and process of drafting the reports are not sufficiently communicated, there is an obvious risk of creating a misleading sense of optimism. In other words, forwarding a message that the present overshoot of emissions will be balanced by large-scale negative emissions CDR technologies in the future is misleading, as such claims only rest on a very limited range of science. If such visions are cemented in mitigation pathways and climate policies, they may carry a performative function if they serve to justify a more relaxed view of the desperately needed near-term measures, which the IPCC also stresses. That phenomenon of IAMs' **performativity** has been raised by several STS researchers. Beck and Mahony (2018, p. 1) clearly illustrate what is meant by the performative function when urging the IPCC to be more self-reflective: "We suggest that pathways and scenarios have a 'world-making' power, potentially shaping the world in their own image and creating new political realities. Assessment bodies like the IPCC need to reflect on this power, and the implications of changing political contexts, in new ways."

Although the foregoing account critically examines how the IPCC scientific process and methods have treated CDR, some clarification is warranted. In recent years, the IPCC has faced more diverse and increasingly incompatible expectations. The IPCC "now more than ever – is mobilized for different and sometimes contradictory political purposes" (Schenuit, 2023, p. 166). Acknowledging this context, the analysis of boundary work is not intended to question the scientific authority of the IPCC. Instead, I suggest that using the concept of boundary work offers an analytical approach to studying how relevant science is established. This approach can hopefully support the communicative efforts of the reports' messages and results while also stressing that science is co-produced by both political actors and different scientific disciplines and perspectives. This is especially the case when it comes to the assessment of the future. What the boundary work analysis also shows is that there is a need to continue striving for not restricting what is considered relevant science to a few scientific fields.

Acknowledgment

This work was funded by the Swedish Research Council Formas (Grant no. 2019-01973). I also want to acknowledge Jonas Anshelm's, Simon Haikola's, and Mathias Fridahl's contributions to the research this chapter rests on.

Further Reading

Beck, S. and Mahony, M. (2018) "The politics of anticipation: the IPCC and the negative emissions technologies experience," *Global Sustainability* (1), pp. 1–8.

Carton, W., Hougaard, I.M., Markusson, N., and Lund, J.F (2023) "Is carbon removal delaying emission reductions?," *Wiley Interdisciplinary Reviews: Climate Change*, e826.

Haikola, S., Hansson, A., and Fridahl, M. (2019) "Map-makers and navigators of politicised terrain: expert understandings of epistemological uncertainty in integrated assessment modelling of bioenergy with carbon capture and storage," *Futures* 114, 102472.

Hansson, A., Anshelm, J., Fridahl, M., and Haikola, S. (2021) "Boundary work and interpretations in the IPCC review process of the role of bioenergy with carbon capture and storage (BECCS) in limiting global warming to 1.5° C," *Frontiers in Climate* 3.

References

Beck, S. and Mahony, M. (2018) "The politics of anticipation: The IPCC and the negative emissions technologies experience," *Global Sustainability* (1), pp. 1–8.

van Beek, L., Oomen, J., Hajer, M., Pelzer, P., and van Vuuren, D. (2022) "Navigating the political: An analysis of political calibration of integrated assessment modelling in light of the 1.5°C goal," *Environmental Science and Policy* 133, pp. 193–202.

Carton, W., Hougaard, I. M., Markusson, N., and Lund, J. F (2023) "Is carbon removal delaying emission reductions?," *Wiley Interdisciplinary Reviews: Climate Change*, e826.

Cointe, B. and Guillemot, H. (2023) "A history of the 1.5°C target," *WIREs: Climate Change* e824.

Geden, O. (2015) "Climate advisers must maintain integrity," *Nature* 521, pp. 27–28.

Gieryn, T. F. (1983) "Boundary-work and the demarcation of science from non-science: strains and interests in professional ideologies of scientists," *American Sociological. Reviews* 48, pp. 781–795.

IPCC (2018). *Global Warming of 1.5°C. An IPCC Special Report on the Impacts of Global Warming of 1.5°C Above Pre-Industrial Levels and Related Global Greenhouse Gas Emission Pathways, in the Context of Strengthening the Global Response to the Threat of Climate Change, Sustainable Development, and Efforts to Eradicate Poverty.* V. P. Masson-Delmotte et al. (eds.) Cambridge: Cambridge University Press.

IPCC (2019a) *IPCC WGI SR15 Second Order Draft Review Comments and Responses - Summary for Policy Makers.* Available at: www.ipcc.ch/site/assets/uploads/sites/2/2019/07/SR15FOD_Summary_for_Policymakers_Comments_and_Responses.pdf. (Accessed: December 17, 2021).

IPCC (2019b) *IPCC WGI SR15 Second Order Draft Review Comments and Responses - Chapter 2.* Available online at: www.ipcc.ch/site/assets/uploads/sites/2/2019/05/SR15SOD_Chapter2_Comments_and_Responses.pdf. (Accessed: December 17, 2021).

Haikola, S., Hansson, A., and Fridahl, M. (2019) "Map-makers and navigators of politicised terrain: expert understandings of epistemological uncertainty in integrated assessment modelling of bioenergy with carbon capture and storage," *Futures* 114, 102472.

Hansson, A., Anshelm, J., Fridahl, M., and Haikola, S. (2021) "Boundary work and interpretations in the IPCC review process of the role of bioenergy with carbon capture and storage (BECCS) in limiting global warming to 1.5° C," *Frontiers in Climate* 3.

Low, S. and Schäfer, S. (2020) "Is bio-energy carbon capture and storage (BECCS) feasible? The contested authority of integrated assessment modeling," *Energy Research & Social Science* 60, pp. 1–9.

Schenuit, F. (2023) "Staging science: dramaturgical politics of the IPCC's Special Report on 1.5°C," *Environmental Science and Policy* 139, pp. 166–176.

34 Boundary Work in Solar Geoengineering Assessment and Experiments

Sean Low

A Focus on Experts

When solar geoengineering[1] is introduced to new audiences, it is common to start with a figure like the one below (see Figure 34.1). They typically show a landscape of *technical* options in *geophysical* locales – and these two emphases shade over important realities. Solar geoengineering is not composed of tidy, finished, and functional systems, and it is not devoid of people, politics, and timelines.

I prefer this second figure (see Figure 34.2): a bibliometric mapping of solar geoengineering research. Bibliometrics is a common method, which allows us to see, describe, and analyse publication and citation networks among experts, in this case, geoengineering researchers. The content (from 2014) is a little dated, but it represents an immature field of techno-science as a contested product of scientific experts. With this in mind, I ask the reader to consider three interlinked points.

First: What is solar geoengineering? It is a creature of the future and the imagination – hypothetical sociotechnical systems that focus on a handful of technical and climatic details (e.g., using modified aircraft for deployment; deployment lengths and locations; overarching changes to temperature and precipitation), but are "thin" in most dimensions. Shading the sun would be a systemic (or complex, or "wicked") issue, impacting lives and livelihoods everywhere. This could be for better or for worse – depending on how and for how long solar geoengineering approaches are deployed, where those affected are, and their local capacities and vulnerabilities to a changing climate. It might even have profound impacts as an idea, without ever being deployed. It might relieve heat stress or change the monsoon, protect sea ice and/or change crop yields, buy time for emissions reductions, or start a war.

Second: How might we know? The systemic, future-oriented nature of solar geoengineering makes it difficult to grasp tangibly. As experts, or publics, or policy-makers, we need to "tame" how the future is imagined – that is, make sense of potentially game-changing technologies that do not exist. To do so, we make use of an evolving, imperfect toolkit of assessment practices: from *quantitative* systems models that highlight climatic and techno-economic dimensions, to *qualitative* methods that inquire better after political and ethical dimensions, such as analogies (comparable problem structures and solutions), mixed-methods surveys, deliberative engagements, scenario exercises, and games (Muidermann et al., 2020; Low, Baum, and Sovacool, 2022b). Each has different epistemologies, or ways of knowing. Each invokes different kinds of expertise (scientific and disciplinary, or sectoral, or traditional knowledge), and each could appeal to different audiences (policy-makers vs. publics). Hence, assessment practices are not neutral tools. The tool's fit-for-purpose is significant: Is a research approach capable of assessing the question posed?

Third: Who says so? The hand that wields the tool is also significant: Who designs the geoengineering study and interprets the results? Solar geoengineering is a technocratic and polarized

DOI: 10.4324/9781003409748-44

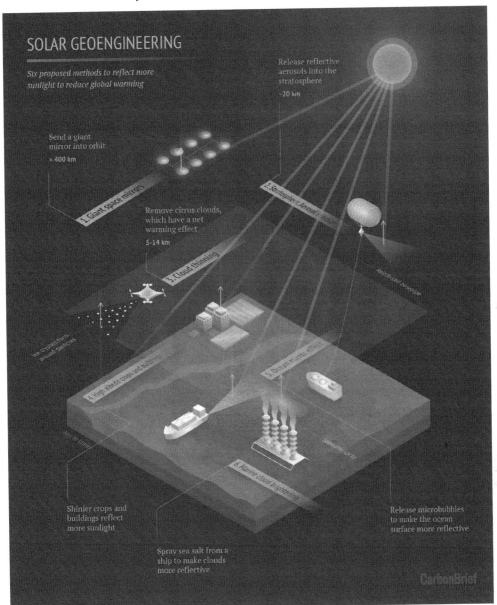

Figure 34.1 A typical introductory figure to solar geoengineering approaches.

Source: Carbon Brief (2018): "Explainer: Six ideas to limit global warming with solar geoengineering", graphic by Rosamund Pearce. Summary graphics for solar geoengineering typically shows a landscape of technical options in geophysical locales, and while useful and commonplace, such depictions do not represent side-effects and harms, politics, and timelines. Reproduced with permission.

conversation. It hovers on the fringes of mainstream climate governance; national policy-makers have not taken a stand; and public awareness is low. But there is a tug-of-war over what can be known about solar geoengineering – its balance of benefits vs. risks, and by extension, whether it should be integrated into climate strategy in the era of the Paris Agreement – between different

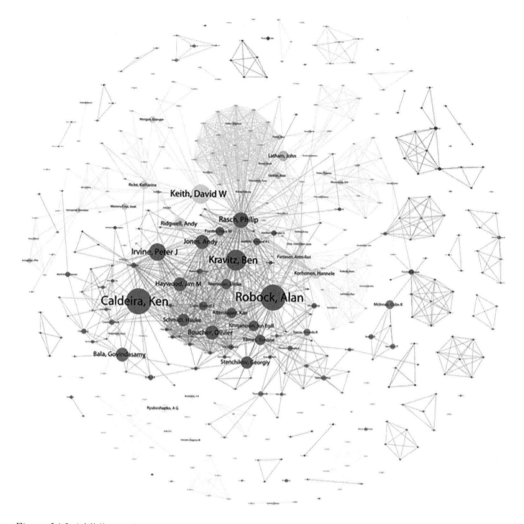

Figure 34.2 A bibliometric mapping of solar geoengineering research.

Source: Oldham et al., 2014. The data points have most certainly changed, but the principle has not: we might usefully view an immature field of techno-science as the contested product of scientific experts. If it interests the reader, the author is (or was) one of the small red dots. Reproduced with permission.

networks of experts and civil society organizations. To link to the previous point, the design and conclusions of assessments – and by extension, the state of knowledge on solar geoengineering that is passed to publics and policy-makers – are contested by different, even antagonistic, groupings of experts.

At the core is this question: Who gets to create authoritative knowledge? Who shapes the bounds of debate that could (re)shape the coming century of climate action?

In what follows, I trace some illustrative practices and politics of this emerging field. Using a central concept – *boundary work* – I ask: how is knowledge and evidence about solar geoengineering created? Moreover, what does this knowledge do, and how does it further shape governance? I do so in two linked spheres of activity: expert assessment and outdoor experimentation.

Throughout, we uncover examples of politics in the conduct of science: in representation (who predominates, and who is missing), procedure (who decides and how), epistemology (how actors know what they know), and outcome (what avenues for action are made prominent, and what alternatives are delegitimized).

Expert Assessment

Research may very well shape whether solar geoengineering is developed and deployed, and how it is governed. To understand how research is unfolding, let us examine three related explanatory concepts. In any nascent field of techno-science, there are strong, competing perspectives in research, innovation, and policy over how to best tame uncertainties. **Boundary work** describes how key terms of debate are shared across competing expert networks – but are defined to better align with different assessment practices, or research or political agendas (Gieryn, 1983; Star and Griesemer, 1989). Terms of debate include terminology (What is "geoengineering"?), research and governance practices (What are climate models, or public engagement, capable of?), and objectives (Who should assessment inform: a local community or the US government?). Stakeholders circle around the same conversations while using them to draw different conclusions, reinforcing their own practices and politics against others in a process called **demarcation** (Gieryn, 1983).

Boundary work is in conversation with **co-production**: actors design fields of techno-science to mirror what they desire in nature and society (Miller and Wyborn, 2018). That is: expert assessments of solar geoengineering are not neutral, but entwined with normative beliefs about the state and direction of climate governance. This is all the more significant because solar geoengineering studies are increasingly geared to inform governmental policy-making (what Gieryn, 1983 calls "mandated" science).

This is not to say that experts lie, or that all research is compromised. The point is more subtle: that scientists grapple with the stakes of doing science-for-policy, and in both tacit and instrumental ways, set up the questions so that they are best suited to give the answers to the audiences they deem crucial.

To illustrate, I pose two competing modes of assessment activity: mutually-reinforcing complexes of intents, expert networks and communities, audiences deemed relevant, assessment methods and epistemologies, projections or imaginings of future implications, and resulting proposals for appropriate governance (Table 34.1).

The first is a "mission-oriented" mode of technical, model-centric, policy-facing assessment that aspires toward the establishment of a governmental research program, and largely takes place in the US. The general direction is to use climate models to simulate best-case scenarios for deployment – broadly, the scheme is "optimally" designed so that global climatic conditions on balance improve – and to depict those scenarios as steering visions for implementation and international cooperation. Assessment is only set up to grapple with technical uncertainties and broad climatic effects (temperature, precipitation, etc.). Complex ethical and societal questions are bracketed – that is, mentioned up front, then set aside (and for all intents and purposes, ignored) as orthogonal to technical study. The argument is that solar geoengineering, sensibly following optimised modelling, can buy time for mitigation while protecting the most vulnerable. Yet, there are no governments, international bodies, or (perverse) geopolitical considerations present – one must assume that in an increasingly multi-polar world, all governments agree to cooperate on solar geoengineering for many decades.

The second is a more "precautionary" mode of qualitative, deliberation-based, society-facing assessment with a variety of more critical leanings. The general direction is to point out the

Table 34.1 Competing modes of expert practice and politics in solar geoengineering

	Mission-oriented assessment and advocacy	*Precautionary or prohibitive governance*
Intent (*mandate*)	• Reduce *technical* uncertainties. • Implicitly *highlight the benefits* of solar geoengineering as a backstop for ambitious targets.	• Highlight *socio-political* implications and concerns, or historically-grounded logics of delay. • Diverse aims, ranging from inclusive *participation* in assessment, to *precaution*, to *prohibition*.
Researchers (*expert networks and communities*)	• Network of research *advocates* • Earth systems modeling use predominant but not exclusive.	• Looser networks typically *critical* of climate interventions. • Social science and governance scholars; civil society organizations.
Users (*audiences deemed relevant*)	• *National decision-makers*: science-policy bodies, national governments and funding bodies.	• *Local-to-regional demographics* & decision-makers; galvanizing new modes of *assessment*. Nascently, science bodies and governments, but for constraining rather than facilitating further activity.
Practices (*methods, epistemologies*)	• *Earth systems modeling* on high-level impacts.	• *Social science*: analogies to comparable governance issues; public and expert engagements.
Outputs (*projections or imaginings of implications*)	• *Technical deployment schemes* depicting high-level benefits/ risks.	• Analyses of *perverse incentives* in business, geopolitics, and global governance.
Recommendations (*appropriate governance*)	• Governmental research program (increase academic and governmental attention, public funding). • Incorporate into IPCC pathways.	• Global, multi-level stakeholder engagement. • Non-Use Agreement (reduce academic attention and public funding, ban deployment).
Illustrative references	• Parson and Keith, 2013; Keith, 2017; MacMartin & Kravitz, 2019; Wanser et al., 2022	• Stilgoe, Owen, and Macnaghten, 2013; McLaren & Corry, 2021; Biermann et al., 2022

Source: adapted from Low, Baum, and Sovacool, 2022b.

Note: Simplified representation of two predominant strands of solar geoengineering, for illustrative purposes

politics that technical study and a drive towards greater policy attention is missing: unequal capacities and vulnerabilities between and within the global North and South, military and national security planning dimensions, and the prospects of fragmented governance in a multipolar world. Moreover, solar geoengineering feeds into a logic of **mitigation deterrence**, or moral hazard, where techno-fixes delay rather than buy time for decarbonization (McLaren, 2016). Indeed, there are widespread political pressures and a long history of doing so – for example, through market mechanisms and offsets, carbon capture and carbon removal, and "bridging fuels" such as natural gas (Low and Boettcher, 2020). For some, if solar geoengineering assessment is to be conducted properly, there must be global public consultation and consent, especially of marginalized and vulnerable peoples: all who stand to be affected deserve a say. For others, the best-case scenario for deploying and governing solar geoengineering is divorced from reality; it is so dangerous a distraction that it should receive no further policy attention. (It is worth pointing out that I fall somewhere within these two points of view.)

Experiments

A related contested area surrounds the conduct of outdoor experiments: small-scale field trials are intended to test deployment mechanics or atmospheric processes at a much-reduced scale (see Table 34.2). Although experiments are usually described and conducted as inquiries into technical issues, their greatest uncertainties and implications have always been political. The physical equipment and impacts of these experiments – in most cases, a balloon attached to aerosol-dispensing and measurement equipment, with negligible, time-limited effects – have been belied by a great deal of expert dispute and civic opposition.

There is a lesson here: however powerful climate modelling (or other kinds of assessment work) is for shaping debate and policy; academic work is difficult for publics and policy-makers to parse and grapple with. This can be true even for academics from different disciplines. **Controversy studies** tell us of instances in which scandal provides an opportunity to peer inside usually hidden, or black-boxed, scientific activities (Nelkin, 1979) – a famous example in climate assessment is "Climategate". In this vein, experiments present a physical, tangible site for questioning and contestation.

Table 34.2 Experiment case studies

Experiment or project	Project leads or host	Year and location	Opposition and key implications
Yuri Izrael's Field Experiment on Studying Solar Radiation Passing through Aerosol Layers	Roshydromet and Russian Academy of Sciences, led by Yuri Izrael.	2008; Saratov, Russia	• Conducted in Russia by climate advisor (Y. Izrael) to Putin; since passed away. • Study blacklisted within mainstream research. • Russian locale exemplifies latent geopolitical dimensions of research.
SPICE "Test bed" – Stratospheric Particle Injection for Climate Engineering	Bristol University, led by Matthew Watson (part of a wider consortium).	Suspended 2012; Norwich, UK	• Detailed governance framework with technical and societal dimensions with "stage gate" process for progress. • Ostensibly cancelled due to an unexpected patenting issue from within research team.
SCoPEx – Stratospheric Controlled Perturbation Experiment	Harvard University, led by Frank Keutsch and David Keith, with Swedish National Space Agency.	Suspended 2021; Kiruna, Sweden	• High NGO opposition due to involvement of leading advocates. • Suspended due to opposition by Saami Council and allied Swedish and international NGOs.
Make Sunsets – Start-up company	Luke Iseman.	2022 and ongoing, Mexico and US	• Introduced commercial intent to solar geoengineering through sale of $10 credits to crowd-fund deployment. • Denounced as rogue science. • Led to Mexico banning solar geoengineering experiments.
SATAN – Stratospheric Aerosol Transport and Nucleation experiment	Andrew Lockley (then, unpaid affiliate with University College London), with European Astrotech (company).	2022, UK	• Introduced clandestine research and development, avoiding peer review and scrutiny. • Denounced by some as rogue science. • Defended by others as making a misguided point about a governance gap in experimentation.

Source: partially adapted from Low, Baum, and Sovacool, 2022a.

For such a small field, experiments have had startlingly diverse dimensions. The first (by Yuri Izrael) was conducted in Russia by a climate advisor to Putin, and has been blacklisted from mainstream conversations due to its unclear technical value and political implications. The next two (SPICE and SCoPEx) were conducted by university-based researchers, inquiring primarily after atmospheric chemistry or engineering questions – but have become a proxy battleground for different research and NGO networks over the strength of research governance and public engagement. And the final two (Make Sunsets and SATAN) were conducted by individuals or companies – and not to test technical issues, but to challenge "establishment' researchers" views on governance or commercialization. Detailed case studies of the Yuri Izrael, SPICE, and SCoPEx experiments, governance frameworks, and scientific/social contestation can be found in Low, Baum, and Sovacool, 2022a. The Make Sunsets and SATAN activities took place or came to light post-publication; the reader can find many online articles on their conduct.

Again, we can observe boundary work and co-production at play: the same themes, with different agendas, dimensions, and conclusions (Table 34.3). Mission-oriented assessment – exemplified by the design and governance of SCoPEx – emphasizes that small-scale experimentation presents no "slippery slope" towards large-scale, sustained deployment – exiting any scale of research is possible. Thus, there must be "allowed zones" with scale-based thresholds of field trials. The objective is to reduce technical uncertainties (as noted: regarding deployment mechanics or atmospheric processes), while ethical and societal concerns are bracketed (that is, set aside as dimensions that technical study should not be asked to cover). Self-policing (e.g., institutional review, expert advisory boards) has been the norm in experimental governance. Public engagement has increasingly become a motif – perhaps in response to social opposition and intra-scientific critique. However, such engagements are envisioned as soliciting local input on small-scale technical and environmental impacts posed by the experiment, rather than on the more complex enterprise of global geoengineering. Indeed, experiments are described as "basic science" – without political and commercial intent, and separate from any political or commercial repercussions that might eventuate. In this view, scientists cannot be held responsible for unknowable, evolving circumstances.

Meanwhile, precautionary assessment – here, see the SPICE "test bed" – warns against a slippery slope, in which research labelled as "basic science" functionally creates space to consider deployment. The precautionary view criticizes "mission-oriented" research advocates for bracketing thorny ethical and societal questions, and bringing only small-scale, time-limited technical risks into focus. This sets in place a benign conception of field experiments, and fosters conditions for progress to later stages and larger scales. Moreover, the dangers of moral hazard and mitigation deterrence are conveniently left aside, and public engagement, to the extent it is pursued, is limited to issues of local concern. In the precautionary view, public engagement should extend local harms to consider open-ended and long-term implications of climate interventions. Such consultation should also go beyond acquiring a "social license to operate", to create meaningful criteria on which public consent and experimental progress – or cancellation – should be based.

These concerns are made all the more urgent by recent experiments. *Make Sunsets* is the first foray by profit-driven actors into solar geoengineering: a pair of Silicon Valley innovators selling cooling credits to crowd-fund aerosol-spraying balloons, but with questionable (or non-existent) assessment and governance criteria. The technically-grounded but amusingly-named *SATAN* experiment maintained a deliberate secrecy – supposedly, to make a misguided and opaque point about the need for proper experimental governance. In short order, longstanding concerns about the commercialization of solar geoengineering, as well as clandestine research and development, were proven plausible. Both trials have been widely denounced by solar geoengineering researchers – especially advocates. We might see these critiques as face-saving examples of

Table 34.3 Competing modes of governing outdoor experiments

	Mission-oriented assessment and advocacy	*Precautionary or prohibitive governance*
Scale	• Emphasize small-scale, negligible environmental impacts.	• Emphasize that small-scale activities implicitly imply progression to larger, eventually global scales and stages.
Slippery slope?	• No. • Separation between stages: research informs but does not lead to development and deployment.	• Yes. • Resonant, optimistic conceptions of deployment are sent in place during research, creating conditions for further development and deployment.
Scope	• Objective is to reduce technical uncertainties. • Bracket ethical and societal concerns.	• Objective is to highlight consultation and consent, local and global dimensions of concern. • Emphasize ethical and societal concerns.
Evaluation process	• Institutional review boards, scientific self-governance. • Increasingly, public engagement.	• Stakeholder and public engagement.
Public engagement	• Targeted: examining technical characteristics and local implications.	• Expansive: examining global and long-term implications.
Political or innovation dimensions	• Separate scientific activity from political decision-making and commercial intent.	• Show that scientific activity shapes, and is shaped by, political decision-making and commercial intent.

Source: adapted from Low, Baum, and Sovacool, 2022a, Stilgoe, Owen, and Macnaghten, 2013, Parker, 2014.

boundary work that seeks to demarcate "rogue" activity from established, legitimate research. We might also see them as a reminder to always consider political implications of research, however remote. Responsibility cannot end at the lab door – or whatever the outdoor equivalent is!

Politics Shapes Science Shapes Politics

Solar geoengineering research and assessment demonstrates how – paraphrasing Hoppe (2005) – *politics is scientized, and science is politicized*. In other words, stakeholder communication and decision-making rely on the authority provided by expert assessment; meanwhile, assessment is crafted to face different audiences and processes, containing epistemological and political biases. Modes of assessment set in play resonant conclusions that actively steer solar geoengineering governance in their image. Each mode, moreover, highlights actions and conclusions that reduce the other's capacity to define what is at stake. These are efforts to shape the norms of scientific conduct (Gupta and Möller, 2018), as a prelude to established processes of (recall: "mandated") science-for-policy.

We can already see that the mission-oriented mode is making headway in high-level science-policy circles in the US, with reports published by the National Academies of Sciences broadly reflecting their points of view (e.g. NASEM, 2021). A federal research programme may soon follow (Voosen, 2023). The precautionary mode has less clear collective aims, but one prominent initiative is a Non-Use Agreement that seeks to ban deployment, as well as refocus public funding and academic attention (Biermann et al., 2022).

Let us take a reading of each mode, in light of their (best) intents. The mission-oriented mode of assessment paces and sequences more traditional movements from technical to societal questioning, and from expert to public deliberations, to ensure that the politics do not outstrip the science. The precautionary mode emphasizes the dangers of technocracy, however well-meaning. Politics always outstrips science, and must be addressed upfront; assessment requires diverse inputs to prevent solar geoengineering from distorting climate governance in the era of the Paris Agreement.

But let us also consider a critical reading of each. The mission-oriented mode creates depictions of solar geoengineering where complex societal implications that cast doubt on solar geoengineering's feasibility are described in technical terms for better incorporation into policy, and preferred policy (say, governmental funding for a federal research programme) is aided by depictions of long-term controllability and near-term necessity. Conversely, the precautionary mode is perhaps driven by a political agenda of resistance, and aims to set such a high bar for assessing and governing solar geoengineering *a priori* that there is no scope for "muddling through" – as global actors arguably do with any number of complex governance issues deemed necessary or entrenched.

I will close with a note on positionality. It should be clear that while I have presented (and streamlined) two modes of activity, I sympathize much more strongly with the precautionary strand. With due consideration, I believe that solar geoengineering is an artificial conversation. I worry about creating a false equivalence between a small and static network of (research) advocates, versus scholars, publics, and stakeholders invested in mainstream climate governance. I am concerned that the history of global climate governance demonstrates that states and industries have structured targets and instruments to delay decarbonization, and their interests in solar geoengineering will be no different (Low and Boettcher, 2020).

But perhaps recognizing these constraints and trends gives us a base of knowledge on which to overcome them. There are many ongoing initiatives to frame the direction of solar geoengineering research (Parson, 2023), and by extension, the shape of global environmental assessment (Low, Baum, and Sovacool, 2022b). If the reader chooses to participate in them, I hope they will recall that experts – and indeed, boundary work, or contestations over who counts as a relevant expert – stand at the centre of novel fields of techno-science. Therein lies responsibility.

Note

1. Solar geoengineering refers to a broad range of envisioned systems (see Figure 34.1). But to better manage content, I will refer only to the assessment of stratospheric aerosol injection (SAI) – the most highly profiled approach, given its resonant characteristics of being planetary, cheap, and fast, but imperfect. I will continue to use the term solar geoengineering, due to the unwieldiness of SAI.

References

STS Concepts

Gieryn, T. F. (1983) "Boundary-work and the demarcation of science from non-science: strains and interests in professional ideologies of scientists". *American Sociological Review*, 48(6), pp. 781–795.

Hoppe, R. (2005) "Rethinking the science-policy nexus: From knowledge utilization and science technology studies to types of boundary arrangements". *Poiesis & Praxis*, 3(3), pp. 199–215.

Miller C. A. and Wyborn, C. (2018) "Co-production in global sustainability: History and theories". *Environmental Science and Policy*, 113, pp. 88–95.

Nelkin, D. (1979) *Controversy: Politics of Technical Decisions.* Beverly Hills: Sage Publications.

Star, S. L. and Griesemer, J. R. (1989) "Institutional ecology, translations and boundary objects: Amateurs and professionals in Berkeley's museum of vertebrate zoology, 1907–39". *Social Studies of Science*, 19(3), pp. 387–420.

STS-Driven Examinations of Solar Geoengineering Assessments and Experiments

Gupta, A. and Möller, I. (2018) "De facto governance: How authoritative assessments construct climate engineering as an object of governance". *Environmental Politics*, 28(3), pp. 480–501.

Low, S. and Boettcher, M. (2020) "Delaying decarbonization: Climate governmentalities and sociotechnical strategies from Copenhagen to Paris". *Earth System Governance*. DOI: 10.1016/j.esg.2020.100073

Low, S., Baum, C. M., and Sovacool, B. K. (2022a) "Taking it outside: Exploring social opposition to 21 early-stage experiments in radical climate interventions". *Energy Research & Social Science*, 90, 102594. DOI: 10.1016/j.erss.2022.102594

Low, S., Baum, C. M., and Sovacool, B. K. (2022b) "Undone science in climate interventions: Contrasting and contesting anticipatory assessments by expert networks". *Environmental Science and Policy*, 137, pp. 249–270. DOI: 10.1016/j.envsci.2022.08.026

McLaren, D. P. and Corry, O. (2020) "The politics and governance of research into solar geoengineering". *WIREs Climate Change*, 12, e707.

Muidermann, K., Gupta, A., Vervoort, V., and Biermann, F. (2020) "Four approaches to anticipatory climate governance: Different conceptions of the future and implications for the present". *WIREs Climate Change*, e673.

Stilgoe, J., Owen, R., and Macnaghten, P. (2013) "Developing a framework for responsible innovation". *Research Policy*, 42, pp. 1568–1580.

Background on Solar Geoengineering Politics

Biermann, F., Oomen, J., Gupta, A., Ali, S. H., Conca, K., Hajer, M. A., and VanDeveer, S. H. (2022) "Solar geoengineering: The case for an international non-use agreement". *WIREs Climate Change*, e754.

Carbon Brief (2018) "Explainer: Six ideas to limit global warming with solar geoengineering". Available at: www.carbonbrief.org/explainer-six-ideas-to-limit-global-warming-with-solar-geoengineering/ (Accessed: 15 March 2023).

Keith, D. W. (2017) "Toward a responsible solar geoengineering research program". *Issues in Science and Technology*, 33(3).

MacMartin, D. G. and Kravitz, B. (2019) "Mission-driven research for stratospheric aerosol geoengineering". *Proceedings of the National Academies of Science*, 116(4), pp. 1089–1094.

McLaren, D. P. (2016) "Mitigation deterrence and the 'moral hazard' of solar radiation management". *Earth's Future*, 4, pp. 596–602. DOI:10.1002/2016EF000445

National Academies of Sciences, Engineering, and Medicine (NASEM) (2021) *Reflecting Sunlight: Recommendations for Solar Geoengineering Research and Research Governance*. Washington, DC: The National Academies Press. https://doi.org/10.17226/25762

McLaren, D. P. and Markusson, N. (2020) "The co-evolution of technological promises, modelling, policies and climate change targets". *Nature Climate Change*, 10, pp. 392–397. DOI:10.1038/s41558-020-0740-1

Oldham, P. Szerczynski, B., Stilgoe., J., Brown, C., Eacott, B., and Yuille A. (2014) "Mapping the landscape of climate engineering". *Philosophical Transactions of the Royal Society A*, 372, 20140065. http://dx.doi.org/10.1098/rsta.2014.0065

Parker, A. (2014) "Governing solar geoengineering research as it leaves the laboratory". *Philosophical Transactions of the Royal Society A*, 372, 20140173.

Parson, E. A. (2023) "Solar geoengineering in the news – again and again". Available online: https://legal-planet.org/2023/03/15/solar-geoengineering-in-the-news-and-again-and-again/ (Accessed: 15 March 2023).

Parson, E. A. and Keith, D. W. (2013) "End the deadlock on governance of geoengineering research". *Science*, 339, pp. 1278–1279.

Wanser, K., Doherty, S. J., Hurrell, J. W., and Wong, A. (2022) "Near-term climate risks and sunlight radiation modification: A roadmap approach for physical sciences research". *Climatic Change*, 174(23).

Voosen, P. (2023) "Could solar geoengineering cool the planet? U.S. gets serious about finding out". *Science*, 379(6633) DOI: 10.1126/science.adh1684

Part XI

Climate Futures

Introduction

Zeke Baker

In 1986, sociologist Ulrich Beck introduced the concept of **risk society** to characterize what more or less amounted to a new historical epoch: one characterized by ubiquitous and thoroughly human-caused hazards. Marked by problems ranging from nuclear arms to chemical pollution of waterways to the disintegration of the atmosphere's ozone layer, Beck argued that risk society gets expressed in a generalized existential anxiety over future risks of society's own making. Examples are everywhere, ranging from the microplastics in the guts of whales to the toxic, but largely unknown, class of chemicals called PFAS (or, per- and polyfluoroalkyl substances) that are likely circulating through your own body, to the melting of polar ice. In many cases, the signs of damage appear not only as reminders of socially produced hazards but also as harbingers of deeply uncertain times ahead.

Risk society, Beck argues, is structured by the distribution of risks, rather than by more traditional modern stratification along lines of economic class or race (although these may very well intersect). If the traditional, age-old economic and political question had long been about who controls the resources (especially the surplus), novel questions around exposure to, responsibility for, and protection from risks now come to the center. Moreover, risk society is politically ordered around fear of the unknown and of problems revolving around a risky collective future.

When it comes to climate change, risk and future uncertainty are everywhere the target of knowledge, policy, and concern. In popular culture, "cli-fi" (climate fiction) is now an established literary genre, and it is generally dystopian in character. Mental health professionals are holding workshops and conferences on "climate anxiety" because of many peoples' deep worry about the environmental future. Climate adaptation policy (see Part VIII, this volume) is held together by the expectation of a rupture between the present and the future. Climate mitigation policy, as addressed throughout this book, is supported by models of future trajectories, with scenarios constructed to visualize how actions or failures to act may plausibly impact future outcomes. (Directing action to align the present with future projections in this way is not unlike public measures taken during COVID-19 to "flatten the curve" of infection rates.) Parallel to future-oriented policy efforts, yet exhibiting more moral and political charge, the climate movement creatively synthesizes more traditional environmental themes of preservationism (framed around the past) and resource conservationism (present and future) with those of catastrophe, intergenerational justice, emergency, and collapse (future).

In climate science and technology, future risks are represented by GHG emission reduction pathways and climate change scenarios that represent impacts based on how much societies successfully decarbonize. Important also are models of how climate change may affect a range of economic dimensions (e.g. land value and agricultural production), meteorological phenomena (e.g. tropical storm intensity and heat waves), and sociopolitical outcomes (e.g. migration, civil conflict over resources). Pathways, scenarios, and models are empirically informed by the observable past and are theoretically related to observable trends. They are fundamentally about the future – future risk, to be precise.

DOI: 10.4324/9781003409748-45

In short, the future is everywhere. Science, politics, civil society, and culture thus come together to construct a range of distinct futures that make some coherent sense, at least to specific audiences or groups. Risky futures may be embraced, avoided, or moderated, but they are in any event amenable to human collective and policy choices. Like the risk of nuclear annihilation that first emerged in the Cold War, climate change in the Anthropocene era therefore beckons an inward gaze on society: what can be done, based on what people have created? Although they rely upon different conventions, all representations of the future are socially constructed (see Lahsen, Chapter 4, this volume). Whether futures are being depicted through aesthetic visualization, projected in a scientific model, or taken into consideration when creating public policy, they are necessarily made in the present. It stands to reason that practices of constructing futures are therefore inherently tied to present interests, assumptions, and concerns.

Herein lies the core premise for Science and Technology Studies (STS) research on climate change futures. Future climate risks are constructed. They are described, evaluated, planned for, avoided, or otherwise envisioned and acted upon through social processes that can be observed, explained, and questioned. The future cannot explain itself, meaning its active construction must be accounted for. If futures are made, or constructed, they can also be critiqued, reimagined, and changed.

The chapters in this Part invite us into concrete spaces where futures are indeed actively being made, constructed as objects of knowledge, and framed as issues for political contestation. Oomen (Chapter 35) draws upon the study of social movement activism and discourse, situated as it is in the midst of governmental, cultural, scientific, and political framings around climate change futures. While Yearley (Chapter 9, this volume) focuses on social movement organizations' use of mainstream climate science, Oomen is more interested in their imaginations of the future. Oomen argues that STS helps us to bridge scientific understandings of risk ("what might happen?") with normative and political approaches to risky futures ("what do we *want* to happen or not happen?"). The future, in this perspective, is an active site of contestation. By bringing attention to what Oomen calls a "politics of the imagination," we can grasp the social processes by which different groups, at times in concert and at times in competition, bring others together around particular visions of risk, consequence, and prescriptions of what should be done about it.

Howe (Chapter 36) invites us to think across space and time, specifically with respect to the interconnections and networks drawn together through novel, climate-impacted ocean dynamics. The world's water systems and oceans are indeed physically interconnected, even if human experience is not always sensitive to such connections and their changes. Howe asks, how might radical ongoing, interconnected changes in the social relationship to water – melting glaciers *here* to rising seas *there*, for example – provide opportunities for "making kin" with water, other people, and other life forms? Drawing from a rich, multi-sited study of "ocean connectivities," Howe provokes that the changing material conditions of life in a changing hydrosphere can indeed be more than risk, tragedy, and disaster at the hands of a ravaged and ravaging nature. Instead, might what Howe calls "hydrological globalization" productively and meaningfully bind together people, places, and cultures brought together by our liquid world?

The Part, and book, conclude with Andy Stirling's incisive account and critique of prevailing constructions of the climate future in mainstream science, policy, activism, and even within some STS research. Stirling argues that past environmental struggles and environmental movements (regarding health, food and pollution, for example) were markedly successful because they centered around horizontal movements channeled by an ideology of *care*. Caring for and living with the dynamic, uncertain, and interdependent Earth was the heart of prior environmentalisms, and it worked. By comparison, Stirling argues that the prevailing tack is to rely upon an ideology of *control*, namely the control of the planetary atmosphere through various means. Stirling finds this ideology even in the heart of the climate movement – a paradox, insofar as environmentalism had long critiqued ideas of control, technical fixes, and top-down management. Thinking with Stirling, might it be possible to reshape knowledge about climate away from control, toward mutual care, and thereby more successfully and democratically address climate change?

35 Futuring in Climate Politics

Activism and the Politics of the Imagination

Jeroen Oomen

Introduction

On 13 December 2022, German police forces raided the apartments of climate activists suspected of forming a criminal organization. Part of an ongoing investigation into an activist attempt to block an oil refinery, these raids targeted people connected to a loosely knit and ill-defined group of climate activists calling itself *Letze Generation* (Last Generation). In the months prior to the raids, Letzte Generation had made a name for itself with hundreds of protests across Germany. They had attempted to block and disrupt key fossil fuel infrastructures and had garnered press attention by staging highly visible blockades and disruptions. Their most emblematic form of protest has become gluing their hands to asphalt to block highways and the runways of the airports of Berlin and Munich. In doing so, Letzte Generation has attracted outsized attention, positive and negative, for a group without any stable funding or political support. Politically too, they have become an influential if divisive actor, attracting both the hero's admiration from sympathizers and the villain's hatred and blind rage from motorists and skeptics. The German chancellor Olaf Scholz, reputed to be aloof, even called their actions "völlig bekloppt" – totally crazy. Across the globe, similar disruptive action groups have sprung up. The most visible exponent of these groups is the Extinction Rebellion, whose chapters have made waves in the United States, England, The Netherlands, and many other countries.

About six months before the raids, thousands of scientists across the globe had been arrested for similar protests. Following the publication of the third part of the Intergovernmental Panel on Climate Change (IPCC)'s sixth assessment report, they had chained themselves to banks in the US and glued themselves to the steps of parliament in Spain and to bridges and roads in Germany. Wearing white coats to symbolize their status as scientists, their protests were the manifestation of anger about political complacency – and of fears about the future. Loosely affiliated with Extinction Rebellion, they called themselves the *Scientist Rebellion*. Peter Kalmus, a NASA scientist acting as a charismatic spokesperson for the group, said, "We've been trying to warn you guys for so many decades that we're heading towards a fucking catastrophe, and we've been ignored. The scientists of the world are being ignored, and it's got to stop. We're not joking. We're not lying. We're not exaggerating."

In the eyes of activists and climate scientists, the future is clear: if we don't "act" quickly, it will be a "fucking catastrophe". The future, in their eyes, appears as a threat, as an imminent warning to change our present ways. This conception of the future is at the heart of climate politics. It has been a central feature of environmental politics for decades. The famous *Limits to Growth* report in 1972 – the first major environmental warning based on computer models – opened with a quote by U Thant, the former UN Secretary General, that the world had "perhaps ten years" to address mounting social and environmental problems. Yet clearly, it is a message

DOI: 10.4324/9781003409748-46

that is not shared widely enough, and that does not resonate deeply enough. Despite these decades of recurring warnings, convincing political action on climate change remains elusive. Why, then, aren't these scientific and activist warnings effective (enough)? It is not that the scientists of the world are being ignored. It is that scientific warnings become part of a political struggle over the future, in which scientists, activists, and politicians all have different readings of the political implications of such warnings.

In this chapter, I address climate change, climate science, and climate activism from the perspective of **futuring**, a novel theoretical approach that investigates how and why images of the future become influential (or not). As a social science approach, futuring helps to understand the **politics of the imagination**, the political struggle through which different groups try to bring people together around *their ideas of a desirable future*. The power to influence how people imagine the future is a key domain of political contestation. In climate politics, the political cockpit and most policy projections present the future as more of the same, just (hopefully) without carbon emissions. At the same time, activist groups attempt to make their ideas about the future matter, fighting for a future that is radically different from the present. This chapter is about that struggle. It is an open-ended investigation of whose ideas about the future prevail and why ideas about the future inspire action.

Futuring, Sociotechnical Imaginaries, and the Politics of the Future

In recent years, fears about the environmental future have become so prominent that climate activists explicitly frame their action in relation to those fears. "Fridays for Future", Last Generation, Extinction Rebellion. Such names make sense, as large parts of society now imagine a certain futurelessness, a catastrophe lurking on the horizon. The imagined future is dire. Yet constructing narratives about the future, imagining climate futures, has always been a key part of climate change science and politics. As such, it is important to understand the ways in which imagined and projected futures influence societies in the present.

In the past two decades, a growing field of social science has emerged around studying the future. Even though it is obvious that ideas about the future influence the present, "the future" has historically been a difficult challenge for the social sciences. For decades, they struggled with incorporating "the future" into their work, because how can you study what has not happened yet? How can one meaningfully study the uncertainty of the future? While early sociologists were explicitly utopian, and actively saw their sociological work as a method to create better societies, such lofty aims receded between the 1950s and 1970s. Utopias, especially deliberately designed ones, were not the purview of social science. Nor was *predicting* the future, for that matter. The social sciences sought to explain how societies were structured and analyse what made people act in certain ways. As a result, they became backward-looking, seeking to explain society by looking at past and present structures.

The first serious attempts to incorporate the future into social science came through ideas around **performativity**, the observation that ideas about the future have social effects in the present. How imagined futures affect the present can be studied. The **sociology of expectations** has convincingly shown that the fictional expectations, as economic sociologist Jens Beckert calls it, that people hold about the future coordinate economic and political action in the present. Expectations aren't simply representations of the future, "they do something: advising, showing direction, creating obligations" (van Lente, 1993, p. 191). For example, the expectation that climate damages will grow immensely in the coming decades has already led insurance companies to reevaluate what they are willing to insure. Beyond credible expectations, however, the future influences the present in many more ways. Hope for a better future may mobilize action. Fear for and anger over an apocalyptic climate future also motivates action, as we saw in the example of the Extinction Rebellion and Letzte Generation. Such emotional relationships with the future are **affective**. Rather than relying on credibility and rationality, as expectations do, affective ideas about the future provide actions in the present with meaning, with a reason for acting. Collectively shared ideas about the future – such as viewing climate change as a fundamental threat to society – animate decision-making, based on both expectations and affective ideas about how the world *ought* to be organized. Science and Technology Studies (STS) and related disciplines have created a good understanding of how the future influences the present. What remains a struggle, however, is *how* such futures come to affect people. How do imagined futures become collectively shared? Through what social processes do imagined futures become believable or effective? How do imagined climate futures become so influential that they can reasonably be expected to alter a whole way of life?

The emerging literature on futuring aims to fill that gap. As an analytical framework, futuring analyzes how people (try to) bring others together around their images of the future. It observes that the activity of futuring is one of the fundamental social practices of Western modernity. The future is everywhere. Scientists construct climate projections, policy projections, and emissions scenarios. Commercials show aspirational futures of fancy cars, far-away vacations, and investment opportunities. Both business and national policy rely fundamentally on cost-benefit analyses: economic projections about whether investments today can be weighed against their benefits in the future. Not a country on Earth makes its public policy without an eye on *economic growth*, the idea that through shrewd economic management the future will be materially better than the present. In her book *The Future of the World*, the historian Jenny Andersson observes that over the course of the 20th century, societies have developed a whole range of practices that interact with the future, such as scenario planning and simulation modelling. According to her, these practices have become "staples of global governance", as they are used to justify political decisions and policy in the present (Andersson, 2018, p. 11). As such, making images of the future is always "an intervention in the present and an attempt to shape coming times" (ibid, p. 4). Futuring is interested in that attempt. It aims to understand how actors, such as climate scientists, politicians, or activists, construct images of the future. Rather than solely examining the performativity of imagined futures in the present, it tries to explain why and how such imagined futures *become* performative in the first place.

Like the literature on sociotechnical imaginaries, futuring sees the **collective imagination** as crucially important. The performativity of imagined futures and imaginaries relies on people and institutions believing them and collectively acting them out. Visions of the future need to capture the collective imagination to steer action. As such, futuring zooms in on how people come to believe in images of the future. It analyzes how actors construct images of the future and try to convince others that these images are a) accurate, or feasible, or desirable, and b) that such images of the future mean something for how one should act, make policy, or do politics in the present. A key term in futuring research is the **techniques of futuring** concept. This concept focuses on "practices bringing together actors around one or more imagined futures and through which actors come to share particular orientations for action" (Hajer and Pelzer, 2018, p. 222). With this approach, futuring research is interested in the construction, performance, and enactment of imagined futures and sociotechnical imaginaries in the present. What are the practices that *construct* images of the future? How do people *perform* those imagined futures in order to convince other people these futures are plausible or desirable? And how do imagined futures come to be *enacted*, responded to politically, personally, or in terms of policy?

In short, futuring is interested in *the act* of imagining climate futures. In the following section, I show how the futuring lens helps us understand the contestation of those projected futures, exemplified by climate activists including those involved with Letzte Generation and Extinction Rebellion.

Imagining Climate Futures: Science and the Political Battleground

Climate politics has always been about the future. For decades, scientists have attempted to place their projections of the climatic future at the heart of political debate. They have produced reports. They have given press conferences. They have published angry letters and memoirs, and appeared in TV shows and on the radio. They have endured the organized denial and orchestrated attacks from vested interests and detractors. Most of all, they have defended their scientific findings vigorously. The science, they say, is clear. Climate change is happening. And climate science and activism together have, with fits and starts, been highly successful at putting climate change and other environmental issues *on the agenda*. Without this hard work, without the incessant futuring commitment of these scientists, that would not have happened. It was the public performance of these findings, by scientists, activists, and politicians alike, that gave these dire projections their extraordinary political authority. The kinds of political decisions that climate change necessitates are unprecedented. That societies would even consider doing so based on the authority of scientific projections of the future is nothing short of amazing. It would have been unthinkable just 70 years ago.

In their book *The Environment: A History of the Idea*, the environmental historians Paul Warde, Libby Robin, and Sverker Sörlin (2018) observe that there are four important dimensions that influenced how "the environment" became a key term of 20th- and 21st-century science and politics. First, the environment carried, from its beginning, a notion of *the future*. The environment, like the climate later, connected to fears about how the present might endanger the future. These fears about the future were addressed through the second dimension: *expertise*. Scientific expertise, Warde and his co-authors stress, was crucial to making environmental futures visible and to adjudicate between different predictions about the future. A deep-seated *trust in numbers*, a belief in numerical expertise to make the future legible and predictable, was the third dimension – the dimension that made it possible to quantify the future and make it appear as plausible and objective. That faith in numerical expertise aided the final dimension of the environment, *scale and scalability*. The term "the environment" made it possible – again, like the climate would later – to connect local phenomena to an interconnected global system.

For climate change, these dimensions are crucial. Climate politics revolves around fears about the future. Quantified numerical expertise is at the heart of climate science, and it is at the heart of drawing political and economic implications too. As the historian Paul Edwards has observed, much of what we know about the climate, we know through models. These days, models are what Edwards called a knee-jerk response: any contentious policy issue, any impactful political decision, is based on projective modelling. In climate politics and climate science, models are the pre-eminent way of constructing images of the future. Yet trust in these models is not self-evident. Models are imperfect, dependent on the quality of the data that goes into them and the assumptions upon which they are based. And, as we have seen in both the COVID-19 pandemic and in climate politics, model projections are always contested.

For futuring scholars, that contestation is precisely the point. Because making images of the future is always also an attempt to influence the present, it is a *political* endeavour. We should analyse it as such. That means that the prime concern is not whether an image of the future is correct. Whether a projection is scientifically accurate, or at least as accurate as it can be, does not in itself make a difference in politics and society. Instead, it is the *authority* that people award to this projection that matters: people need to act *as if* that projection were true. A futuring analysis does not ask whether a prediction is true, nor whether a car commercial presents a future that *is* desirable. Instead, it investigates why such images of the future become performative in the first place. A futuring approach to climate politics, then, would ask how and why we come to believe in climate futures, how modellers try to bring people together around their model projections, and how such projections come to have enough weight to shift a whole political system.

To understand the ways in which the future animates climate politics, we need to look backwards. After decades of increasing scientific worries about climate change, the political recognition that something might be amiss grew in the 1980s. As a response, in 1988, the IPCC was formed to assess climate change based on the best available science. Set-up in the interface between climate science and governments, the IPCC has a unique mandate. Its assessments come with an official recognition from governments as being accurate and policy-relevant. Since the IPCC's formation, climate change has become the defining threat for the future. Politically, anthropogenic climate change arrived in the 2020s as a principal battleground, implicitly coding for a wide array of sociocultural convictions, economic positions, and private interests – a battleground over what the future should be. Political parties in Europe debate whether to steer towards technological solutions or systemic reductions of consumption and production. In the United States and Australia, former frontier nations with deeply embedded fossil fuel industries, disagreements continue to rage around whether climate policy is needed at all. Even there, however, public opinion roundly supports decisive political action on the issue. The scientific message, outlined convincingly in the IPCC's 6th assessment report, is clear: climate change presents unprecedented and potentially catastrophic risks for the future if carbon emissions are not curbed immediately. Droughts, floods, storms, and heat waves are already disrupting life all over the globe. In the scientific and popular imagination, the climatic future appears almost apocalyptic. Yet decisive climate policies remain elusive and contested.

Despite their incessant warnings, climate projections have not yet managed to break through established routines of policymaking, nor have they seriously threatened embedded ideas about economic growth and material wellbeing. Although climate politics nominally accepts the premise "if we do not act, the future will be dire", it holds a rather limited idea of what it means to "act". For decades, it has been "almost too late to act". Newspapers and politicians have reiterated deadlines time and again. Yet global emissions do not decline. Economic growth continues to be the main political story about the future, which climate policy cannot be allowed to threaten. Via technological innovations and mostly painless policy interventions, political cockpits attempt to simultaneously safeguard the environmental *and* economic future. This story about the future is so dominant that it has also pervaded scientific model projections. Integrated assessment models, the key models used to gauge how climate targets might be reached, all assume economic growth and technological development. As Lisette van Beek et al. (2022) show, they are **politically calibrated**, constructed in relationship to dominant assumptions about feasible and desirable policies. In the cockpit of climate politics, the dominant story is still one in which the future is not meaningfully different from the present.

It is in this context that Extinction Rebellion, Scientist Rebellion, and Letzte Generation place their interventions. The dominant political story about the future, they feel, is not adequate. It is not honest. Despite decades of scientific warnings, of increasingly elaborate scientific projections about the state of the world, they say, political leaders still do not (or cannot) implement the necessary policies and restrictions. Beyond climate change, multiple planetary systems, such as biodiversity and nutrient flows, are already imminently threatened. The discourse of these activist communities is laced with quotes from prominent climate scientists who say that "the coming twenty years will be worse than the last 20 years" or that "a 4°C future is probably incompatible with an equitable international community". Groups like Extinction Rebellion (XR) and Letzte Generation present the apocalyptic future as a self-evident truth. These activists are continually (re)producing and circulating images of the future. Given that

294   *Jeroen Oomen*

governments cannot be trusted to safeguard the climatic future, they bring the future into the present. Whether they hold die-ins – in which they symbolize their futurelessness by lying like the dead on the streets surrounded by protest signs saying "dying for change" – or yell "tell the truth" and "listen to the science", climate activists mobilize particular ideas about the future. As with Extinction Rebellion, so with Letzte Generation: futurelessness is even represented in their names. Where early climate scientists were blindsided by the opposition and vitriol their climate projections evoked, a new generation of activists (including the Scientist Rebellion, Fridays for Future, and decolonial activists around the world) treats the future as a political battleground, a battleground of the collective social and political imagination. Where scientists historically hid behind objectivity, a new generation of activist scientists choose openly theatrical methods and symbolically reject the legitimacy of climate politics. Through their activism, they try to bring people around their imagined futures. The futures they draw on are still constructed by climate scientists. They still come from the IPCC and other climate projections. The narrative about the future is also still the same: if we don't act, the future will be "a fucking catastrophe". Yet these activists choose different ways to perform and enact these futures. The Fridays for Future movement's school strikes centred on the rights of the young, the generations that will bear the brunt of the changes to come. The German Letzte Generation pioneers new, more disruptive, ways of blocking infrastructure, such as *cementing* their hands to the asphalt, to demand better investment in sustainable mobility and other climate measures. Their politics of the imagination is openly combative – and deeply divisive. Extinction Rebellion in the Netherlands, which hasn't shied away from civil disobedience and controversy either, increasingly focuses on expanding its support. When a group of protest organizers were arrested by the Dutch government to public outcry, they used the state's judicial overreach to reach out to academics, civilians, and celebrities. More and more, their periodic actions demanding an end to fossil subsidies are attended by support groups, happily singing, clapping, and chanting along.

Conclusion: Imagining Climate Futures as a Politics of the Imagination

It is a common observation that in climate politics, it is always five minutes to midnight. It is always, almost, too late to act. Yet the deadline always moves, with every report, with every negotiation, and with every press conference. To understand the institutional and cultural inadequacy of climate politics, it is necessary to come to understand the ways in which ideas about the future animate, structure, and guide our ideas about climate change, as well as our attempts to govern it. In that context, we should see climate activism as an attempt to bring people together around their reading of the future, to hammer home the threat and the deadlines. Imaginaries and imagined futures are always stories we tell to understand the relationship between past, present, and future. In this light, the future is a battleground in the **politics of the imagination**, the political struggle over shaping norms and shared visions. What parts of the present and the past have no place in the future? What aspects of the present do we want to retain?

For climate activists, the answer to those questions is very different than for those currently making political decisions. Where climate activists and scientists proclaim that catastrophic consequences of climate change are now almost unavoidable, the political centres continue to act towards ideas about the future that aren't meaningfully different from the present. Climate change is widely accepted as a key concern, but it is still one of many concerns, one that can still be addressed by projecting the present into the future, by doing the same things that politics have always done, just in a more "sustainable" way. The underlying sociotechnical imaginary, the collectively shared and institutionally stabilized way of imagining and acting on the future, still maintains 20th-century measures of success and progress, most notably economic growth.

For activist groups, this is more than a misreading of the issue at stake. It is fundamentally dishonest. The future, in their eyes, will not be the same. Extinction Rebellion's "die-ins", Letzte Generation's highway blockades, and Scientist Rebellion's donning of white coats at protests are recognizable performances, **techniques of futuring**, aimed to bring people around *their* reading of what is at stake. To make people act.

Imagining climate futures is always a political act, an ideological battle over the collective imagination. For social scientists, analysing those politics of the imagination should be a prime concern.

Further Reading

Hajer, M. and Pelzer, P. (2018) "2050 – An energetic odyssey: Understanding 'techniques of futuring' in the transition towards renewable energy", *Energy Research & Social Science*, 44, pp. 222–231. DOI: 10.1016/j.erss.2018.01.013

Oomen J., Hoffman J., and Hajer M. A. (2021) "Techniques of futuring: On how imagined futures become socially performative", *European Journal of Social Theory*, 25(2), pp. 252–270. DOI: 10.1177/1368431020988826

Warde, P., Robin, L., and Sörlin, S. (2018) *The Environment: A History of the Idea*. Baltimore, MD: Johns Hopkins University Press.

References

Andersson, J. (2018) *The Future of the World: Futurology, Futurists, and the Struggle for the Post Cold War Imagination*. New York: Oxford University Press.

Hajer, M. and Pelzer, P. (2018) "2050 – An energetic odyssey: Understanding 'techniques of futuring' in the transition towards renewable energy", *Energy Research & Social Science*, 44, pp. 222–231. DOI: 10.1016/j.erss.2018.01.013

van Beek, L., Oomen, J., Hajer, M. et al. (2022) "Navigating the political: An analysis of political calibration of integrated assessment modelling in light of the 1.5°C goal", *Environmental Science & Policy*, 133, pp. 193–202. DOI: 10.1016/j.envsci.2022.03.024

van Lente, H. (1993) *Promising technology: The dynamics of expectations in technological developments*. [PhD thesis, University of Twente] Delft: Eburon.

Warde, P., Robin, L., and Sörlin, S. (2018) *The Environment: A History of the Idea*. Baltimore, MD: Johns Hopkins University Press.

36 The World Ocean and Climate Connectivities

Cymene Howe

Bodies of water have always been fundamental to the human **imaginary**. Water is a life *force* that sustains all living beings, and it is a life *space* creating environments and ecosystems for organisms to thrive. In many cosmological systems, bodies of water are also life *forms*, sentient beings with powers uniquely their own and forever interacting with other beings around them. But bodies of water in our times of environmental distress are now becoming infused with new meanings and new potentials as they are impacted by human-created heat channeled into the world's waters through rising atmospheric temperatures and a warming Earth. Melting ice, sea level rise, drought, desertification, shifting weather systems and superstorms—in addition to the perils of water pollution, acidification, and toxicity—all indicate that both the meaning and the effects of water are being transformed.

As we consider how the human relationship to water is changing, some insights from STS, especially regarding the relationality between humans and nonhumans, are particularly relevant. Beginning in the 1980s, STS scholars including Bruno Latour, John Law, and Michel Callon were interested in the interconnectedness of human and nonhuman **actors** (or "actants")—entities that would influence one another in unexpected ways creating webs of interaction that were both social and material in form. This became known and developed over time into **actor network theory** or ANT—where social and natural worlds always condition each other in shifting networks of relationships. Importantly, from the perspective of ANT—there is no "outside" of the sociomaterial worlds we inhabit and with which we interact.

The relationships between people and nonhuman others—whether technological devices like infrastructures and machines or other species and stones—has been a critical point of analysis within **feminist STS**, a subfield that recognizes how inequalities affect all the dimensions of both science and society. Donna Haraway (2016), a feminist STS philosopher, for instance, has very famously proposed that we **make kin** with animals and plants, as well as with ecosystems that we collectively share with them. For Haraway this means recognizing that we are part of larger cycles and life systems, and importantly, that there is a politics to this proposition. She writes,

> Making kin seems to me the thing that we most need to be doing in a world that rips us apart from each other, in a world that has already more than seven and a half billion human beings with very unequal and unjust patterns of suffering and well-being. By kin I mean those who have an enduring mutual, obligatory, non-optional, you-can't-just-cast-that-away-when-it-gets-inconvenient, enduring relatedness that carries consequences.
>
> (Qtd in Paulson, 2019)

DOI: 10.4324/9781003409748-47

Stacy Alaimo, furthering aspects of Haraway's interconnective, kin-centered thinking, has created the concept of **transcorporeality**—a term that envisions all living beings (including humans) as intermeshed with a material world that crosses through and into bodies and beings, transforming them.

When thinking about water, or the world ocean, or melting ice and rising seas, why are these perspectives important? First off, each of these STS positions provide an important set of tools for understanding the interconnectedness of us all—in the broadest sense—that is, not only among human beings, but beyond the human as well. For many thinkers concerned with climate change, including me, relationality is one of the most important dimensions for us to recognize. Global environmental impacts can now be seen everywhere because of human-caused greenhouse gasses and the warming of the Earth's atmosphere that has caused. If, in our research and interactions more generally, we can see the connections we have to each other and to the Earth system itself, then perhaps we might make better decisions about our practices now and in the future, including reducing our collective environmental impact. Not all people are equally responsible for the new condition that has been called the **Anthropocene**—an epoch of significant human impact across key geological markers and the Earth system—including the hydrosphere, cryosphere, lithosphere, biosphere, and atmosphere. Some populations around the world will be (and already are) more affected by Anthropocene damage, particularly people living in places with fewer resources to respond to extreme weather, drought, sea level rise, and other climate-driven disasters.

Water, especially when it is so thoroughly impacted by and connected to human systems, leads us to a few important principles. First, it is important to think across the social and natural sciences. Second, environmental harm, including that embedded in the hydrosphere, is unequally distributed across the world. Finally, working in collaboration with one another (and "making kin" with nonhumans) may be essential to solving water-related issues. Nowhere is this more apparent than in the dynamics of the planet's oceans.

Finding the World Ocean

What if we could see ourselves connected by water, related to each other through a world ocean? For ocean scientists, water encircles the Earth, creating a world hydrosphere. The seas of the planet are one body, never quite neatly captured in a name, like Pacific or Indian or Atlantic. And this world ocean does not just lay there inert and subdued. Its motion is constant in tides, currents, and circulation patterns. Its movement is also skyward, rising up into the atmosphere evaporatively becoming clouds, and returning as rain and snow, sleet, and storm. A hydrological cycle of watery interplay. As Hi'ilei Julia Hobart (2020) puts it, "Water's nature is diasporic … Often, it misbehaves in such a way that reminds humans of the limits of their own power and control." Its disobedience may be its magic.

Let's think of the water we call ice. At the southern pole of the world, the frozen continent of Antarctica has lost almost 6,500,000,000,000 tons of ice in the last 30 years. In the Arctic region, heat is turning glaciers and ice sheets into meltwater faster than anywhere else on Earth. We know that these sites of melt are pouring vast quantities of water into the world oceans. Arctic glaciers, being transformed from solid to liquid, now represent the source of more sea level rise than any other: more than the distending effects of oceanic thermal expansion and more than the disintegration of the icy continent of Antarctica.

In my research I have been trying to understand the social consequences of a melting world and, in turn, how meltwater becomes rising seas that also impact the physical and social dynamics of coastal cities. I am not a glaciologist or an oceanographer. I am a cultural anthropologist interested in the social dimensions of climate change—how it is reshaping our encounters with the world and with each other. For me, the tools of STS have been invaluable in this work because STS prioritizes the fusion of the social and natural sciences. Meltwater and rising seas present both a "physical" problem but also a "cultural" problem because of how people interpret, deal with, adapt to, and envision their future relationship to water.

For several years I have been studying the loss of ice in places where it has long been a part of lives and landscapes. I focus on how people understand their environments and all that is sustained by them, and how that is being transformed by climate change. In the course of this work, my research partner and I came across Okjökull—the first of Iceland's major glaciers to be destroyed due to climate change. Scant attention had been given to Okjökull's passing and so we, enlisting a handful of Icelandic colleagues, held a funeral for the little glacier called Ok; it was the world's first memorial for a glacier. As our friend Andri Snær Magnason said at the time, "never in the history of humanity, has there been a need for a glacier funeral." Now, there is. While glaciers are not in fact "alive" in the scientific sense, by recognizing their "passing"— their breakdown because of human-caused climate change—we found ourselves enacting what Haraway calls "making kin" with nonhuman others. Nonhuman others, in the broadest sense, might be animals or mountains, insects or rivers, plants or glaciers: entities that we share the world with, our cohabitational partners on Earth.

In trying to understand the social impact of melting ice, I have also been keenly aware of the other end of this geohydrological event, namely, rising seas brought on by melting ice. I began to ask: what if we could follow the water from our melting glaciers to the coasts of the world where glaciers now manifest as sea level rise? Where would we go and what stories would we find along the way? As our planet's ice becomes ocean, its ice now made water, may it create

Figure 36.1 Installing the glacier memorial to Okjökull in Iceland, August 2019.

Source: Photo by Sigtryggur Ari Jóhannsson. Used by permission.

routes and passages of connectivity between otherwise distant places? Put another way: how might melting ice, transformed into rising seas, tie us together across continents through the world ocean? And, if we could come to recognize this togetherness, might that provide new possibilities for making kin in a melting world?

The Social Life of Ice and Water

Mary Douglas, a cultural anthropologist writing in the mid-20th century, thought quite a lot about what she called "matter out-of-place." She wrote specifically, and quite eloquently, about dirt. For her, dirt was an idea. But it was also a substance that might be just fine over there, on what we consider the "outside" of our social spaces, but that was never welcome here, on the "inside" of those places we deem to be interior, social, and sometimes sacred. Indeed, some dirt, when out of place, could even rise to the level of taboo.

Water moving outside of its usual boundaries is, of course, much more than a philosophical question. It can be a very destructive force, even deadly—as we see in the case of extreme rain events or tsunamis. But water out-of-place—that is "outside" of the places we expect it or want it to be—is a growing reality in our times of sea level rise, floods, and superstorms. Douglas's meditation on dirt, therefore, offers us an interpretation of attitudes that many people continue to hold about how water can be made to remain under human control. Like dirt, we may *want* water to stay in its place and we may engineer many ways to do that. However, this human conceit of control over water is being increasingly challenged by climate change and global conditions where there is a lot of matter—in fact many *matters*—out-of-place.

The causes of global warming are not equally shared (historically and in the present the "Global North" has been the source of many more greenhouse gas emissions than countries of the "Global South"). And we know that the impacts of climate change will not be equally distributed either; countries with less economic resources will have less capacity to combat the effects that they have done little to produce in the first place. In a global picture, however, we recognize that momentous environmental changes are upon us all. We are collectively struggling to manage all that is now out-of-place, including water in the form of sea level rise, storms, and floods. Douglas's notion of matter out-of-place is therefore helpful in thinking through these considerable challenges. But, perhaps more useful still is Hi'ilei Julia Hobart's formulation: namely, that water is *diasporic*. The definition of diaspora is to be "dispersed from one's traditional homeland." Ice that once inhabited land, now melted into the ocean to become rising seas, feels very much this way: dispersed, dislocated, detached from where it has traditionally resided. This water, put into motion through human actions is much more than matter out-of-place: it is a diaspora.

As ice continues to transform around the world, I have started to follow the water from once-frozen places to others, where the ocean swells and rises. To follow the water, I begin in the Arctic region—Greenland and Iceland—and then travel to coastal cities further south—Honolulu and Cape Town—where the meltwater from those countries is most pronounced as sea level rise. Part of this work involves trying to unravel how water in motion is being charted and graphed by scientists, a key method of STS. Another part of this work requires asking questions about how communities, industries, government officials, and others are coping with ice loss or, alternately, bracing for sea level surges. I am interested in how these responses, or "adaptations," might be similar or different across and between sites. Adaptation to a changing waterscape may look very different in places impacted by ice melt and those being affected by sea level rise. For example, cities facing sea level rise may choose to install seawalls whereas places experiencing ice loss may build new infrastructures to control overflowing glacial rivers (see Boyer, Chapter 28, this volume). More broadly, and in a comparative mode, I research how these different responses might actually be similarly scaled over time, for example, by looking at the policy and planning timeframes being used by coastal engineers or local governments. I am also curious about what drives the political and economic commitments behind adaptations, now and into the future. Is it governments or industries that are leading the charge, or is it affected residents that are leveraging the most compelling demands for adaptation? Critically important too is how local communities are reckoning with these tremendous changes to their home environments and how this may result in a sense of loss, outrage, or even hope for new possibilities.

To be able to follow the water from sites of melt to sites of rise is, I discovered, a very challenging and complex scientific task. For this, I was lucky to find a group of physicists who knew just how to do so.

From Ice to Ocean

In elementary school, most of us learn about world geography through maps and images. Cartographic pictures of the world neatly designate Earth's oceans: the Arctic, Atlantic, Indian, Pacific, and Southern. Each has its proper name and place. But for oceanographers and other scientists these are mutable waters, interconnected and moving swiftly and fluidly across lines marked on maps. This is the world ocean. One ocean. One that covers about 70 percent of the Earth's surface.

The relationship between a changing cryosphere—Earth's ice—and its precise impact on sea levels along the world's coasts has been very challenging for scientists to track. In part, that is because the "bathtub model" predominates in many climate projections—indicating the volume of meltwater being added to the world ocean, but not how it acts very differently in coastal locations according to a host of criteria including local currents, subsidence, and erosion, to name a few. However, a group of physicists at National Aeronautics and Space Administration (NASA) has recently created a way to map this relationship. Their model, called the Gradient Fingerprint Map (or GFM), can calculate where specific sites of melting ice will appear as sea level rise in 293 of the world's coastal cities. By pinpointing mass and meltwater contributions from each major glacial basin on Earth—from the Himalayas to Antarctica and from the Arctic to the Southern Andes—the model demonstrates a precise relationship between lost ice and sea level gain. In short, it shows how Earth's gravitational and rotational processes, as well as the redistribution of mass, are contorted by the melting of the world's ice and how that influences the world ocean on a planetary scale.

When we try to follow meltwater to where it becomes sea level rise, we learn that unlike the borders drawn on maps, the movement of the world's water does not conform to a linear path. In a sense, there is no path at all, but instead a contortion of Earth's "liquid envelope." As meltwater from glaciers and ice sheets joins with the world ocean, the planet's gravity, as well as its rotation, are shifted, creating different spatial patterns of sea level rise around the world. In one of our conversations together, Eric Larour, a lead physicist in the development of the GFM tool, outlined how the tool itself worked through a series of complex calculations, drawing data from satellites and tide gauge readings. Drawing upon the physics of shifting ice and water, the GFM tool allows users to see with precision where melting ice results in specific amounts of sea level rise in particular cities the world over. It shows, for instance, that Cape Town, South Africa, is more affected by Icelandic melt than any other city in the world and that Greenlandic melt impacts sea level rise in Honolulu, Hawai'i more than any other city in the United States.

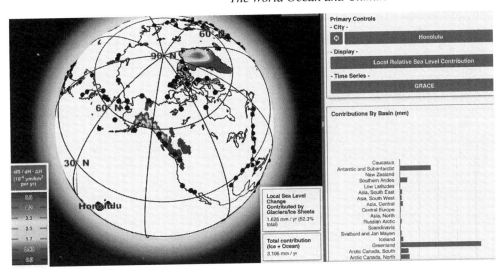

Figure 36.2 The Gradient Fingerprint Map tool developed by NASA, here showing the contribution of Greenlandic ice loss to sea level rise in Honolulu, Hawai'i.

Source: Image created by the US National Aeronautics and Space Administration (NASA) https://vesl.jpl.nasa.gov/sea-level/slr-gfm/. Used with permission.

Are We Living in a Time of Hydrological Globalization?

The world ocean is not simply water in motion, it is also in concert with ice, connecting points on the globe in rather counterintuitive ways. The consequences of a melting world and its effects on the greater ocean and those of us—indeed all of us—who are impacted by it, cannot focus only on the physics of displaced water as an Anthropocene fugitive. Instead, we ought to look to the multiple (and multiplying) human-nonhuman interconnections that are made through these processes. As a social scientist I am interested in what these watery relationships might show us. Can we begin to think of this liquid diaspora as a kind of hydrological globalization—a set of planetary relationships that may not have been uniquely created by a warming climate, but which are made more apparent, and perhaps urgent, by the rapid transformation of the world's hydrosphere?

As our planetary water becomes differently distributed under the impact of human forces—through a collapsing cryosphere and sea level rise, extreme rain events, and groundwater depletion—I believe that the concept of **hydrological globalization** can help us to signal the emergent connections between water in motion and responses to it. Hydrological globalization emphasizes global connectivities not primarily through links of trade, migration, finance, or media, but through physical relationships created as our hydrosphere is reshaped. Hydrological globalization is global in one sense, but it also offers a possibility for more local engagements in the process of "making kin" with one another, and potentially, for a more intimate understanding of the world's water in the context of rapid hydrological change. Conceptually, hydrological globalization allows us to follow the impacts and uptakes of watery rearrangements—connecting geographical sites, both materially and socially, by literally connecting them through their (which is really *our*) water.

Further Reading

Haraway, D. (2016) *Staying with the Trouble: Making Kin in the Chthulucene*. Durham, NC: Duke University Press.

Howe, C. (2022) "Melt/Rise: Climate Change and the Global Interconnectedness of Water," *The Berlin Journal*. DOI: www.americanacademy.de/melt-rise/

Kuznetski, J. and Alaimo, S. (2020) "Transcorporeality: an interview with Stacy Alaimo," *Ecozon@: European Journal of Literature, Culture and Environment*, 11(2), pp. 137–146. DOI: https://ecozona.eu/article/view/3478

Thomas, J. A., ed. (2022) *Altered Earth: Getting the Anthropocene Right*. Cambridge: Cambridge University Press.

References

Paulson, S. (2019) "Making kin: an interview with Donna Haraway," *Los Angeles Review of Books*. Available at: https://lareviewofbooks.org/article/making-kin-an-interview-with-donna-haraway/ (Accessed: June 27, 2023).

Hobart, H.J. (2020) "On oceanic fugitivity," *Items* (Insights from the Social Sciences, Social Science Research Council). Available at: https://items.ssrc.org/ways-of-water/on-oceanic-fugitivity/ (Accessed: June 27, 2023).

37 From Controlling Global Mean Temperature to Caring for a Flourishing Climate

Andy Stirling

Introduction

Actions to prevent disruption of Earth's climate have proven tragically slow and insufficient. Notwithstanding some progress, gaps between aspiration and realisation seem larger than ever. This is despite climate concerns being among the most high-profile and active fields of environmentalism over the past few decades. The contrast is stark with other areas of continuing ecological struggle, which have in different areas yielded greater (though still incomplete) degrees of past success, often against comparably major interests. The puzzle is further compounded in that persistent relative failure on climate issues has unfolded at the same time as seemingly unprecedented levels of environmentalist 'policy impact' – at least if judged by the clamour in elite circles around 'evidence-based decisions' driven by 'sound science'.

Why is this? How can progress be so slow on climate action, when the acknowledged basis in values, evidence and the shared long-run self-interest of worldwide societies are all so compelling? By what practical means might this impasse be remedied? It is here that the diverse interdisciplinary field of STS may offer some crucial but under-appreciated contributions. For it is in STS that academic attention arguably turns most distinctively to the ways politics unfold *inside* (as well as around) science – and knowledge in general. And it is here that some particularly distinctive insights also begin to emerge. Unlike many other kinds of policy-relevant study scrutinising how interests and privilege bear on how *truth speaks to power*, STS has also pioneered attention to ways and degrees in which (even in supposedly *pure* science) *power shapes truth* (Wilsdon and Doubleday, 2015).

So, what does this key strand in STS research have to say about why climate action has so far been so unsuccessful? How might some of the most emphatically stated 'truths' spoken by climate science and environmentalism already contain the imprints of the specific patterns of power and privilege, which are arguably strongly implicated in the problem itself? How might these warping effects on understanding the challenges compromise efforts at remedies? The answers to these kinds of queries may hold some profound but neglected practical implications.

General Background

Illuminating these insights about the importance of politics *inside* knowledge (as well as politics *around* knowledge) involves close critical attention to the contentions raised above. Is it really the case that a worldwide environmental movement currently expressing well-founded regrets over failures to ameliorate climate disruption has in the past been more successful in comparably challenging struggles? By what means might success be conceived, let alone compared? What are the repercussions for future climate action? Here, the broad field of STS offers some compelling lenses through which to view the complexities.

DOI: 10.4324/9781003409748-48

To see this, it is first necessary to consider a crucial feature of past decades of environmental struggle, as seen from this 'politics of knowledge' angle. Past successes that can be looked at in this light include workplace hazards (e.g., around the banning of asbestos, heavy metals and benzene); conservation threats (e.g., curtailing of whaling, coastal overfishing, soil-destroying agriculture); environmental pollution (e.g., reductions in use of pesticides, carcinogens, neuro-toxins); food safety challenges (e.g., regulating contaminants, additives, pathogens, allergens); or energy pathways (e.g., away from nuclear power and towards energy efficiency and renewa-bles) (Harremoës et al., 2001). Although only ever partial – and still leaving much unfinished business – major green successes in all these fields have been achieved against very strong inter-ests in business, academia and government, often of an overall magnitude broadly comparable to those around fossil fuels (Gee et al., 2013).

Around many household and industrial products (including those mentioned above) aggres-sively promoted at the time to be societal necessities (just like fossil fuels), the massive interna-tional chemical industry has been in continuous retreat in the face of successful environmentalist action over the past few decades. Likewise, despite weighing in with degrees of power and privilege rarely matched in other sectors, concentrated transnational food commodity firms (e.g., around grain and sugar) have suffered their own striking reversals driven by collective action in key strategic areas (e.g., around hormone-fed meat or genetically modified foods). In each case, what was assumed and asserted by industry to be 'the' inevitable new 'way forward' for progress in their sector has been actively substituted with an alternative path (IAASTD, 2009).

There are also other cases where powerful elite interests in a trajectory that is challenged by environmentalism seem at least as entrenched as around fossil fuels. Perhaps most striking in this regard is the transnational civil-military nuclear industrial complex. This entrenched nexus of incumbent power has combined a dominant role in official imaginations of electricity futures with largely hidden added force from even more concentrated security and geopolitical inter-ests. Commanding immense financial flows, this dominatingly powerful corporate-state cartel spans ostensibly opposing national sides in various axes of 'nuclear deterrence'. Such aspira-tions to control through concentration of energy or projection of violent force, is a hallmark of colonial modernity. We see this crystallised in infrastructures like nuclear systems. Indeed, it is this global infrastructure more than any other that governs access to the top table of interna-tional affairs. Associated nuclear interests have over the years championed exactly the tactics deployed by the fossil fuel industries around research obfuscation, false denial, political capture, and suppression of dissent. Yet even this unrivalled combination of global power and privilege has not prevented environmental movements from inflicting an unprecedented worldwide rever-sal in the fortunes of nuclear power (Elliott, 2007).

An STS-Informed Approach

So what is it about the ways in which so much progress was made in past environmental con-troversies that distinguishes them from how the struggle against climate disruption has been fought more recently? How does politics *inside* the knowledge driving climate struggle differ from the knowledge that helped realise success in other areas? This is where STS points towards a distinctive set of insights. In short, the 40-year struggle for climate action has seen environ-mentalism tend to move *away* from a bottom-up politics mobilised around **care** and *towards* elite policies driven more by aspirations to **control**. Not only has this entailed a retreat from the kinds of caring understandings associated with so much past success, but this previously fruitful caring approach is being substituted by precisely the kind of controlling idiom that was so often ineffectively deployed *against* green action.

Before looking more closely at this shift in environmental politics, it is worth briefly reviewing the underlying general distinction prominently made in STS-related work between 'care' and 'control'. Roots here can be traced back through Ivan Illich's notions of 'conviviality' – about *caring for and about others*, rather than *manipulation and **control*** (Illich, 1972). Owing a great debt to feminism (Fisher and Tronto, 1991), STS authors like Annemarie Mol, Vicky Singleton, Chris Groves, Maria Puig de la Bellacasa, Luigi Pellizzoni, Saurabh Arora, Sarah Davies and Maia Horst have all approached this broad contrast between 'care' and 'control' as being more about imaginations than materialised practice alone (Felt et al., 2013).

In my related characterisation, the salient difference between control and care echoes that between equally familiar kinds of agency people encounter in everyday life. On one hand, people experience functioning machines, like a light switch, computer or car. On the other hand, people have experiences involving friends, family or comparably empathised with others, like fellow citizens, cherished places or other living beings. The former machine model of control assumes a singular reified intention, with impacts realised on an objectified focus in perfect one-to-one correspondence with aims (with negligible ambiguities, uncertainties, shortfalls, collateral effects or feedbacks): when the switch works, the light turns on, and nothing else happens. The latter empathetic pattern of care is about multiple, contingent and contending intentions involving affordance for diverse subjectivities (including on the part of what is being related to), with implications embraced as unfolding in messy, open-ended and often indeterminate ways. Whether either is 'good' or 'bad' will typically depend largely on context and perspective (Stirling, 2019a). With this characterisation in mind, key dimensions of what might variously count as 'control' or 'care' are shown in the framework summarised in Table 37.1.

In contemplating this contrast between care and control, it is worthwhile to consider that the political and cultural context within which this chapter is being written (and, most likely, read) – like the context for climate struggle itself – is not neutral between the two. It is widely recognised across diverse traditions in contemporary thought that the encompassing formation of 'colonial modernity' (referred to earlier in discussing its iconic crystallisations in nuclear power and weapons) is constituted by metastasized imaginations of control. Centred on exactly the machine paradigm described above, it is the associated fantasies, fallacies and fixations around control that tellingly account for the otherwise diverse canonical distinguishing features of modernity – like individualization (control of life courses), rationalization (control of reason), nation-forming (control of territory), industrialization (control of production), capitalization (control of value), bureaucratization (control of organisations) and democratization (control of polities).

The dominant context for the unfolding of all these historic processes was arguably the unfolding of European colonialism, involving an especially extreme and brutal assertion of controlling imaginations. So it is in this sense that the contrast between care and control summarised in Table 37.1 is not symmetrical in any contemporary real-world context. Across diverse manifestations of currently globalising colonial modernity, it is the features described in the left-hand column that tend to be most strongly expressed in the public domain. Although not absent, the qualities shown on the right-hand side tend institutionally to be more recessive and relegated to private life.

Taken together, the summary picture in Table 37.1 might be called a **polythetic distinction** between key features of 'ideologies of care' and 'ideologies of control' that typically underlie (respectively) earlier environmentalism and later climate action. This relational contrast is 'polythetic' in the sense of being transparent and accountable about the many dimensions under which it is defined, not all of which are involved in every instance (Schneider and Wagemann, 2012). Varying differently across these dimensions, real-world cases will be diverse and ambiguous, with many exceptions. So, the resulting broad contrast is not a categorical binary:

Table 37.1 A relational distinction between caring and controlling styles of politics in knowledge

Politics in Knowledge	
Controlling	*Caring*
singular intentions & imaginations	**multiple** intentions & imaginations
objectified focus of attention	focus is afforded its own **agency**
knowledge & action separated & **sequenced**	knowledge & action deeply **entangled**
closed deterministic static **categories**	open dynamic uncertain **relations**
hubris about perfectly realised aims	**humility** about gaps, side-effects, feedbacks
visionary individualised **leadership**	resonating collective **mutualistic** values
emphasis on **technical** solutions	**social** innovations, cultural creativity
nudged trust for **incumbent** structures	uninvited sceptical **subaltern** dissent
driven especially by overbearing **fears**	motivated mostly by emancipating **hopes**
disciplined **technocratic** expertise	**democratically** plural understandings
projecting force, asserting **authority**	bearing witness, exercising **persuasion**

control and care are always inter-related, typically normatively ambiguous, often intimately co-occurring and fundamentally mutually constitutive. But the key point is that – as in 'family resemblances' – the overall set of dimensions represented above may often offer a meaningful basis for recognising significant commonalities and distinguishing wider patterns relevant for action.

Such a distinction between care and control may be useful when it comes to comparing styles of politics in knowledge that underlie contemporary climate action with those that characterised earlier waves of environmentalism. Here, I argue that there is a broad tendency in which the kinds of understandings driving past ecological struggles have tended to lie more on the right side of Table 37.1, with those driving more recent climate action tending to be better described on the left. It is in these terms that it seems contemporary climate struggle is being captured by the dominating ideology of what it is contending with.

Care and Control: Some Broad Patterns

To test this argument, let us consider existing STS-related work on environmental struggles. In short, past successes against toxic chemicals, ecological devastation, species eradication, tainted food and nuclear threats, for instance, have all clearly been associated more with democratic struggle for values-based *care* for the wellbeing of people and their environments than with technocratic authority making science-based claims to be able to *control* the threats in focus. Indeed, it was technocratic forms of 'risk-based', 'cost-benefit', 'dose-response', 'critical loads' and 'dilute and disperse' regulatory modelling that were often used by adversaries to *attack* environmentalist positions on these issues. Examples of this dominant controlling idiom still saturate public discourse around, for instance: risks of economic crashes, nuclear reactor accidents, scenarios for machine intelligence, vaccination programmes, pandemic lab-leaks and the calibrating of security threats (Scoones & Stirling, 2020).

This broad pattern of preponderance of an idiom of control over imaginaries of care can be discerned in more detail on climate issues, by considering each line in Table 37.1 in turn. To an extent greater than virtually any other environmental issue, the global scientific-governmental institutions routinely credited with leading climate action (e.g. IPCC) tend to assert a single, comprehensively integrated picture of both the problem and prescribed responses. By contrast with the more horizontal, messy diverse alliances in previous ecological struggles, the intentions and imaginations behind the climate problem and responses tend to be represented as unitary rather than plural. By contrast with the formative agency earlier afforded to 'Gaia' – conceiving of environmentalists as 'friends' of an agency-endowed Earth – contemporary more singularised visions of *planetary management* objectify the Earth itself and its climate as a thing, to be acted upon by means of what are openly called *control variables*.

Even critical circles enthusiastically herald variants of a supposed *Anthropocene* epoch, in which the former ineffable alter-agency of Gaia is terminologically suppressed into the supposed 'human dominion' of an objectified Earth. The current stratigraphically-fleeting horizon of threatened harm is scientistically projected into a notional multi-million-year geological epoch named (in advance!) after the enthusiasms of its authors. So has environmentalism morphed from seeking to protect the Earth from the devastations of colonial modernity, towards the engineering of a *safe operating space* for a humanity typically conceived in the singularised image of exactly this previously-resisted formation.

In keeping with this colonially modern controlling machine-like model of change, knowledge is treated as objective, separate from more overtly normative action, rather than always partly subjective and also itself conditional on action (see Schäfer, Chapter 32, this volume). Whereas past ecological discourses and mobilisations were built mainly around values, current climate discourse centres more on *evidence based policy*, reinforcing longstanding political pressures for justification with academic conceits that expertise alone can be an unambiguous driver for action. While past environmental struggles emphasised action itself (more than science) as a basis for change, frequent current climate campaigning slogans now urge, *do what the science says*. Such categorical framings of action can obscure the ever-present uncertainties concerning open-ended relational dynamics that were once emphasised in environmentalism. Instead, the global climate is increasingly framed as if it were naturally static. Asserted with hubristic precision, it is often treated as self-evident that a categorically defined *global mean temperature* should be maintained in the *holocene optimum* state.

In this, the above Anthropocene contradiction is further hardwired in the supposedly impeccable authority of intergovernmental modelling. The allegedly optimal global climate target chosen for fixing is that preferred by precisely the extant globalised formations of colonial modernity that are actually causing the climate problem. It is damagingly left to deniers of climate disruption to remind us of the many neglected uncertainties and of the fact that the Earth's climate has always been naturally highly variable. As a result, the undoubted severity of disruptive impacts on a myriad of people, other beings and ecologies are attributed more to the property of 'change' in itself, than to the political, economic and cultural iniquities and marginalisations that constitute vulnerabilities to particular rates and forms of disruption in a world rendered brittle by colonial modernity.

Across all these tendencies, climate action is undertaken increasingly as technical *science-based* control, as if over a deterministic machine. This sidelines more overtly-political, values-based and democracy-affirming struggles centred around care. The chance thereby diminishes to resist climate disruption in ways that acknowledge the uncertainties, but which (like previous environmental struggles) rest on more readily-accountable charges of moral unacceptability.

This is not abstract criticism. Mainstream climate modelling projections are increasingly committed to euphemistically-named *negative emissions technologies*, which also quietly further entrench incumbent interests. Climate geoengineering infrastructures include those pioneered in the military imaginations that are so central to colonial modernity. With geoengineering, the machine-like model of climate control takes a very material form, assuming singularised aims in perfect one-to-one mappings onto defined intended states of the world. In a betrayal of past environmental sensibilities, this also neglects possibilities of significant gaps, side-effects or unintended feedbacks (see Low, Chapter 34, this volume).

An expedient machine-like subordination to existing structures of privilege is reinforced by the rise of a plethora of technocratic *dashboard* and *cockpit* metaphors of kinds that used to arise against, rather than for, ecological struggles. Misleading technocratic idioms regarding quantitative precision using 'risk-based' metrics were deployed again and again against past environmental struggles across diverse chemical hazards, nuclear risks, biodiversity impacts and food safety threats. Such metrics were perennially revealed to neglect the crucial uncertainties, ambiguities, variabilities and sensitivities that form grounds for more caring precaution. Yet, it is a sign of the obduracy in the entrenched interests in play that repeated refutation of this calculating style by practical experience has done little to disturb the underlying authoritarian imagination. And it goes beyond mere obduracy, for this same failed controlling idiom to be increasingly adopted in climate debates by exactly the environmentalist interests who so often in the past revealed its manifest flaws.

A further feature of this shift from ideologies of care to control is also in keeping with a set of essentially colonial preoccupations with the dispensing of privilege and patronage. This is the origin, for instance, of the caricature 'take me to your leader' greeting. In this vision, 'leadership' at lower levels in an enacted hierarchy may expediently be framed as representations *upwards* towards overarching power, but they are more often quietly realised in practice as assertions *downwards* of the stratified interests to which such leadership has become a client. It is in keeping with this tendency that key roles for previously more anonymously-led social movements (like Greenpeace or Friends of the Earth) have, in mainstream climate policy arenas, tended to become more individualised and personified in notionally heroic visionaries, charismatic influencers and distinguished leaders.

The kinds of economic, institutional, cultural, behavioural and broader societal transformations advocated in earlier waves of environmentalism are, of course, still present – interlinking with progressive social justice aims in movements like degrowth, Buen Vivir, Pachamama, uBuntu and so on (Perkins, 2019). Even so, within climate activism and policy making, the tone tends again to be more about control than care – with a greater emphasis on technical 'solutions' than on social innovation, cultural expression or political transformation as ends in themselves. Similarly, issues beyond climate tend increasingly to be disciplined under a climate lens – with peace disproportionately addressed through *climate conflict* and emancipation increasingly seen in the lens of *climate justice*. In this process of further domestication and subsumption of earlier more direct forms of progressive dissent, emphasis has also shifted: from actively encouraged scepticism towards the tacit nudging of *behaviour change*; from uninvited mobilisations towards invited forms of *public engagement*; and towards trust by ordinary people for powerful interests, more than the other way around.

In place of the unruly forms of subaltern agency through which environmentalism and other progressive movements helped give rise to new institutional orders of *global civil society* amid the cultural revolutions of the late 1960s, climate action tends to place a greater weight on existing incumbent structures. Unlike earlier hopeful motivations towards diverse kinds of emancipatory politics, drivers of climate action tend to be framed more exclusively in terms of – often apocalyptic – fear. In contemporary climate discourse, we see a marked departure from the multiplicity of concerns, values, identities and understandings that had flowed through early environmentalism into the plurality of sustainable development goals emphasising the importance of democratic struggle.

Whilst much past environmental mobilisation was driven by bearing witness and exercising persuasion, current forms of climate activism tend to rest more on projecting force and asserting the authority of elite expertise. Counterproductive efforts are even made in social media, for example, towards banning statements from public discourse if they are judged by some incumbent authority to be incorrect. In such a view, the problem lies in the alleged *stupidity* of ordinary people (to be remedied by assertive education), rather than in entrenched interests (against which the agency of ordinary people is a crucial resource). Indeed, democracy itself is increasingly portrayed as an *enemy of nature* that should be *put on hold* (Stirling and Scoones, 2020). This further risks eroding the precious democratic political spaces that have in the past proven so essential to environmentalism – self-destructively conceding ground to more authoritarian forces against which ecological struggle was in the past opposed.

One area in which these contrasts have arguably become most momentously materialised is around nuclear technologies. In both their civilian energy and strategic military manifestations, STS-related work has long revealed how the imaginaries that constitute these interlinked nuclear infrastructures are canonical expressions of the constituting controlling imagination of colonial modernity discussed earlier. In its unsurpassed embodiment of technocratic hubris, concentrated power and force projection, the worldwide nuclear complex formed, from the outset, an iconic target for early green activism. As a result of decades of struggle (despite nuclear-driven efforts to suppress and retard alternative, low-carbon energy sources), nuclear power is now widely acknowledged in specialist energy circles to be more expensive, less rapid, less secure, less job-intensive, less safe and more problematic in other ways than renewable energy and energy efficiency (Scrase and MacKerron, 2009).

Yet these problems with nuclear energy are strangely sidelined in contemporary climate debates. The immense interests at work help support an explosion of commentary on social media seeking not only to ignore the overwhelming practical difficulties, but to portray nuclear power as somehow essential for climate action. Cost, speed, impact and alternatives are neglected. Specific features of nuclear power that have in the past proven problematic (like its inflexibility) are now noisily celebrated in notions (like baseload) that are acknowledged by the electricity industry to be outdated. Individual climate scientists (with no background in energy) acquire high profiles on many media platforms in nuclear-committed countries for often highly-charged advocacy of a hitherto untested new generation of small modular designs – still at the stage of being 'powerpoint reactors'. These proposals forego economies of scale that were previously crucial to nuclear viability and without any clear rationale for how this disadvantage may be addressed.

Especially in countries with nuclear weapons, organisations formerly campaigning most actively about the connection between ecology and peace have become oddly muted on nuclear power. Patronage pressures lead mainstream energy policy analysis in these countries simply to take official nuclear commitments for granted. The question posed is *how* nuclear renewal can be achieved and how fast, rather than *why* it should be pursued at all. In all the name-calling about irrational anti-nuclear sentiment, it is often oddly invisible that nuclear power is a manifestly less effective, slower and more expensive way to achieve climate action (Johnstone and Stirling, 2022). Despite economic and industrial trends driving the opposite direction – and despite past green movements being vindicated in their formerly vilified advocacy of alternative strategies – it is more telling in the case of nuclear power than anywhere else, that much contemporary environmentalism is now dancing to its former opponents' tune.

Qualifications, Conclusions and Onward Implications

As in much STS-inspired thinking, the kind of analysis offered here may at first seem daunting in its apparent critical abstraction. Yet it is also a general feature of STS that such challenges are often inherent to asking the most important – and therefore difficult – questions. The peril in taking prevailing ways of thinking for granted is to forego a basis for resisting existing patterns of action. Here, the concrete implications for driving imaginations around contemporary climate struggles could hardly be more pragmatic. By turning the focus onto how structures of power and privilege leave their imprints *inside* knowledge in this field, it can be recognised how – around climate issues, as elsewhere – modernity has arguably colonised its own critique (see Callison, Chapter 3, this volume). If climate action is successfully to defend the autonomy of the Earth and its people – rather than usher in a new military-derived 'Anthropocene' geoengineering-based global control regime – then it is difficult to imagine higher stakes (see Schubert, Chapter 31, this volume).

As a core strand in STS has always argued, the issue at the heart of controlling vs. caring politics in knowledge lies with democracy. Democracy in this case concerns not only making paths for *truth to speak to power*, but also promoting everyday political recognition for how *power shapes truth*. This latter injunction is at no time more important than when the truths involved are those propounded by self-consciously critical interventions. Such a view of power does not need to be seen as intrinsically pejorative. It is just that power is – under virtually any view – more likely to become negative when it is not challenged. So, conceived far more broadly than just stale parliamentary procedures, or a concentrated media, or increasingly-managed voting systems, an STS-inspired vision of democracy is much wider. In the terms referred to here, it is about *access by the least powerful to the capacities for challenging power* (Stirling, 2019b). Perhaps most crucially, this vision of democracy is as much about the knowledge interlinked with actions, as about the actions themselves. In the climate struggle – as in science, technology, innovation and politics more widely – an analysis that actively interrogates the contrasts between controlling and caring modes of knowledge can illuminate a much greater diversity of possible future pathways for change compared to those that are convenient for entrenched interests to accept or promote.

As with previous (still unfinished) progressive struggles beyond environmentalism – including against slavery, colonialism, racism, patriarchy, class snobbery, sexual bigotry and many other injustices – the most effective means to suppress dissent is to instil the imagination even among critics that *there is no alternative*. And (despite all the radical rhetoric) it is this incipient fatalistic indifference to underlying structures that growingly characterises much contemporary climate action. Overbearing pressures are increasingly conditioning acceptance that the appropriate reaction is *panic* leaving *no time for democracy*. It seems forgotten that all of the above progressive mobilisations depended centrally on care for democratic struggle. None were driven primarily by aspirations to assert control.

Also apparently forgotten is that it was precisely on occasions when a controlling idiom did come to the fore in past progressive struggles (as in some violent revolutions), that it was exactly this condition that arguably led to the betrayal of the overarching, more caring motivations. In a complex and turbulent world, significant roles were obviously played across all these historic fields by policy styles described on the left-hand side of Table 37.1. But the key point here is that these rarely took the form of major drivers. What primarily motivated successful progressive struggles was not so much aspirations to control, but rather more pluralistic, values-driven, horizontally-organised, relational political qualities of care: entangling knowledge and action towards their diverse hoped-for futures. So, the main question raised here is: is a move from a caring to a controlling politics in knowledge a syndrome that is currently impeding policy success on climate?

Of course (as in any analysis), many complexities, conditionalities, qualifications and limitations beset the present argument. The current confined format restricts attention to a somewhat stylised summary of central points. Especially in a spirit of caring attention, there are many sources of uncertainty, variability, contingency and countervailing meaning that might productively be explored. So, the point in the end is less to assert a single central narrative than to open up space for considering the depth and scope of critique that is too often foreclosed by the urgency and normative weight of climate struggle. For instance, if the controlling counterforce of authoritarian populism is invoked as grounds to hold back on this kind of more caring politics, then this too can be addressed according to relational insights around care. In an important and growing analysis, it is exactly the kinds of globalising elite technocratic control often urged in climate action itself, that may most strongly provoke this remarkably globally coordinated worldwide backlash of authoritarian populism. Here the point is that – no matter how well motivated – imaginaries of control are themselves authoritarian. Relationally, control engenders control – one kind of authoritarianism often breeds another. Authoritarian populism is technocratic paternalism in the mirror. So – both in enacting climate protection and resisting its current contrarian reactions – *less can be more*. The adoption of a more caring than controlling idiom may address both dilemmas together. Around care as so much else, the medium is the message.

In conclusion, the scope of the political transformations needed to tackle climate disruption extends far further than can be reached purely through routine policy. Expanding as they do beyond capitalism alone, the driving problems must be diagnosed in terms of a more expansively encompassing political formation – here termed colonial modernity. When this deeper and older politics is recognised in pan-continental histories, political economies and cultures, we can appreciate that conventional controlling climate governance practices around centralised coordination and vertical delivery are inadequate to be rapidly effective. What is under any view clearly needed more than just a 'policy control' idiom are the kinds of broad-scale political movement engaging disruptively with entire political economies, cultures and histories, of kinds which have characterised the previous progressive transformations discussed above.

Yet this is arguably the greatest dilemma of all for a control-based vision of climate action. For it is precisely distinctive of these most fundamental and momentous of societal movements that they are well beyond even the most fevered aspirations to machine-like control. In this sense, it is not just that a controlling imagination is relatively overplayed on climate issues, but that it is intrinsically inappropriate. For those for whom colonial modernity renders control the only imaginable kind of agency, the despair engendered in this kind of understanding, is further disabling. To recognise care as an alternative possibility for political agency, can overcome this barrier.

Here at the end there emerges potentially the most important cue for onward – critically caring – thought and action. This hinges on the engaging familiarity of what is arguably the most obvious and fertile alternative to a control model, in imagining the kinds of large-scale cultural, political and historical dynamic necessary to properly address climate disruption. For it is well understood that the exquisitely choreographed emergent flocking motions observable among birds and fishes do not depend on any kind of vertical, individualised categorical control from any individual visionary leader or 'cockpit'. Albeit not fully understood, these motions are evidently achieved instead by much more horizontally-coordinated, mutually-calibrated relational processes spanning the assemblage as a whole – in other words, they are more about care. The microdynamics involved are more about 'thriving out', than 'scaling up' or 'cascading down'. And when the time is right, it is one of the most striking features of these 'murmurations', as they are called (recalling the cultural revolutions of the 1960s) that lead to radical reorientations with a rapidity that would be the envy of central control.

It is perhaps telling then (not least of the caring relational wisdom in language?) that the word *murmuration* refers both to this kind of relational coordination as well as to the power of distributed dissent? Resulting questions are for further consideration and experiment. But what may be concluded with confidence is that, in an ailing climate movement struggling for renewal, it may be in imaginations and interventions around care (as summarised in Table 37.1) that the most fruitful prospects lie for urgently needed actions to reverse disruptions to a flourishing climate.

Further Reading

Leach, M., Scoones, I. and Stirling, A. (2010) *Dynamic Sustainabilities: Technology, Environment, Social Justice*. London: Earthscan.

Pryck, K. De and Hulme, M. (eds) (2022) *A Critical Assessment of the Intergovernmental Panel on Climate Change*. Cambridge: Cambridge University Press.

Hilgartner, S., Miller, C. A. and Hagendijk, R. (eds) (2015) *Science and Democracy: Making Knowledge and Making Power in the Biosciences and Beyond*. London: Routledge.

Bellacasa, M. P. de la (2017) *Matters of Care: Speculative Ethics in More than Human Worlds*. Minneapolis: University of Minnesota Press.

Scoones, I. and Stirling, A. (eds) (2020) *The Politics of Uncertainty: Challenges of Transformation*. London: Routledge.

References

Elliott, D. (ed.) (2007) *Nuclear or Not? Does Nuclear Power Have a Place in a Sustainable Energy Future*. London: Palgrave MacMillan.

Felt, U., Barben, D. Irwin, A., Pierre-Benoît, J., Rip, A., Stirling, A. and Stöckelová, T. (2013) *Science in Society: Caring for Our Futures in Turbulent Times*. Strasbourg: European Science Foundation.

Fisher, B. and Tronto, J. (1991) 'Toward a feminist theory of care', in Able, E. and Nelson, M. (eds) *Circles of Care: Work and Identity in Women's Lives*. Albany, NY: SUNY Press.

Gee, D., Grandjean, P., Hansen, S. F., van den Hove, S., MacGarvin, M., Martin, J., Nielsen, G., Quist, D. and Stanners, D. (eds) (2013) *Late Lessons from Early Warnings: Science, Precaution, Innovation*. Copenhagen: European Environment Agency.

Harremoës, P., Gee, D., MacGarvin, M., Stirling, A., Keys, J., Wynne, B. and Vaz, S. G. (eds) (2001) *Late Lessons from Early Warnings: The Precautionary Principle 1896–2000*. Copenhagen: European Environment Agency.

IAASTD (2009) *Agriculture at a Crossroads: International Assessment of Agricultural Knowledge Science and Technology for Development (IAASTD)*. Washington, DC: Island Press.

Illich, I. (1972) *Tools for Conviviality*. New York: Harper and Row.

Johnstone, P. and Stirling, A. (2022) 'Beyond and beneath megaprojects: Exploring submerged drivers of nuclear infrastructures', *Journal of Mega Infrastructure & Sustainable Development*, pp. 1–22. doi: 10.1080/24724718.2021.2012351.

Perkins, P. E. E. (2019) 'Climate justice, commons, and degrowth', *Ecological Economics*, 160, pp. 183–190. doi: 10.1016/j.ecolecon.2019.02.005.

Schneider, C. Q. and Wagemann, C. (2012) *Set Theoretic Methods for the Social Sciences: A Guide to Qualitative Comparative Analysis*. Cambridge: Cambridge University Press.

Scrase, I. and MacKerron, G. (eds) (2009) *Energy for the Future: A New Agenda*. London: Palgrave Macmillan. Available at: http://books.google.co.uk/books?id=Wm4ZOwAACAAJ.

Stirling, A. (2019a) 'Engineering and sustainability: Control and care in unfoldings of modernity', in Michelfelder, D. P. and Doorn, N. (eds) *Routledge Companion to the Philosophy of Engineering*. London: Routledge, pp. 461–481.

Stirling, A. (2019b) 'How deep is incumbency? A "configuring fields" approach to redistributing and reorienting power in socio-material change', *Energy Research & Social Science*, 58, 101239. doi: 10.1016/j.erss.2019.101239.

Stirling, A. and Scoones, I. A. N. (2020) 'COVID-19 and the futility of control in the modern world', *Issues in Science and Technology*, 36(4), pp. 25–27.

Wilsdon, J. and Doubleday, R. (eds) (2015) *Future Directions for Scientific Advice in Europe*. Cambridge: Cambridge University Press.

Index

scientific controversy 2, 68; and climate science methods 261–2; disregarding scientific literature within 165; and inauthenticity 164–5; manufacturing doubt or pseudo-facts within 165; studies of 30, 281
Scientist Rebellion 79–80, 288
Scientists for the Future (S4F) 80
scientized politics 283; *see also* co-production
sea level rise 105, 224, 229, 231–2, 299–300
seawalls 196, 224–5, 299
smart energy technologies 180-4
social construction: of climate change xviii, 91; of technology 41, 187–8; of users in technology design 183–4
social constructionist theory 27–8, 29–37, 43, 187–8
social life of science 2
social movements xix; anti-environmental movement 30; climate change denial countermovements 75; climate justice movement 76, 308; *see also* climate activism; environmental non-governmental organizations; environmentalism
sociology of expectations 290; *see also* climate change futures
sociotechnical development: consumption/user practices 168; innovation space 167; landscape 167; markets 168; niche opening 167; political support for 167–8; and regime change 167; and shaping futures 168, 188–94; social network support 167; solution for climate change; STS tools for 168, 188–94; university investment in schools for 194
sociotechnical imaginaries xxii, 68, 168, 190–2; alternative imaginaries 190–1; public and stakeholder engagement in 191; water in relation to 290–1; *see also* renewable energy transition
sociotechnical process 195, 209; definition of 212
sociotechnical systems 45, 113, 170, 187, 193
sociotechnical transition xx-xxi, 167–8, 180; and bridge technologies 264
sociotechnologies 51, 188; democratization of 193; pathways for development 190; redesign in equitable ways 192–3; and users 189; *see also* renewable energy transition
soil health 219
solar energy 105, 115, 170

Solar Radiation Management (SRM) 257–8, 263, 266, 268; *see also* geoengineering
standpoint theory 4
state: history in relation to climate change 7, 11–12, 256
Strong Programme in the sociology of science 29–30
STS theory: ontology of 27–8; epistemology of 27–8
Sustainable Livelihoods Approach 203
symmetrical inquiry 30, 33–4; asymmetrical studies of environmental science 31–2; *see also* methodological relativism

technical fix xx–xxi, 101, 113, 211, 288
technocracy 284
technological determinism: critique of 167
technological innovation 39–40, 44
technology: assemblages 188; diffusion 181; emerging (and precautionary assessment of) 253, 279–80; speculative 254
technosphere 189
thermal energy systems 114
Thunberg, Greta xx, 63, 70–1, 83
tipping points 147, 257
tragedy of the commons 130
treadmill of production theory 28, 42–3; *see also* political economy of climate change theory
trust: and expertise 162; in modernity 161; in science policy 162–4, 166; as sociocultural product 166

uncertainty 155–6, 246–7, 307; in geoengineering 268–9, 279, 281; social dynamics as a source of 288; *see also* climate change knowledge
United Nations Framework Convention on Climate Change (UNFCCC) 135
upstream engagement 121, 125

visions: of climate mitigation 258, 269; role of science in informing 257

water: STS approaches to 232, 236, 296; the world ocean and 297
watershed 119
Watt, James 40–1, 44
weather-world 210
wildfire 23–4

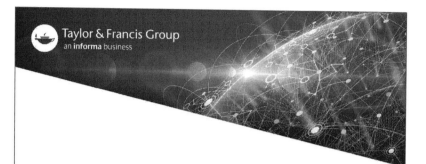

9781032530178